Felix Sharipov

Rarefied Gas Dynamics

Related Titles

Bergman, T.L., Lavine, A.S., Incropera, F.P., DeWitt, D.P.

Fundamentals of Heat and Mass Transfer

Seventh Edition

2011
ISBN: 978-0-470-50197-9, also available in electronic formats

Jorisch, W. (ed.)

Vacuum Technology in the Chemical Industry

2014
ISBN: 978-3-527-31834-6, also available in electronic formats

Nikrityuk, P.A.

Computational Thermo-Fluid Dynamics

In Materials Science and Engineering

2011
ISBN: 978-3-527-33101-7, also available in electronic formats

Grigoriev, R., Schuster, H. (eds.)

Transport and Mixing in Laminar Flows

From Microfluidics to Oceanic Currents

2012
ISBN: 978-3-527-41011-8, also available in electronic formats

Felix Sharipov

Rarefied Gas Dynamics

Fundamentals for Research and Practice

WILEY-VCH
Verlag GmbH & Co. KGaA

Author

Prof. Dr. Felix Sharipov
Departameto de Física
Universidade Federal do Paraná
CaixaPostal 19044
81531-990, Curitiba
Brazil

Library of Congress Card No.: applied for

British Library Cataloguing-in-Publication Data
A catalogue record for this book is available from the British Library.

Bibliographic information published by the Deutsche Nationalbibliothek
The Deutsche Nationalbibliothek lists this publication in the Deutsche Nationalbibliografie; detailed bibliographic data are available on the Internet at <http://dnb.d-nb.de>.

Print ISBN: 978-3-527-41326-3
ePDF ISBN: 978-3-527-68507-3
ePub ISBN: 978-3-527-68553-0
Mobi ISBN: 978-3-527-68554-7
oBook ISBN: 978-3-527-68552-3

Cover Design Adam Design, Weinheim, Germany
Typesetting SPi Global, Chennai, India
Printing and Binding Markono Print Media Pte Ltd., Singapore

Printed on acid-free paper

Contents

Preface

During lectures, seminars, and mini-courses given by me in universities, research institutes, and schools, I was frequently asked to suggest a book for beginners in order to learn quickly fundamentals and main results of rarefied gas dynamics. During last decades, the interest to this area drastically increased due to a necessity to model gas flows in many technologies related to rarefied gas flows. For instance, many technological processes take place in vacuum chambers under low pressure conditions where the continuous medium mechanics is not valid anymore. A further optimization of such processes requires more detailed information about gas flows in complex geometrical configurations. Another example of application of rarefied gas dynamics is the rapid miniaturization of electronic and mechanical equipment, which led to the necessity to take into account the gas rarefaction. In fact, the molecular mean free path became close to characteristic sizes of the miniaturized equipment even under the atmospheric pressure. Thus, the number of researchers and engineers dealing with rarefied gases drastically increased. In spite of many excellent books in this area, it was not so easy to suggest one of them that would describe the fundamentals of rarefied gas flows in concise and easily acceptable form. The present textbook intends to fill this lacuna. It is addressed to students, researchers, and engineers who wish to learn fundamentals and main results of rarefied gas dynamics and then to apply this knowledge to their practice.

In the first part of the present book, the main concepts related to velocity distribution function, Boltzmann equation, gas–surface interaction are given in an easily acceptable form. The main techniques to model rarefied gases such as discrete velocity method and direct simulation Monte Carlo method are described. Most of the results are given without hard mathematical derivations, but many references are suggested to those readers who want to study this field deeper. In the second part of the book, the classical problems of fluid dynamics, namely, Couette flow, heat transfer between solid surfaces, flows through various kinds of pipes, wave propagation, and so on, are solved analytically and numerically. Both linear and nonlinear transport phenomena are considered. For the sake of simplicity, only a single monatomic gas is considered, but some recommendations about the applicability of these results to polyatomic gases and gaseous mixtures are given. The book draws more attention to deterministic approaches such as the discrete velocity method. Indeed, in many technological processes, the Mach number is so small that the probabilistic methods widely used in aerothermodynamics become time consuming because of the statistical scatter. In situations when the Mach number is extremely small, the deterministic methods based on numerical solution of the kinetic equation become unique tools for a modeling of rarefied gas flows.

Most of the numerical solutions given in this book are provided by numerical codes based on the discrete velocity method with recommended input data, allowing the readers to obtain new results which are not reported in papers. The codes are neither optimized nor parallelized, but they require modest computational effort and can be run in an ordinary ultrabook. The readers can modify the code and solve new more complicated problems. Each chapter ends by exercises helping the readers to understand better the chapter matter and to apply it to some practical situations.

It is hoped that the manner to describe the deterministic method will enable many students, researchers, and engineers to learn easily the main concepts and results of rarefied gas dynamics. In the future, this knowledge will make easier the study of other books and papers in this field.

The manuscript of this book has been used in a course that I teach at the Post-Graduation in Physics of Federal University of Paraná. I wish to thank my students and colleagues for comments on the present manuscript.

Curitiba, Brazil
April, 2015

Felix Sharipov

List of Symbols

a	characteristic size
$\hat{A}$	linearized scattering operator, Eq. (5.51)
$A_1,\ A_2,\ A_3$	coefficients in model equations, Eqs. (7.44), (7.45)
A_{ij}	discretized scattering operator, Eq. (9.10)
$\hat{A}_n$	linearized scattering operators in power expansion, Eq. (5.80)
A_Ψ	amplitude of Ψ, Eq. (16.5)
b	impact parameter, Fig. 1.2
b_M	cut-off impact parameter
B	tensor in ellipsoidal model, Eq. (7.32)
$\boldsymbol{c}$	dimensionless molecular velocity, Eq. (5.2)
c_p	specific heat per particle at constant pressure, Eq. (6.4)
c_p	magnitude in polar coordinates, Eq. (9.18)
C_{k_1}, C_{k_2}	coefficients of finite difference scheme, Eqs. (9.37), (9.47)
C^x_{kj}, C^θ_j	coefficients of finite difference scheme, Eq. (9.76)
d	potential zero distance, Eq. (1.19)
e	specific internal energy, Eqs. (1.13), (2.24)
E	internal energy, Eq. (1.12)
E_r	kinetic energy of relative motion, Eq. (1.22)
$\dot{E}$	energy flow rate, Eqs. (11.75), (12.78), (13.3),
f	velocity distribution function, Eq. (2.1)
f_i	set of functions, Eq. (9.1)
f^M	Maxwellian, Eq. (2.37)
f^M_0	global Maxwellian, Eq. (5.1)
f^M_R	reference Maxwellian, Eq. (5.28)
f^M_w	surface Maxwellian, impermeable surface, Eq. (5.56)
f^M_w	surface Maxwellian, permeable surface, Eq. (5.58)
F	cumulative function, Eq. (8.3)
F_N	representation of model particles, Eq. (8.26)
g	bulk source term, Eqs. (5.34)
$\tilde{g}$	dimensionless bulk source term, Eqs. (7.49)
$\boldsymbol{g}_r$	relative velocity, Eq. (1.20)
$\bar{g}_r$	average relative speed, Eq. (3.34)

$g_{r,\max}$	maximum relative speed
$\boldsymbol{G}$	center mass velocity, Eq. (8.30)
G	dimensionless flow rate, Eqs. (15.3), (15.20)
G_{P}	Poiseuille coefficient for short channel and tube, Eqs. (13.56), (14.16)
G_{P}^{*}	Poiseuille coefficient for infinite channel and tube, Eqs. (11.81), (12.80), (13.5)
$\mathcal{G}_{\mathrm{P}}$	Poiseuille coefficient for short channel and tube, Eqs. (13.60), (14.16)
G_{T}	thermal creep coefficient for short channel, Eq. (13.56)
G_{T}^{*}	thermal creep coefficient for infinite channel and tube, Eqs. (11.81), (12.80), (13.6)
$\mathcal{G}_{\mathrm{T}}$	thermal creep coefficient for short channel, Eq. (13.60)
h	perturbation function, Eq. (5.6)
h_0	split part of perturbation function, Eq. (11.118)
h_i	set of perturbation functions, Eq. (9.6)
h_{R}	reference perturbation, Eq. (5.30)
h_w	surface source term, Eq. (5.59)
H	term of model equation, Eq. (7.48)
$\mathcal{H}$	coefficient, Eq. (13.11)
i	imaginary unit
I	unit tensor, Eq. (A.9)
$\mathcal{I}_0$	Bessel function, Eq. (A.27)
I_n	function, Eq. (A.29)
$\Im$	imaginary part of complex number, Eq. (A.39)
$\boldsymbol{J}_e$	flow vector of energy, Eq. (2.20)
J_k	thermodynamic flux, Eq. (5.81)
$\boldsymbol{J}_m$	flow vector of mass, Eq. (2.19)
$\boldsymbol{J}_N$	flow vector of particles, Eq. (2.18)
$\boldsymbol{J}_s$	flow vector of entropy, Eq. (2.36)
$\boldsymbol{J}_\psi$	flux of property ψ, Eq. (2.17)
k	wave number, Eq. (16.1)
k_{B}	Boltzmann constant, Table A.1
k_{I}	attenuation coefficient, Eq. (16.4)
Kn	Knudsen number, Eq. (1.33)
ℓ	equivalent free path, Eq. (1.32)
L	length-to-height ratio, Figure 13.3
L	length-to-radius ratio, Figure 14.1
L	dimensionless distance, Eq. (16.13)
$\hat{L}$	linearized collision operator, Eq. (5.21)
$\hat{L}_{\mathrm{B}}$	linearized BGK collision operator, Eq. (7.8)
$\hat{L}_{\mathrm{E}}$	linearized ellipsoidal collision operator, Eq. (7.34)
$\hat{L}_{\mathrm{S}}$	linearized S-model collision operator, Eq. (7.19)
m	mass of one particle, Eq. (1.1)
M	mass of gas

$\mathcal{M}$	atomic weight, Eq. (1.2)
M	momentum flux tensor, Eq. (2.21)
$\dot{M}$	mass flow rate, Eqs. (11.75), (12.78), (13.2), (13.39), (14.5)
n	number density, Eq. (1.5)
n_R	reference number density, Eq. (5.28)
N	number of particles
N_A	Avogadro number, Table A.1
N_{coll}	number of collisions in cell, Eq. (8.24)
N_m	number of model particles
N_{mc}	number of model collisions in cell, Eq. (8.27)
N_x	number of nodes for x variable
N_c	number of nodes for velocity
N_θ	number of nodes for θ variable
N_L	Loschmidt number, Table A.1
p	pressure, Eq. (1.7)
P	pressure tensor, Eq. (2.28)
Pr	Prandtl number, Eq. (6.4)
$\boldsymbol{q}$	heat flow vector, Eq. (2.32)
Q_P^*	mechanocaloric coefficient for infinite channel and tube, Eqs. (11.82), (12.81), (13.7)
Q_T^*	heat flow coefficient for infinite channel and tube, Eqs. (11.82), (12.81), (13.8)
$\tilde{\boldsymbol{q}}$	dimensionless heat flow vector, Eq. (5.16)
Q	collision integral, Eq. (3.5)
Q_B	BGK collision integral, Eq. (7.1)
Q_E	ellipsoidal model of collision integral, Eq. (7.31)
Q_S	S-model collision integral, Eq. (7.17)
$Q^{(u)}$	heat flow rate due to velocity gradient, Eq. (10.15)
$\boldsymbol{r}$	position vector
r	intermolecular distance
r	radial coordinate
R	scattering kernel, Eq. (4.4)
R_{CL}	Cercignani–Lampis scattering kernel, Eq. (4.17)
R_d	diffuse scattering kernel, Eq. (4.10)
R_{ds}	diffuse–specular scattering kernel, Eq. (4.15)
R_g	molar gas constant, Table A.1
R_f	random number
R_{ij}	discretized kernel, Eq. (9.5)
R_n, R_{t1}, R_{t2}	components of scattering kernel, Eq. (8.7)
R_s	specular scattering kernel, Eq. (4.13)
$\Re$	real part of complex number, Eq. (A.39)
s	entropy density, Eq. (2.35)
S	entropy, Eq. (4.41)
S	rate-of-shear tensor, Eq. (6.1)
$\mathcal{S}$	coefficient, Eq. (13.11)

t	time
T	temperature, Eq. (2.26)
$\hat{T}$	time reversion operator, Eq. (5.38)
T^*	reduced temperature, Eq. (6.30)
T_R	reference temperature, Eq. (5.28)
T_w	wall temperature
$\boldsymbol{u}$	mean (bulk or hydrodynamic) velocity, Eq. (2.16)
$\tilde{\boldsymbol{u}}$	dimensionless bulk velocity, Eq. (5.13)
u_m	speed amplitude, Eqs. (16.18), (16.59)
u_R	reference bulk velocity, Eq. (5.28)
$\tilde{u}_R$	reference dimensionless bulk velocity, Eq. (5.29)
$\boldsymbol{u}_w$	wall velocity
$\tilde{\boldsymbol{u}}_w$	dimensionless wall velocity, Eq. (5.53)
U	intermolecular potential, Eq. (1.15)
$\boldsymbol{v}$	molecular velocity
v_0	most probable speed at T_0, Eq. (5.2)
v_m	most probable speed, Eq. (2.40)
v_w	most probable speed at T_w, Eq. (4.9)
$\mathcal{V}$	volume of gas
$\mathcal{V}_C$	volume of cell
$\boldsymbol{V}$	peculiar velocity, Eq. (2.22)
w	function in collision integral, Eq. (3.5)
W	dimensionless flow rate, Eqs. (13.76), (14.24)
W'	dimensionless flow rate, Eq. (13.80)
W_0	transmission probability, Eq. (13.45)
W_i	weight of ith node, Eq. (9.2)
α	accommodation coefficient, Eq. (4.18)
α_c	condensation coefficient, Eq. (4.32)
α_d	diffuse part of scattering, Eq. (4.15)
α_n	accommodation coefficient of energy of normal motion, Eq. (4.17)
α_t	tangential momentum accommodation coefficient, Eq. (4.17)
β	radii ratio, Figure 12.1
β	aspect ratio, Figure 13.1
γ	dimensionless wave number, Eq. (16.74)
δ	rarefaction parameter, Eqs. (1.34), (7.46)
δ'	coefficient proportional to δ, Eq. (9.24)
ε	potential well depth
ϵ	azimuthal impact parameter, Eq. (3.7)
ζ_T	temperature jump coefficient, Eq. (10.39)
θ	angle of polar coordinates, Eq. (9.18)
θ	frequency parameter, Eq. (16.11)
κ	heat conductivity
$\tilde{\kappa}$	dimensionless heat conductivity, Eq. (6.29)
Λ_{kn}	kinetic coefficient, Eq. (5.84)
Λ^t_{kn}	time-reversed kinetic coefficient, Eq. (5.85)

μ	dynamic viscosity
$\tilde{\mu}$	dimensionless viscosity, Eq. (6.29)
μ_w	chemical potential, Eq. (4.39)
ν	number of moles, Eq. (1.3)
ν	part of collision integral, Eq. (3.13)
ν_B	parameter of BGK model, Eqs. (7.15), (7.16)
ν_c	intermolecular collision frequency, Eq. (3.33)
ν_E	parameter of ellipsoidal model, Eq. (7.42)
ν_S	parameter of S model, Eq. (7.30)
ξ	small parameter
ξ_n	thermodynamic force, Eq. (5.81)
ξ_P	small parameter related to pressure, Eqs. (11.72), (13.50), (13.82), (14.12)
ξ_T	small parameter related to temperature, Eqs. (10.40), (11.72), (13.50), (16.61)
ξ_u	small parameter related to velocity, Eqs. (10.3), (16.61)
ρ	mass density, Eq. (1.6)
ϱ	density deviation, Eq. (5.12)
ϱ_R	reference density deviation, Eq. (5.29)
Π	pressure tensor deviation, Eq. (5.15)
σ	total entropy production, Eq. (4.45)
σ_{coll}	entropy production due to intermolecular collisions, Eq. (4.43)
σ_t	total cross section, Eq. (1.26)
σ_d	differential cross section, Eq. (1.23)
σ_P	viscous slip coefficient, Eq. (10.2)
σ_T	thermal slip coefficient, Eq. (10.17)
σ_w	entropy production due to gas–surface interaction, Eq. (4.36)
Σ_g	imaginary surface in gas
Σ_w	wall surface
ω	oscillation frequency
τ	temperature deviation, Eq. (5.14)
τ_R	reference temperature deviation, Eq. (5.29)
τ_w	temperature deviation of wall, Eq. (5.49)
φ_Ψ	phase of Ψ, Eq. (16.5)
Φ	perturbation, Eqs. (11.19), (11.53), (11.54), (11.97), (11.98)
χ	deflection angle, Fig. 1.2
$(.,.)$	inner product of two functions, Eq. (5.3)
$((.,.))$	inner product of two functions, Eq. (5.5)
$(.,.)_B$	boundary inner product, Eq. (5.64)
$\langle \dots \rangle$	average value per volume unity, Eq. (2.11)

List of Acronyms

AC	accommodation coefficient, Section 4.4
AI	*ab initio* (potential), Section 1.4
BE	Boltzmann equation, Section 3.2
BGK	Bhatnagar, Gross, and Krook model, Section 7.1
CL	Cercignani–Lampis (scattering kernel), Section 4.3
EFP	equivalent free path, Section 1.8
DCS	differential cross section, Section 1.6
DVM	discrete velocity method, Chapter 9
DSMC	direct simulation Monte Carlo (method), Section 8
HS	hard sphere (model), Section 1.4
LJ	Lennard-Jones (potential), Section 1.4
MFP	mean free path, Section 1.8
MPS	most probable speed, Section 2.4
TCS	total cross section, Section 1.6
TSC	thermal slip coefficient, Section 10.3
TJC	temperature jump coefficient, Section 10.5
VDF	velocity distribution function, Section 2.1
VSC	viscous slip coefficient, Section 10.2

1
Molecular Description

1.1
Mechanics of Continuous Media and Its Restriction

Rarefied gas flows can be modeled by many methods dependent on the flow regime. Under certain conditions, a gas can be considered as a continuous medium and then the hydrodynamic equations (see Section 6.1) are successfully applied. These equations provide a description of gas flows in terms of the so-called macroscopic quantities such as density, pressure, temperature, and velocity. Analytical and numerical methods of the mechanics of continuous media are well elaborated and described in numerous books, handbooks, and textbooks so that they are widely used in practical calculations. However, the consideration of gas as a continuous medium imposes some conditions and, consequently, restricts the application of the hydrodynamic equations. What are these restrictions?

The first restriction is based on the assumption that a characteristic size of gas flow (or macroscopic size) must be significantly larger than the so-called molecular mean-free-path (or microscopic size), that is, a path that a gaseous particle flies between two successive intermolecular collisions. What is the characteristic size? When macroscopic variables change slowly and smoothly, the characteristic size is a typical distance between boundaries of the gas flow. However, the macroscopic variables can change significantly over the mean-free-path, and then the assumption on the continuousness of the medium is broken at least in the region of the significant change. For instance, if a gas has a stepwise temperature distribution, continuous mechanics does not work near the temperature step.

The second restriction is related to nonstationary flows, namely, the continuous mechanics is valid if a significant change of macroscopic variable happens during a time interval significantly larger than the mean-free-time of gaseous particles. For instance, if an oscillation frequency of solid surface is close to the intermolecular collision frequency, a gas flow near this surface cannot be described by continuous mechanics even if the first assumption is fulfilled. Another example when the second assumption is not fulfilled is a sudden motion of a solid surface or sudden variation of its temperature.

Rarefied Gas Dynamics: Fundamentals for Research and Practice, First Edition. Felix Sharipov.

Thus, if at least one of the above-mentioned assumptions is not fulfilled, continuous mechanics cannot be applied, but a modeling should be done at a microscopic level, that is, a gas must be considered as rarefied. This branch of fluid mechanics is called rarefied gas dynamics. The aim of this field is to obtain macroscopic characteristics based on microscopic behavior of gaseous particles. Such a behavior involves two kinds of interactions, namely, intermolecular collisions and gas–surface interaction. In this chapter, basic information about the intermolecular interaction is given.

1.2
Macroscopic State Variables

In this section, we will define main macroscopic quantities used in fluid mechanics. Consider a portion of gas occupying a volume $\mathcal{V}$. The amount of the gas can be measured by its mass M or by the number of particles N. The mass of one particle is given as

$$m := \frac{M}{N}. \tag{1.1}$$

If the number of particles is equal to the Avogadro number N_A (see Table A.1), then the amount of gas is one mole. The mass of one mole is called atomic weight and is calculated as

$$\mathcal{M} := mN_A. \tag{1.2}$$

The atomic weight is usually given in grams per mole. Its values for the noble gases are given in Table A.2. The gas amount can also be measured in the number of moles as

$$\nu := \frac{N}{N_A} = \frac{M}{\mathcal{M}}. \tag{1.3}$$

The quantities characterizing the gas amount, for example, volume $\mathcal{V}$, mass M, the number of particles N, or the number of moles, are called extensive. It is common to write equations in terms of specific variable given as a ratio of two extensive quantities. For instance, the specific mass, or mass density, is defined as

$$\rho := \frac{M}{\mathcal{V}}. \tag{1.4}$$

The number of particles per gas volume, or number density, is defined as

$$n := \frac{N}{\mathcal{V}}. \tag{1.5}$$

The mass density and number density are related via the molecular mass as

$$\rho = mn, \tag{1.6}$$

which follows from Eqs. (1.1), (1.4), and (1.5).

Any gas occupying a container of volume $\mathcal{V}$ produces a pressure p on its walls, which is defined as a force F acting on an area A unity

$$p := \frac{F}{A}. \tag{1.7}$$

The unit of pressure in the International System of Units (SI) is given by newtons (N) per square meter (m^2) and is called pascal (Pa).

The pressure of gas depends on its amount, volume, and also on its temperature T, which is measured in kelvins (K). The definition of this quantity in the frame of equilibrium thermodynamics can be found in many textbooks, for example, Ref. [1]. Later the temperature definition will be given via the velocity distribution function.

1.3
Dilute Gas

A relation of gas pressure to its amount, volume, and temperature is called state equation. This book is restricted only to the so-called diluted gas obeying the following state equation:

$$p\mathcal{V} = \nu R_g T, \tag{1.8}$$

where R_g is the molar (or universal) gas constant (see Table A.1). The state equation in the form (1.8) is given in most of textbooks on thermodynamics, but in statistical physics it is usually written in terms of the number density n or mass density ρ as

$$p = nk_{\mathrm{B}}T, \quad p = \frac{\rho}{m}k_{\mathrm{B}}T, \tag{1.9}$$

where k_{B} is the Boltzmann constant (see Table A.1), which is related to the molar gas constant R_g and Avogadro number N_A as

$$k_{\mathrm{B}} = \frac{R_g}{N_A}. \tag{1.10}$$

The combination of this equation with (1.2) yields the relation

$$\frac{k_{\mathrm{B}}}{m} = \frac{R_g}{\mathcal{M}}, \tag{1.11}$$

which is quite useful because the kinetic theory deals with the molecular mass m, while in many practical fields, the atomic weight $\mathcal{M}$ is more preferable.

A relation of internal energy E of gas to its state variables is called the energy equation. For a dilute monatomic gas, the internal energy is proportional to its temperature and is given as

$$E = \frac{3}{2}Nk_{\mathrm{B}}T, \quad \text{or} \quad E = \frac{3}{2}\nu R_g T. \tag{1.12}$$

The specific internal energy e, or energy per mass unity, is defined as

$$e := \frac{E}{M} \tag{1.13}$$

so by using (1.1) the energy equation (1.12) is transformed to

$$e = \frac{3}{2}\frac{k_B T}{m}. \tag{1.14}$$

The state equation (1.9) and energy equation (1.14) work well for the atmospheric pressure and for any lower pressure. Consequently, all results of the book are valid for this range of the pressure.

1.4 Intermolecular Potential

1.4.1 Definition of Potential

In case of dilute gas, the intermolecular interactions do not affect the state and energy equations, but they are important to describe the transport phenomena in gases, like mass, heat, and momentum transfers. In this section, main models of intermolecular interactions are given.

The intermolecular potential $U(r)$ is defined so that the potential energy of two particles separated by a distance r is equal to U. If this potential is known, then the interaction force F between these two particles is calculated as

$$F(r) = -\frac{dU(r)}{dr}. \tag{1.15}$$

An exact calculation of the potential $U(r)$ is a very hard task, that is why manysimplified models were proposed.

1.4.2 Hard Sphere Potential

The most simple potential is the hard sphere (HS) model given as

$$U(r) = \begin{cases} \infty & \text{at} \quad r < d, \\ 0 & \text{at} \quad r > d, \end{cases} \tag{1.16}$$

where d is the sphere diameter. Physically, it means that two particles cannot be closer than their diameter, but when they are separated by a distance $r > d$ then the interaction force is zero. In many applications, this model works pretty well, that is why it is widely used in practical calculations. The molecular diameter can be extracted from the transport coefficients such as viscosity and thermal conductivity. However, if one calculates the diameter d from the gas viscosity at two different temperatures, one obtains two different values of the diameter. It means that we cannot calculate the diameter once at one temperature and then use it for any other temperature.

1.4.3 Lennard-Jones Potential

As will be shown later, see Section 6.4.1, the dynamic viscosity μ of gas composed of hard spheres is proportional to the square root of the temperature T, $\mu \propto \sqrt{T}$, while empirical data, see for example, the review by Kestin *et al.* [2], indicate a different dependence of viscosity on the temperature. Such a discrepancy is explained by the neglect of attractive force between particles when they are separated by a distance $r > d$. Moreover, the repulsive force arising at a short distance is really large but not infinite as for the HS potential. These two factors are taken into account by the Lennard-Jones (LJ) potential given as

$$U(r) = 4\varepsilon \left[\left(\frac{d}{r} \right)^{12} - \left(\frac{d}{r} \right)^{6} \right], \tag{1.17}$$

where ϵ is the potential well depth. This potential contains two fitting parameters d and ϵ that allow us to describe better the dependence of transport coefficients on the temperature. The values of d and ε vary from one bibliography source to another because they are extracted from different coefficients. Some numerical values of the LJ potential parameters extracted from viscosity experimental data reported in the book by Hirschfelder *et al.* [3] are given in Table 1.1. The books in [4, 5] provide very similar data.

1.4.4 *Ab initio* Potential

Recently, a calculation of the potential *ab initio* (AI) became possible. A technique to calculate the AI potentials is well elaborated and the corresponding data for

Table 1.1 Parameters ε/k_B and d for LJ and AI potentials.

	ε/k_B (K)		d (nm)	
Gas	*a*	*b*	*a*	*b*
He	10.22	10.997898	0.2576	0.2640950
Ne	35.7	41.152521	0.2789	0.27612487
Ar	124.0	143.123	0.3418	0.335741
Kr	190	193	0.361	0.363
Xe	229		0.4055	

a) LJ, extracted from viscosity, Ref. [3].
b) AI, Refs [6–9].

Table 1.2 Parameters of *ab initio* potential given by (1.18).

Param.	Unit	Value	
		Ne, Ref. [7]	Ar, Ref. [8]
A	K	$4.02915058383 \times 10^{7}$	4.61330146×10^{7}
a_1	$(\text{nm})^{-1}$	$-4.28654039586 \times 10^{1}$	$-2.98337630 \times 10^{1}$
a_2	$(\text{nm})^{-2}$	-3.33818674327	-9.71208881
a_3	nm	$-5.34644860719 \times 10^{-2}$	$2.75206827 \times 10^{-2}$
a_4	$(\text{nm})^{2}$	$5.01774999419 \times 10^{-3}$	$-1.01489050 \times 10^{-2}$
b	$(\text{nm})^{-1}$	$4.92438731676 \times 10^{1}$	4.02517211×10^{1}
C_6	K (nm)^{6}	$4.40676750157 \times 10^{-2}$	$4.42812017 \times 10^{-1}$
C_8	K (nm)^{8}	$1.64892507701 \times 10^{-3}$	$3.26707684 \times 10^{-2}$
C_{10}	K (nm)^{10}	$7.90473640524 \times 10^{-5}$	$2.45656537 \times 10^{-3}$
C_{12}	K (nm)^{12}	$4.85489170103 \times 10^{-6}$	$1.88246247 \times 10^{-4}$
C_{14}	K (nm)^{14}	$3.82012334054 \times 10^{-7}$	$1.47012192 \times 10^{-5}$
C_{16}	K (nm)^{16}	$3.85106552963 \times 10^{-8}$	$1.17006343 \times 10^{-6}$

many gases and mixtures can be found in numerous works, see for example, Refs [6–14]. The main advantage of the AI potentials is the absence of any parameters to be extracted from experimental data. Numerical results on the AI potential usually are given in terms of interpolating formulas. One of them has a form: [7, 8]

$$\frac{U(r)}{k_B} = A \exp\left[a_1 r + a_2 r^2 + \frac{a_3}{r} + \frac{a_4}{r^2}\right] - \sum_{n=3}^{8} \frac{C_{2n}}{r^{2n}}\left(1 - e^{-br}\sum_{k=0}^{2n}\frac{(br)^k}{k!}\right). \tag{1.18}$$

The numerical values of the interpolating coefficients for neon and argon reported in Refs [7, 8] are reproduced in Table 1.2. The expressions of the potential $U(r)$ for helium and krypton have slightly different form and can be found in Refs [6] and [9], respectively.

The well depth ε for the AI potentials and the distance corresponding to the zero point

$$U(d) = 0, \tag{1.19}$$

are compared with those for the LJ potential extracted from viscosity in Table 1.1. As is expected, the values obtained by two quite different methodologies are close to each other. Both LJ and AI potentials are plotted on Figure 1.1, which shows that they are only slightly different, but qualitatively they are quite similar to each other. Thus, the LJ potential with the parameters ϵ and d extracted from viscosity provides very reliable description of the intermolecular interaction.

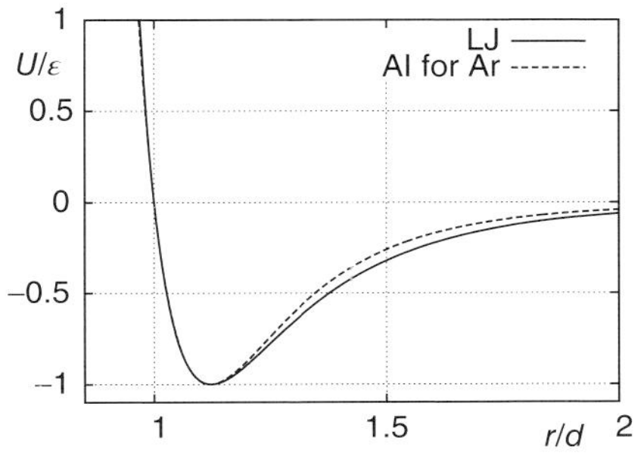

Figure 1.1 Potential U versus intermolecular distance r.

1.5 Deflection Angle

Once the intermolecular potential is known, the binary collision problem can be solved. Let us consider two particles of the same mass m moving with velocities $\boldsymbol{v}$ and $\boldsymbol{v}_*$ before collisions and changing their velocities to $\boldsymbol{v}'$ and $\boldsymbol{v}'_*$ after the collision. The relative velocities before and after the collision are denoted as

$$\boldsymbol{g}_r = \boldsymbol{v} - \boldsymbol{v}_*, \boldsymbol{g}'_r = \boldsymbol{v}' - \boldsymbol{v}'_*, \tag{1.20}$$

respectively. If the collision is elastic, the vector of their relative velocity changes only its direction, that is, $|\boldsymbol{g}'_r| = |\boldsymbol{g}_r|$. The angle χ between the vectors $\boldsymbol{g}_r$ and $\boldsymbol{g}'_r$ is called the deflection angle. Thus, when this angle is known, the post-collision velocities $\boldsymbol{v}'$ and $\boldsymbol{v}'_*$ are easily related to the pre-collisions velocities $\boldsymbol{v}$ and $\boldsymbol{v}_*$.

To calculate the angle χ, a binary collision is considered in the reference frame related to one of the particles, while the other particle moves with a velocity $\boldsymbol{g}_r$ as is shown in Figure 1.2. From this scheme, one can easily calculate the deflection angle χ for the HS potential determined by the impact parameter b as

$$\chi = 2\arccos(b/d). \tag{1.21}$$

For an arbitrary potential $U(r)$, the deflection angle is determined not only by the parameter b but also by the kinetic energy of the relative motion related to the relative motion speed g_r as

$$E_r = \frac{mg_r^2}{4}. \tag{1.22}$$

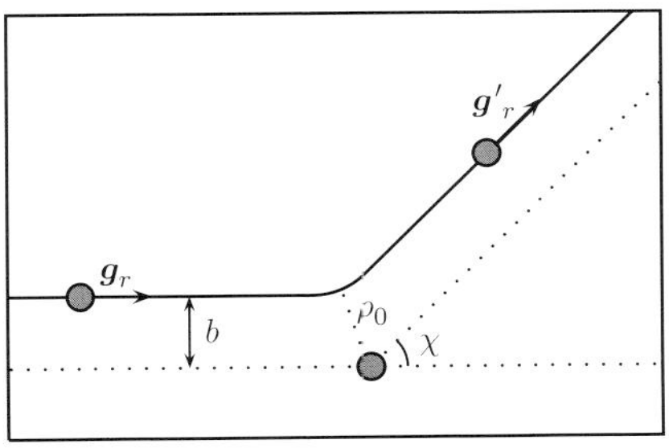

Figure 1.2 Scheme of binary collision.

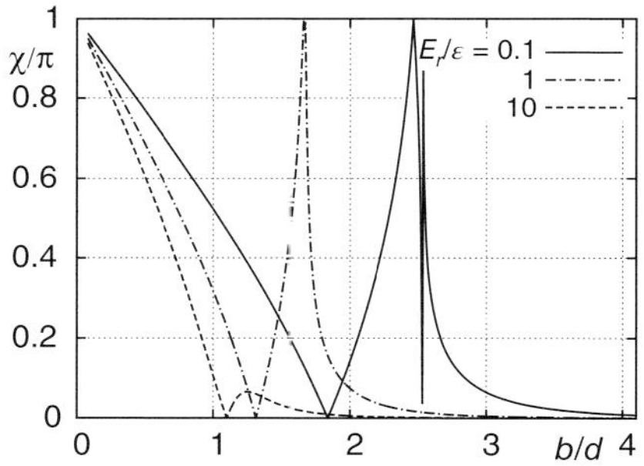

Figure 1.3 Deflection angle χ versus impact parameter b for LJ potential.

The calculation of χ can be performed numerically following a procedure described in many textbooks on classical mechanics, see for example, Ref. [15], and those on kinetic theory [3–5, 16–20]. Some details on the numerical scheme is given in Ref. [21].

In case of the LJ potential, the deflection angle χ is usually calculated in terms of the dimensionless quantities E_r/ε and b/d because the function $\chi(E_r/\varepsilon, b/d)$ once calculated can be used for the LJ potential with any numerical values of the parameters d and ϵ. It is not possible to do the same with the AI potential, which requires calculations of the deflection angle χ for each combination of E_r and b. Typical plots of χ on b/d based on the data reported in Ref. [21] are depicted on Figure 1.3 for three values of the energy ratio $E_r/\varepsilon = 0.1$, 1, and 10. In contrast to the hard spheres (1.21), the dependence of χ on b/d for the LJ potential is not monotone, but it reaches zero, then it increases and then vanishes at $\beta/d \to \infty$.

1.6 Differential Cross Section

The kinetic equation will be given on terms of the so-called differential cross section (DCS) $\sigma_d(E_r, \chi)$. If we imagine a beam of particles uniformly distributed in the space and moving with the same velocity equal to g_r relative to a fixed target particle, then the number of the beam particles scattered into the angle interval $[\chi, \chi + \mathrm{d}\chi]$ will be proportional to $\sigma_d(E_r, \chi) \sin\chi \,\mathrm{d}\chi$. On the other hand, for a given energy E_r, the deflection angle is determined only by the impact parameter b. Since the particles are distributed uniformly, the number of particles having the impact parameter in the interval $[b, b + \mathrm{d}b]$ is proportional to b so that the DCS can be defined as

$$\sigma_d(E_r, \chi) \sin\chi \,\mathrm{d}\chi = b \,\mathrm{d}b. \tag{1.23}$$

If the parameter b is expressed in terms of χ, the DCS can be calculated from (1.23). For the HS potential, the function $b = d\cos(\chi/2)$ is obtained from (1.21) and the DCS takes the form

$$\sigma_d(E_r, \chi) = \frac{b}{\sin\chi} \frac{\mathrm{d}b}{\mathrm{d}\chi} = \frac{d^2}{4} \quad \text{for} \quad \text{HS}. \tag{1.24}$$

For a potential such as LJ, several values of b can correspond to the same value of χ, see Figure 1.3, so that all these values must be included in the expressions of the DCS, that is, more generally it is defined as

$$\sigma_d(E_r, \chi) := \frac{1}{\sin\chi} \sum_i b_i \left| \frac{\partial b_i}{\partial \chi} \right|, \tag{1.25}$$

where b_i are all values of the impact parameter corresponding to the deflection angle χ. A detailed information about the DCS and its calculation for the LJ potential is given in the paper [21].

1.7 Total Cross Section

The total cross section (TCS) σ_t is defined via the integration of the DCS over the whole interval of the deflection angle variation

$$\sigma_t := 2\pi \int_0^{\pi} \sigma_d(E_r, \chi) \sin\chi \, d\chi. \tag{1.26}$$

Substituting (1.24) into (1.26), the TCS for the hard spheres is obtained as the area of circle with a radius d

$$\sigma_t = \pi d^2 \quad \text{for} \quad \text{HS}. \tag{1.27}$$

The TCS also can be treated as the area of the circle around the target particle containing only those beam particles that undergo a collision.

If a potential is not zero at any distance, the TCS is infinite because all beam particles undergo collision with the target particle. In order to avoid such a nonsense result, an angle cut-off should be done, that is, we assume that a deflection within the small interval $[0, \chi_0]$ ($\chi_0 \ll 1$) is not considered as a collision, then the TCS becomes a finite value given as

$$\sigma_t(E_r) = 2\pi \int_{\chi_0}^{\pi} \sigma_d(E_r, \chi) \sin\chi \, d\chi, \tag{1.28}$$

which depends on the energy E_r of the relative motion.

Using the definition (1.23), the TCS can be calculated by integrating with respect to the impact parameter b. Again, to avoid the infinite TCS, the potential should be cut-off. Let us assume that $U(r) = 0$ if $r > b_M$, that is, a collision happens only if the impact parameter is smaller than b_M, then the TCS is given as

$$\sigma_t = 2\pi \int_0^{b_M} b \, db = \pi b_M^2 \tag{1.29}$$

that is the area of circle of a radius b_M. In this case, the TCS is independent of the collision energy E_r.

1.8 Equivalent Free Path

The concept of mean-free-path (MFP) is quite important in rarefied gas dynamics. It is defined as a mean distance traveled by a particle between two successive collisions and inversely proportional to the number density n and TCS σ_t so that

$$\text{MFP} \propto (n\sigma_t)^{-1}. \tag{1.30}$$

A more exact calculation of the MFP made for a gas being in equilibrium leads to the expression

$$\text{MFP} = \left(\sqrt{2}n\sigma_t\right)^{-1}. \tag{1.31}$$

The derivation of this formula can be found, for example, in Section 2.4 of the book [4] or in Section 4.3 of the book [17].

According to Eq. (1.27), the TCS of hard sphere is determined by their diameter d so that the MFP is a quite definite quantity for this potential. However, for a potential without a cut-off, the TCS is infinite and hence the MFP calculated by (1.31) is zero. Evidently, if a potential is nonzero at any distance theoretically, all particles are always colliding with each other. In other words, they never move freely. If a potential is cut-off at some distance b_M, then according to Eqs. (1.29) and (1.31), the MFP will be inversely proportional to b_M^2. On the other hand, macroscopic properties of gas should not depend on the cut-off distance b_M, but they should converge by increasing b_M. Thus, the use of the MFP concept to describe nonequilibrium phenomena in gases is not appropriate, but some equivalent free path (EFP) independent of the cut-off should be used.

The gas viscosity μ when calculated for the LJ potential or for some similar one converges to a fixed value by increasing the cut-off distance b_M. Therefore, the EFP defined via the viscosity coefficient μ as

$$\ell := \frac{\mu v_m}{p} \tag{1.32}$$

is not affected by the cut-off distance and is suitable to treat transport phenomena calculated for different intermolecular potentials. Here, v_m is the most probable molecular speed defined in Section 2.4.

1.9 Rarefaction Parameter and Knudsen Number

In order to choose an appropriate numerical or analytical method to calculate rarefied gas flows, the rarefaction regime should be checked. Let us denote a characteristic size of gas flow as a, then the regime of flow is determined by the ratio

$$\text{Kn} := \frac{\text{MFP}}{a}, \tag{1.33}$$

which is called Knudsen number. However, it is very usual to use the rarefaction parameter to characterize the gas rarefaction defined via the EFP as

$$\delta := \frac{a}{\ell} = \frac{ap}{\mu v_m}. \tag{1.34}$$

The rarefaction parameter is inversely proportional to the Knudsen number because EFP is proportional to MFP. Since most of theoretical papers report the results in terms of the rarefaction parameter, it will be used in this book to characterize the gas rarefaction.

It is possible to distinguish several regimes of gas flows with respect to the rarefaction parameter. When $\delta \to 0$, the EFP becomes significantly larger than the characteristic size, $\ell \gg a$, and molecules move without any collision between them. This regime is called free-molecular or collisionless. It is relatively easy for numerical or analytical calculations. The book by Saksaganskii [22] contains a lot of examples of solutions for many problems of rarefied gases in this regime. Few examples of free-molecular flows will be considered in Chapters 11–16.

In the opposite limit $\delta \to \infty$, the EFP is quite smaller than a characteristic size of gas flow, $\ell \ll a$. Under this condition, the continuous mechanics equations are applicable that is why this regime is usually called continuous or hydrodynamic. There are numerous books dedicated to this regime, see for example, Refs [23, 24], so that this regime will be described very briefly in this book. Several examples of hydrodynamic flows given in Chapters 11–16 demonstrate the use of the so-called velocity slip and temperature jump boundary conditions, which extend an application of the hydrodynamic equations to lower values of the rarefaction δ.

Our main interest will be the so-called transitional regime when the rarefaction parameter is not small to neglect the intermolecular collisions, but it is not so large in order to apply the hydrodynamic equations, that is, when $\delta \sim 1$.

Exercises

1.1 Verify that Eq. (1.9) is equivalent to Eq. (1.8).
Hint: Use Eqs. (1.3), (1.5), and (1.10).

1.2 Calculate the mass of one atom of helium and xenon.
Solution: Using (1.2), we obtain $m = \mathcal{M}/N_A$. Then $m = 6.646 \times 10^{-27}$ kg for helium and $m = 2.180 \times 10^{-25}$ kg for xenon.

1.3 Estimate the EFP ℓ of helium ($v_m = 1.1 \times 10^3$ m/s, $\mu = 19$ μPa s) and xenon ($v_m = 1.9 \times 10^2$ m/s, $\mu = 21$ μPa s) under standard conditions, that is, $T = 273.15$ K and $p = 1$ atm (see Table A.1).
Solution: Substituting the given data into Eq. (1.32), we obtain $\ell = 0.21$ μm for helium and $\ell = 0.039$ μm for xenon.
Comments: Two different species of gas can have quite different EFPs under the same conditions. If one deals with a macroscopic size of flow of about few centimeters, the mechanics of continuous medium is applicable, but for a microflow, when its size is about few microns, the flow is transitional.

1.4 Estimate the EFP ℓ for helium under typical conditions of rough vacuum $p = 1$ Pa and high vacuum $p = 10^{-7}$ Pa. Assume $T = 273.15$ K.
Solution: Substituting the given data into Eq. (1.32), we obtain $\ell = 2.1$ cm at $p = 1$ Pa; $\ell = 2.1 \times 10^5$ m at $p = 10^{-7}$ Pa.
Comments: Assuming a typical size in vacuum systems to be about few centimeters, it can be said that the flow regime is transitional for rough vacuum and it is free molecular for high vacuum.

1.5 Calculate the Loschmidt constant, that is, the number of particles per cubic meter under standard conditions, using the state equation (1.9). Compare it with the value given in Table A.1.
Solution: $N_L = 2.69 \times 10^{25}$ m^{-3}.
Comments: The number is huge so that the statistical approach is justified.

1.6 Calculate the number of particles contained in 1 m^3 under the ultrahigh vacuum condition $p = 10^{-10}$ Pa at $T = 300$ K.
Solution: $N = 2.4 \times 10^{10}$.
Comments: Even for the high vacuum conditions, the number of particles is still huge.

2
Velocity Distribution Function

2.1
Definition of Distribution Function

As has been shown in Exercise 1.6, even under the ultrahigh vacuum conditions, there are a huge number of particles in a cubic meter of gas so that it is possible to use a statistical approach in order to describe a state of such a gas. A state of each particle is determined by its position $\boldsymbol{r}$ and its velocity $\boldsymbol{v}$. For polyatomic molecules, other variables are necessary to describe their state, but in this book we will ignore internal degree of freedom. Does it restrict the theory only by monatomic gases? As will be shown later, many phenomena are not affected by the internal molecular structure. When it is possible, an estimation of the influence of this structure on phenomena will be done. The kinetic theory of polyatomic gases is well described in the book by McCourt *et al.* [25] where the reader can find many details how to take into account the internal structure of gaseous molecules.

Thus, describing a particle motion, we deal with two 3D spaces: physical space of particle positions $\boldsymbol{r}$ and velocity space $\boldsymbol{v}$. Formally, we may consider the 6D space of position and velocity $(\boldsymbol{r}, \boldsymbol{v})$. Let us denote as $\mathrm{d}^6 N$ the number of particles expected in the elementary 6D volume $\mathrm{d}^3\boldsymbol{r}\,\mathrm{d}^3\boldsymbol{v}$ being near the point $(\boldsymbol{r}, \boldsymbol{v})$ at a time moment t. The velocity distribution function (VDF) $f(t, \boldsymbol{r}, \boldsymbol{v})$ is defined as

$$f(t, \boldsymbol{r}, \boldsymbol{v}) := \frac{\mathrm{d}^6 N}{\mathrm{d}^3\boldsymbol{r}\,\mathrm{d}^3\boldsymbol{v}}. \tag{2.1}$$

Apart from the assumption on the huge number of particles, the above given definition requires one more assumption, namely, the absence of correlation between moving particles. In other words, the particles are in a free motion independent of each other during practically all time. A particle undergoes an interaction with another one only during a very short time. Such a condition can be fulfilled if an average distance between particles is much larger than their size. Under the standard conditions, the average distance can be estimated via the Loschmidt constant N_L (see Table A.1) as $N_L^{-1/3} \approx 30$ nm, while the molecular size is two orders smaller, namely, 0.4 nm according to Table 1.1. Thus, the description based on the VDF defined by Eq. (2.1) works well for the atmospheric pressure and any lower pressure. This question is discussed more rigorously in Section 3.2 of the book [4].

Rarefied Gas Dynamics: Fundamentals for Research and Practice, First Edition. Felix Sharipov.

This VDF contains all information about a gas flow. For instance, the integration of the VDF over the whole velocity space

$$\int f(t, \boldsymbol{r}, \boldsymbol{v}) \, \mathrm{d}^3\boldsymbol{v} = \frac{\mathrm{d}^3 N}{\mathrm{d}^3 \boldsymbol{r}} \tag{2.2}$$

gives the ratio of the total number of particles d^3N in the elementary volume $\mathrm{d}\mathcal{V} = \mathrm{d}^3\boldsymbol{r}$ of the physical space, which represents the number density according to its definition (1.5). Here, the integration over the whole velocity domain means

$$\int [*] \, \mathrm{d}^3\boldsymbol{v} = \int_{-\infty}^{\infty} \int_{-\infty}^{\infty} \int_{-\infty}^{\infty} [*] \, \mathrm{d}v_1 \, \mathrm{d}v_2 \, \mathrm{d}v_3. \tag{2.3}$$

Saying in the mathematical terminology, the VDF is normalized by the number density

$$\int f(t, \boldsymbol{r}, \boldsymbol{v}) \, \mathrm{d}^3\boldsymbol{v} = n(t, \boldsymbol{r}). \tag{2.4}$$

If one integrates the VDF only with respect to two velocity components, for example, v_2 and v_3

$$f_1(t, \boldsymbol{r}, v_1) := \int_{-\infty}^{\infty} \int_{-\infty}^{\infty} f(t, \boldsymbol{r}, \boldsymbol{v}) \, \mathrm{d}v_2 \, \mathrm{d}v_3, \tag{2.5}$$

one obtains the distribution of the first velocity component, that is, the quantity $f(t, \boldsymbol{r}, v_1) \, \mathrm{d}^3\boldsymbol{r} \, \mathrm{d}v_1$ means the number of particles in the 4D volume $\mathrm{d}^3\boldsymbol{r} \, \mathrm{d}v_1$. With the help of Eqs. (2.4) and (2.5), it is verified that

$$\int_{-\infty}^{\infty} f_1(t, \boldsymbol{r}, v_1) \, \mathrm{d}v_1 = n(t, \boldsymbol{r}). \tag{2.6}$$

The velocity vector $\boldsymbol{v}$ can be expressed in the spherical coordinates as

$$v_1 = v \cos \vartheta, \quad v_2 = v \sin \theta \cos \varphi, \quad v_3 = v \sin \theta \sin \varphi, \tag{2.7}$$

where $0 \leq \theta \leq \pi$ and $0 \leq \varphi \leq 2\pi$ are the angles determining the velocity direction, v is the molecular speed varying from zero to infinity, $0 \leq v < \infty$. Then the elementary volume in the velocity space is given as

$$\mathrm{d}^3\boldsymbol{v} = v^2 \, \mathrm{d}v \sin \theta \, \mathrm{d}\theta \, \mathrm{d}\varphi. \tag{2.8}$$

An integration of the VDF with respect to the angles θ and φ gives us the speed distribution function

$$f_v(t, \boldsymbol{r}, v) := v^2 \int_0^{2\pi} \int_0^{\pi} f(t, \boldsymbol{r}, \boldsymbol{v}) \, \sin \theta \, \mathrm{d}\theta \, \mathrm{d}\varphi \tag{2.9}$$

The quantity $f_v(t, \boldsymbol{r}, v) \, \mathrm{d}^3\boldsymbol{r} \, \mathrm{d}v$ is equal to the number of particles being in the volume $\mathrm{d}^3\boldsymbol{r}$ and having the speed in the range $[v, v + \mathrm{d}v]$. Using Eqs. (2.4) and (2.9), one verifies

$$\int_0^{\infty} f_v(t, \boldsymbol{r}, v) \, \mathrm{d}v = n(t, \boldsymbol{r}). \tag{2.10}$$

2.2 Moments of Distribution Function

Let us introduce the notation for the average of quantity $\psi(\boldsymbol{v})$ per volume unity as

$$\langle\psi\rangle := \int \psi(\boldsymbol{v}) f(t, \boldsymbol{r}, \boldsymbol{v})\, \mathrm{d}^3\boldsymbol{v}. \tag{2.11}$$

Substituting $\psi = 1$ into this definition, the number of particles per volume unity (or number density) is calculated

$$n = \langle 1\rangle. \tag{2.12}$$

Substituting $\psi = m$ into (2.11), the average mass per volume unity (or mass density) is obtained

$$\rho = \langle m\rangle. \tag{2.13}$$

If the average per particle and per mass unity are needed, the expression (2.11) is divided by the number density n and mass density ρ, that is,

$$\text{average of } \psi \text{ per particle:} = \frac{1}{n}\langle\psi\rangle \tag{2.14}$$

and

$$\text{average of } \psi \text{ per mass unity:} = \frac{1}{\rho}\langle\psi\rangle, \tag{2.15}$$

respectively. When the function $\psi(\boldsymbol{v})$ is a polynomial with respect to the molecular velocity $\boldsymbol{v}$, the integral (2.11) is called the moment of the VDF. For instance, the number density n calculated by Eq. (2.4) with $\psi = 1$ represents the moment of the zero order.

The bulk velocity of gas is calculated as the average of $\psi = \boldsymbol{v}$ per particle

$$\boldsymbol{u}(t, \boldsymbol{r}) := \frac{1}{n}\langle\boldsymbol{v}\rangle, \tag{2.16}$$

so that the bulk velocity is the polynomial of the first order. In hydrodynamics, one deals with exactly this velocity $\boldsymbol{u}$, that is why it is also called hydrodynamic velocity.

A net flux of some property $\psi(\boldsymbol{v})$ can be defined as the average of the expression $\boldsymbol{v}\psi(\boldsymbol{v})$

$$\boldsymbol{J}_\psi := \langle\boldsymbol{v}\psi\rangle = \int \boldsymbol{v}\psi(\boldsymbol{v}) f(t, \boldsymbol{r}, \boldsymbol{v})\, \mathrm{d}^3\boldsymbol{v}. \tag{2.17}$$

For instance, if the functions $\psi = 1$ and $\psi = m$ are substituted into (2.17), then the corresponding fluxes are given as

$$\boldsymbol{J}_N(t, \boldsymbol{r}) = \langle\boldsymbol{v}\rangle = n\boldsymbol{u} \tag{2.18}$$

and

$$\boldsymbol{J}_m(t, \boldsymbol{r}) = \langle m\boldsymbol{v}\rangle = \rho\boldsymbol{u} \tag{2.19}$$

representing the flow vectors of particle and mass, respectively. In the case of kinetic energy, $\psi = mv^2/2$, the corresponding flux reads

$$\boldsymbol{J}_e(t, \boldsymbol{r}) = \left\langle \boldsymbol{v} \frac{mv^2}{2} \right\rangle \tag{2.20}$$

representing the energy flow vector , which is the moment of the third order.

A substitution of a vector property ψ into Eq. (2.17) transforms the flux into a tensor of rank two, for example, if $\psi = m\boldsymbol{v}$, then we have the following tensor:

$$\mathsf{M}(t, \boldsymbol{r}) = \langle m\boldsymbol{v}\boldsymbol{v} \rangle \tag{2.21}$$

representing the momentum flux tensor M. Note that the expression $\boldsymbol{v}\boldsymbol{v}$ is not a product of two vectors, but it means the tensor (see (A.7)). Each element of the tensor M is the second-order moment of the VDF.

The average properties (2.11) and their fluxes Eq. (2.17) are defined in the standard reference frame, but it is usual to calculate averages and fluxes in the frame associated with the gas velocity $\boldsymbol{u}$. For this purpose, the so-called peculiar velocity is defined as

$$\boldsymbol{V} := \boldsymbol{v} - \boldsymbol{u}. \tag{2.22}$$

It characterizes the molecular motion only due to the thermal agitation disregarding the motion of gas as whole so that the average peculiar velocity is zero

$$\langle \boldsymbol{V} \rangle = 0. \tag{2.23}$$

The specific internal energy introduced by (1.13) can be calculated by averaging the kinetic energy of the thermal agitation per mass unity

$$e(t, \boldsymbol{r}) = \frac{1}{\rho} \left\langle \frac{mV^2}{2} \right\rangle. \tag{2.24}$$

The average of the full kinetic energy can be expressed via the internal specific energy as

$$\left\langle \frac{mv^2}{2} \right\rangle = \left\langle \frac{m(\boldsymbol{V} + \boldsymbol{u})^2}{2} \right\rangle = \rho \left(e + \frac{u^2}{2} \right), \tag{2.25}$$

where (2.13) and (2.23) have been taken into account. Physically, Eq. (2.25) means that the full energy of gas is composed of its internal energy ρe and the kinetic energy of its motion as whole $\rho u^2/2$.

Combining the relation (2.24) with Eq. (1.14), the kinetic definition of the temperature can be formulated as

$$T(t, \boldsymbol{r}) := \frac{m}{3nk_B} \langle V^2 \rangle. \tag{2.26}$$

Thus, the temperature represents the second-order moment of the VDF with respect to the peculiar velocity $\boldsymbol{V}$.

The flux $\boldsymbol{J}'$ of some property $\psi(\boldsymbol{V})$ due to the thermal agitation is calculated via the VDF as

$$\boldsymbol{J}'_{\psi}(t,\boldsymbol{r}) := \langle \boldsymbol{V}\psi(\boldsymbol{V})\rangle = \int \boldsymbol{V}\psi(\boldsymbol{V})f(t,\boldsymbol{r},\boldsymbol{v})\,\mathrm{d}^3\boldsymbol{v}. \tag{2.27}$$

In case of the momentum $\psi = m\boldsymbol{V}$, the flux $\boldsymbol{J}'$ becomes a tensor P of rank two defined via the tensor $\boldsymbol{V}\boldsymbol{V}$ as

$$\mathsf{P}(t,\boldsymbol{r}) := \langle m\boldsymbol{V}\boldsymbol{V}\rangle, \tag{2.28}$$

which is called the pressure tensor. From the microscopic viewpoint, its physical meaning is a flux of the i-component of momentum in the j-direction. In the macroscopic level, this flux represents a force of gas acting in the direction i on a unit segment of area with a normal vector directed along the axis j. If $i = j$, then P_{ii} represents a normal pressure along the i-axis. The local pressure p is defined as an average over all three directions

$$p(t,\boldsymbol{r}) := \frac{1}{3}(P_{11} + P_{22} + P_{33}) = \frac{1}{3}\langle mV^2\rangle. \tag{2.29}$$

In general case, the diagonal components are different, that is, the normal pressure depends on the orientation of area segment. If $i \neq j$, P_{ij} represents a shear stress, or a tangential force acting on a unit segment of area. According to the definition (2.28), the pressure tensor is symmetric $P_{ij} = P_{ji}$. Similar to the temperature, the pressure tensor is the second-order moment of the VDF. The pressure tensor P and momentum flux tensor M are related to each other as

$$\mathsf{P} = \mathsf{M} - \rho\boldsymbol{u}\boldsymbol{u}, \tag{2.30}$$

where (2.22) and (2.23) have been used. The pressure tensor should be not confused with the so-called viscous stress tensor $\hat{\sigma}$ used frequently in hydrodynamics. It is related to the pressure tensor as

$$\hat{\sigma} = \mathsf{I}p - \mathsf{P}, \tag{2.31}$$

where I is the unit tensor (see Eq. (A.9)). This tensor is traceless because of Eq. (2.29).

Combining Eqs. (2.4), (2.26), and (2.29), the state equation (1.9) is derived. Thus, if the temperature is defined by Eq. (2.26) and the pressure is given by Eq. (2.29), the state equation (1.9) is valid for a dilute gas even if it is out of equilibrium.

A substitution of the kinetic energy $\psi = \frac{1}{2}mV^2$ of the thermal agitation into (2.27) leads to the definition of heat flow vector

$$\boldsymbol{q}(t,\boldsymbol{r}) = \left\langle \frac{mV^2}{2}\boldsymbol{V}\right\rangle, \tag{2.32}$$

which is the third-order moment of the VDF. It is very important to distinguish this quantity from the energy flow vector $\boldsymbol{J}_e$ given by (2.20). The heat flow vector is associated with an energy transfer only due to the thermal agitation, while the flow $\boldsymbol{J}_e$ represents the total energy transfer including the energy convection. Using

Eqs. (2.16), (2.22)–(2.24), and (2.28), the energy flow vector can be written down as

$$\boldsymbol{J}_e = \boldsymbol{q} + \rho \boldsymbol{u} \left(\frac{1}{2} u^2 + e \right) + \boldsymbol{u} \cdot \mathbf{P}, \tag{2.33}$$

where the last term represents the simple product, see Eq. (A.5), between the vector $\boldsymbol{u}$ and the tensor $\mathbf{P}$.

2.3 Entropy and Its Flow Vector

The function ψ could be not necessarily a polynomial of the velocity $\boldsymbol{v}$. For instance, if the function

$$\psi(t, \boldsymbol{r}, \boldsymbol{v}) = k_B(1 - \ln f) \tag{2.34}$$

is substituted into (2.11), then the entropy density, or entropy per volume unity, is obtained

$$s(t, \boldsymbol{r}) := k_B \langle (1 - \ln f) \rangle. \tag{2.35}$$

The entropy flow vector is defined by substituting the same function (2.34) into (2.17)

$$\boldsymbol{J}_s(t, \boldsymbol{r}) = k_B \langle \boldsymbol{v}(1 - \ln f) \rangle. \tag{2.36}$$

2.4 Global Maxwellian

Let us imagine a gas confined in some reservoir maintained at a temperature T and being in a complete (or global) equilibrium, that is, no macroscopic motion of one portion of the gas relatively another, no heat exchange between different parts of the gas and between the gas and reservoir, no chemical reactions occur. In this state, the VDF is given by the global Maxwellian

$$f^{M}(\boldsymbol{v}) = n \left(\frac{m}{2\pi k_B T} \right)^{3/2} \exp\left(-\frac{mv^2}{2k_B T} \right), \tag{2.37}$$

where the number density n and temperature T are constant over the whole reservoir. In some situations, for example, strong gravitational field or system as whole being in rotation, the number density can vary, but such situations are not considered here.

The Maxwellian distribution (2.37) is a rigorous result obtained in the frame of statistical physics from the basic principles. However, most of the textbooks on statistical physics present the equilibrium distribution in terms of the molecular energy and nominate it as the Maxwell–Boltzmann distribution (see example, Section 11.B [26], §37 in Ref. [27], Section 6.3 in Ref. [28]). A detailed derivation of Eq. (2.37) from the Maxwell–Boltzmann distribution can be found, for example, in Section 12-2 of the textbook [1] or in Section 6.5 of Ref. [29].

There are two mechanisms to reach the global equilibrium, namely, intermolecular collisions and gas–surface interaction. Thus, a system can reach the global equilibrium at any gas rarefaction even in the collisionless regime when the second mechanism becomes unique to establish the equilibrium.

The distribution of one velocity component, for example, v_1, is calculated by the definition (2.5) as

$$f_1^{\mathrm{M}}(v_1) = \int_{-\infty}^{\infty}\int_{-\infty}^{\infty} f^{\mathrm{M}}(\boldsymbol{v})\,\mathrm{d}v_2\,\mathrm{d}v_3 = n\left(\frac{m}{2\pi k_{\mathrm{B}}T}\right)^{1/2}\exp\left(-\frac{mv_1^2}{2k_{\mathrm{B}}T}\right), \tag{2.38}$$

that is a typical Gaussian distribution with the maximum at $v_1 = 0$.

The equilibrium speed distribution function $f_v(v)$ defined by Eq. (2.9) is obtained as

$$\begin{aligned} f_v^{\mathrm{M}}(v) &= v^2\int_0^{2\pi}\int_0^{\pi} f^{\mathrm{M}}(\boldsymbol{v})\sin\theta\,\mathrm{d}\theta\,\mathrm{d}\varphi \\ &= 4\pi n\left(\frac{m}{2\pi k_{\mathrm{B}}T}\right)^{3/2} v^2\exp\left(-\frac{mv^2}{2k_{\mathrm{B}}T}\right). \end{aligned} \tag{2.39}$$

It is easily verified that the function (2.39) reaches its maximum at

$$v_m = \sqrt{\frac{2k_{\mathrm{B}}T}{m}} \tag{2.40}$$

that is why this quantity is called the most probable speed (MPS). The functions (2.37)–(2.39) can be written down in terms of the MPS as

$$f^{\mathrm{M}}(\boldsymbol{v}) = \frac{n}{\left(\sqrt{\pi}v_m\right)^3}\exp\left(-\frac{v^2}{v_m^2}\right), \qquad \boldsymbol{v} = [v_1, v_2, v_3], \tag{2.41}$$

$$f_1^{\mathrm{M}}(v_1) = \frac{n}{\sqrt{\pi}v_m}\exp\left(-\frac{v_1^2}{v_m^2}\right), \qquad -\infty < v_1 < \infty, \tag{2.42}$$

$$f_v^{\mathrm{M}}(v) = \frac{4n}{\sqrt{\pi}v_m^3}v^2\exp\left(-\frac{v^2}{v_m^2}\right), \qquad 0 \le v < \infty, \tag{2.43}$$

respectively.

Using the function f_v^{M}, the mean speed $\langle v\rangle$ is calculated as

$$\langle v\rangle = \frac{1}{n}\int_0^{\infty} v f_v^{\mathrm{M}}(v)\,\mathrm{d}v = \frac{2}{\sqrt{\pi}}v_m = \sqrt{\frac{8k_{\mathrm{B}}T}{\pi m}}. \tag{2.44}$$

This quantity is also called thermal speed.

2.5
Local Maxwellian

Let us consider a gas with the density n, temperature T, and bulk velocity $\boldsymbol{u}$ being functions of the time t and coordinates $\boldsymbol{r}$

$$n = n(t, \boldsymbol{r}), \quad T = T(t, \boldsymbol{r}), \quad \boldsymbol{u} = \boldsymbol{u}(t, \boldsymbol{r}). \tag{2.45}$$

Formally, we may define the Maxwellian with these moments as

$$f^{\mathrm{M}}(t, \boldsymbol{r}, \boldsymbol{v}) = \frac{n}{\left(\sqrt{\pi} v_m\right)^3} \exp\left[-\frac{(\boldsymbol{v} - \boldsymbol{u})^2}{v_m^2}\right], \tag{2.46}$$

which is called local Maxwellian. In this case, the MPS defined by (2.40) depends on the time t and position $\boldsymbol{r}$ via the temperature $T(t, \boldsymbol{r})$. In a particular case when the density and temperature are constant and the bulk velocity is zero, the local Maxwellian (2.46) becomes the global one (2.37).

Substituting the local Maxwellian (2.46) into the definitions of the heat flux vector $\boldsymbol{q}$ (2.32) and pressure tensor $\mathbf{P}$ (2.28), it is easy to show that

$$\boldsymbol{q} = 0, \quad \mathbf{P} = p\mathbf{I}, \tag{2.47}$$

that is, no heat flux and shear stress for the VDF given by Eq. (2.46). The absence of the heat flux and shear stress in a gas with a variable temperature and bulk velocity seems a nonsense. In reality, it means that the VDF never takes the form of the local Maxwellian. According to the Chapman–Enskog method [3–5], we may only claim that the VDF tends to the local Maxwellian by increasing the rarefaction parameter δ (decreasing Kn). Nevertheless, the concept of the local Maxwellian is extremely useful because it represents a zero approximation in many problems.

Exercises

2.1 Check that the local Maxwellian (2.46) is normalized by the number density (2.4) and that its substitution into the definitions (2.16) and (2.26) leads to the identities.
Hint: Introduce the new variable $\boldsymbol{c} = (m/2k_{\mathrm{B}}T)^{1/2}(\boldsymbol{v} - \boldsymbol{u})$ and then use the integrals (A.18).

2.2 Obtain (2.33).
Hint: Use Eqs. (2.16), (2.20), (2.22)–(2.24), and (2.28).

2.3 Estimate the number N and fraction f_N of particles having the speed in the range $[Cv_m, \infty)$ for a gas being in equilibrium and occupying a volume of 1 m^3 at the standard conditions. Calculate this number for (a) $C = 2$, (b) $C = 4$, and (c) $C = 8$.
Solution: The number N is calculated via the speed distribution function (2.43)

$$N = \frac{N_L}{n} \int_{Cv_m}^{\infty} f_v^{\mathrm{M}}(v)\, \mathrm{d}v = N_L \left[\frac{2}{\sqrt{\pi}} C e^{-C^2} + \operatorname{erfc}(C)\right], \tag{2.48}$$

where N_L is the Loschmidt constant (see Table A.1) and the function erfc(C) is defined by (A.24). Using its asymptotic series (A.25), the expression (2.48) is simplified

$$N = \frac{N_L}{\sqrt{\pi}} e^{-C^2} \left(2C + \frac{1}{C} - \frac{1}{2C^3} \right). \tag{2.49}$$

The fraction is calculated as $f_N = N/N_L$. Substituting $C = 2$, 4, and 8 into (2.49), the following values of N and f_N are obtained: (a) $N = 1.2 \times 10^{24}$, $f_N = 0.045$; (b) $N = 1.4 \times 10^{19}, f_N = 5.2 \times 10^{-7}$; (c) $N = 4 \times 10^{-2}, f_N = 1.5 \times 10^{-27}$.

Comments: The number N decreases rapidly by increasing C and practically none of particles have the speed larger than $8v_m$. Even if the number of particles N having the speed larger than $4v_m$ is large, their fraction f_N is very small.

2.4 Obtain (2.47).

Solution: Substituting (2.46) into (2.32) and considering that $\mathrm{d}\boldsymbol{v} = \mathrm{d}\boldsymbol{V}$, we obtain

$$q_i = \frac{mn}{2\left(\sqrt{\pi} v_m\right)^3} \int V^2 V_i \exp\left(-\frac{V^2}{v_m^2}\right) \mathrm{d}\boldsymbol{V}. \tag{2.50}$$

The integrand is odd with respect to one velocity component V_i so that the integral is zero. Substituting (2.46) into (2.28), we obtain

$$P_{ij} = \frac{mn}{\left(\sqrt{\pi} v_m\right)^3} \int V_i V_j \exp\left(-\frac{V^2}{v_m^2}\right) \mathrm{d}\boldsymbol{V}. \tag{2.51}$$

In case $i \neq j$, the integral is zero. If $i = j$, the integral reads

$$\int V_i^2 \exp\left(-\frac{V^2}{v_m^2}\right) \mathrm{d}\boldsymbol{V} = \frac{\pi^{3/2}}{2} v_m^5. \tag{2.52}$$

Then with help of Eqs. (1.9) and (2.40), we obtain $P_{ii} = p$.

2.5 Obtain (2.40).

Solution: The derivative of (2.39) yields

$$\frac{\mathrm{d}f_v^{\mathrm{M}}}{\mathrm{d}v} = 4\pi n \left(\frac{m}{2\pi k_{\mathrm{B}} T}\right)^{3/2} \exp\left(-\frac{mv^2}{2k_{\mathrm{B}}T}\right) v \left(2 - v\frac{m}{k_{\mathrm{B}}T}\right). \tag{2.53}$$

Equating it to zero, Eq. (2.40) is obtained.

2.6 Calculate the mean speed $\langle v \rangle$ for helium and xenon at $T = 300$ K.

Solution: Using (2.44), we obtain $\langle v \rangle = 1260$ m/s and $\langle v \rangle = 220$ m/s for helium and xenon, respectively.

Comments: The heavier the gas, the smaller the speed.

3
Boltzmann Equation

3.1
Assumptions to Derive the Boltzmann Equation

The VDF in an equilibrium state of gas is given by the global Maxwellian (2.37). When a gas is out of equilibrium, an evolution of its VDF is determined by the Boltzmann equation (BE). Its derivation is well described in many textbooks, see for example, Refs [3–5, 18–20], so that we restrict ourselves by mentioning just the main assumptions adopted to derive the BE: (i)only binary collisions are taken into account disregarding collisions between three and more particles; and (ii) molecular chaos, that is, we deal with a statistical distribution of gaseous particles. In fact, both assumptions are fulfilled when we assumed that the time of intermolecular interaction is much smaller than their free motion time and when we consider a huge number of particles. Thus, we already have implicitly adopted these assumptions to define the VDF by Eq. (2.1).

The above-mentioned textbooks take into account an external force acting on particles during their free fly between two collisions. If the particles are electrically neutral, two types of the external forces exist: gravitational and inertial. The first force is negligible, while the second force arises in noninertial systems, which is out of the scope of this book, so that the BE will be given without considering external forces.

3.2
General Form of the Boltzmann Equation

Under the assumptions given earlier, the BE reads as

$$\frac{\partial f}{\partial t} + \boldsymbol{v} \cdot \nabla_r f = Q(f, \boldsymbol{v}), \tag{3.1}$$

where the gradient vector ∇_r is defined by (A.11) and the notation $\boldsymbol{v} \cdot \nabla_r$ is given by Eq. (A.12). The symbol $Q(f, \boldsymbol{v})$ denotes the integral describing binary intermolecular collisions. It is determined by the VDF f and it also depends on the molecular velocity $\boldsymbol{v}$ for a given f.

Rarefied Gas Dynamics: Fundamentals for Research and Practice, First Edition. Felix Sharipov.

To write down its explicit expression in a general form, let us define the function $w = w(\boldsymbol{v}', \boldsymbol{v}'_*; \boldsymbol{v}, \boldsymbol{v}_*)$ so that the quantity

$$w(\boldsymbol{v}', \boldsymbol{v}'_*; \boldsymbol{v}, \boldsymbol{v}_*) f(\boldsymbol{v}) f(\boldsymbol{v}_*)\, \mathrm{d}^3\boldsymbol{v}\, \mathrm{d}^3\boldsymbol{v}_*\, \mathrm{d}^3\boldsymbol{v}'\, \mathrm{d}^3\boldsymbol{v}'_* \tag{3.2}$$

would be the number of collisions per unit time and unit volume with pre-collision velocities $\boldsymbol{v}, \boldsymbol{v}_*$ and post-collision ones $\boldsymbol{v}', \boldsymbol{v}'_*$. This function is determined by the intermolecular potential (see Section 1.4), but some general properties can be obtained from fundamental principles. The first property is a consequence of the reversibility of intermolecular collisions and reads

$$w(\boldsymbol{v}', \boldsymbol{v}'_*; \boldsymbol{v}, \boldsymbol{v}_*) = w(-\boldsymbol{v}, -\boldsymbol{v}_*; -\boldsymbol{v}', -\boldsymbol{v}'_*). \tag{3.3}$$

In other words, if the velocities $\boldsymbol{v}'$ and $\boldsymbol{v}'_*$ of two particles after their collision are replaced by $-\boldsymbol{v}'$ and $-\boldsymbol{v}'_*$, the particles will repeat their trajectory in the reverse direction and will get their initial velocities with the opposite signs $-\boldsymbol{v}$ and $-\boldsymbol{v}_*$. The second property of the function w is its unitarity written as

$$\int w(\boldsymbol{v}', \boldsymbol{v}'_*; \boldsymbol{v}, \boldsymbol{v}_*)\, \mathrm{d}^3\boldsymbol{v}'\, \mathrm{d}^3\boldsymbol{v}'_* = \int w(\boldsymbol{v}, \boldsymbol{v}_*; \boldsymbol{v}', \boldsymbol{v}'_*)\, \mathrm{d}^3\boldsymbol{v}'\, \mathrm{d}^3\boldsymbol{v}'_*. \tag{3.4}$$

More detailed proofs of properties (3.3) and (3.4) of the function w can be found in the textbook [19].

Thus, in terms of the function w, the collision integral reads

$$Q(f, \boldsymbol{v}) = \iiint (f' f'_* - f f_*) w(\boldsymbol{v}', \boldsymbol{v}'_*; \boldsymbol{v}, \boldsymbol{v}_*)\, \mathrm{d}^3\boldsymbol{v}'\, \mathrm{d}^3\boldsymbol{v}'_*\, \mathrm{d}^3\boldsymbol{v}_*, \tag{3.5}$$

where the notations

$$f = f(\boldsymbol{v}), \quad f_* = f(\boldsymbol{v}_*), \quad f' = f(\boldsymbol{v}'), \quad f'_* = f(\boldsymbol{v}'_*) \tag{3.6}$$

have been used. To derive the expression (3.5), the properties (3.3) and (3.4) are used. The function w is related to the DCS defined by Eq. (1.23) as

$$\sigma \sin\chi\, \mathrm{d}\chi\, \mathrm{d}\epsilon = \frac{w(\boldsymbol{v}', \boldsymbol{v}'_*; \boldsymbol{v}, \boldsymbol{v}_*)}{g_r}\, \mathrm{d}^3\boldsymbol{v}'\, \mathrm{d}^3\boldsymbol{v}'_*, \tag{3.7}$$

where g_r is the relative speed

$$g_r = |\boldsymbol{v}_* - \boldsymbol{v}|, \tag{3.8}$$

while the angle ϵ is an azimuthal impact parameter determining the orientation of the collision plane. Using the definition (1.23), this relation takes the form

$$g_r b\, \mathrm{d}b\, \mathrm{d}\epsilon = w(\boldsymbol{v}', \boldsymbol{v}'_*; \boldsymbol{v}, \boldsymbol{v}_*)\, \mathrm{d}^3\boldsymbol{v}'\, \mathrm{d}^3\boldsymbol{v}'_*. \tag{3.9}$$

Then the collision integral can be written as

$$Q(f, \boldsymbol{v}) = \iiint (f' f'_* - f f_*) g_r b\, \mathrm{d}b\, \mathrm{d}\epsilon\, \mathrm{d}^3\boldsymbol{v}_*. \tag{3.10}$$

Usually, this form of the collision integral is given in the textbooks, see for example, Refs [3–5, 18]. In case of a cut-off potential, the collision integral can be presented by two terms as

$$Q(f, \boldsymbol{v}) = \Omega(f, \boldsymbol{v}) - \nu(f, \boldsymbol{v}) f(\boldsymbol{v}), \tag{3.11}$$

where

$$\Omega(f, \boldsymbol{v}) = \iiint f' f'_* g_r b \, \mathrm{d}b \, \mathrm{d}\epsilon \, \mathrm{d}^3\boldsymbol{v}_*. \tag{3.12}$$

$$\nu(f, \boldsymbol{v}) = \sigma_t \int f(\boldsymbol{v}_*) g_r \, \mathrm{d}^3\boldsymbol{v}_* \tag{3.13}$$

with the TCS σ_t defined by (1.29).

A variation of some molecular property $\psi(\boldsymbol{v})$ because of the intermolecular collisions is calculated as a moment of the collision integral

$$\begin{aligned}\left(\frac{\partial \psi}{\partial t}\right)_{\text{coll}} &= \int \psi(\boldsymbol{v}) Q \, \mathrm{d}^3\boldsymbol{v} \\ &= \iiiint \psi(\boldsymbol{v})(f'f'_* - ff_*) w \, \mathrm{d}^3\boldsymbol{v}' \, \mathrm{d}^3\boldsymbol{v}'_* \, \mathrm{d}^3\boldsymbol{v}_* \, \mathrm{d}^3\boldsymbol{v}. \end{aligned} \tag{3.14}$$

After some mathematical manipulations taking into account the properties (3.3) and (3.4), the variation of ψ is written as

$$\begin{aligned}\left(\frac{\partial \psi}{\partial t}\right)_{\text{coll}} &= \frac{1}{4} \iiiint (\psi + \psi_* - \psi' - \psi'_*) \\ &\quad \times (f'_* f'_* - ff_*) w \, \mathrm{d}^3\boldsymbol{v}' \, \mathrm{d}^3\boldsymbol{v}'_* \, \mathrm{d}^3\boldsymbol{v}_* \, \mathrm{d}^3\boldsymbol{v}, \end{aligned} \tag{3.15}$$

where the notation

$$\psi = \psi(\boldsymbol{v}), \ \psi_* = \psi(\boldsymbol{v}_*), \ \psi' = \psi(\boldsymbol{v}'), \ \psi'_* = \psi(\boldsymbol{v}'_*) \tag{3.16}$$

have been used.

If we multiply Eq. (3.1) by a function $\psi(\boldsymbol{v})$ and integrate it with respect to the molecular velocity $\boldsymbol{v}$, then the balance equation of this property is obtained as

$$\frac{\partial \langle \psi \rangle}{\partial t} + \nabla_r \cdot \boldsymbol{J}_\psi = \left(\frac{\partial \psi}{\partial t}\right)_{\text{coll}}, \tag{3.17}$$

where the notations (2.11), (2.17), and (A.13) are used.

3.3
Conservation Laws

If a property ψ satisfies the condition

$$\psi' + \psi'_* = \psi + \psi_*, \tag{3.18}$$

then it is called collision invariant. Due to the fact that the mass m, three components of momentum $m\boldsymbol{v}$, and kinetic energy $mv^2/2$ are conserved in each collision, the following five functions

$$\psi = m, \ \psi = m\boldsymbol{v}, \ \psi = \frac{1}{2}mv^2, \tag{3.19}$$

are collision invariants. In other words, the collision invariants (3.19) yield

$$\left(\frac{\partial\psi}{\partial t}\right)_{\text{coll}} = \int \psi(\boldsymbol{v})Q \, \mathrm{d}^3\boldsymbol{v} = 0. \tag{3.20}$$

Then the general balance equation (3.17) is reduced to

$$\frac{\partial\langle\psi\rangle}{\partial t} + \nabla_r \cdot \boldsymbol{J}_\psi = 0. \tag{3.21}$$

Let us specify this equation for each invariant. The mass conservation law ($\psi = m$) reads

$$\frac{\partial\rho}{\partial t} + \nabla_r \cdot \rho\boldsymbol{u} = 0, \tag{3.22}$$

where Eqs. (2.13) and (2.19) have been used. The momentum conservation law ($\psi = m\boldsymbol{v}$) with help of Eqs. (2.19) and (2.21) takes the form

$$\frac{\partial\rho\boldsymbol{u}}{\partial t} + \nabla_r \cdot \mathbf{M} = 0. \tag{3.23}$$

Finally, the energy conservation law ($\psi = mv^2/2$) is obtained as

$$\frac{\partial\rho(e + u^2/2)}{\partial t} + \nabla_r \cdot \boldsymbol{J}_e = 0, \tag{3.24}$$

where Eqs. (2.20) and (2.25) have been used.

The conservation laws (3.22)–(3.24) are usually written down in terms of the substantial derivative

$$\frac{\mathrm{d}}{\mathrm{d}t} = \frac{\partial}{\partial t} + \boldsymbol{u} \cdot \nabla_r, \tag{3.25}$$

the pressure tensor $\mathbf{P}$ and heat flow vector $\boldsymbol{q}$ as

$$\frac{1}{\rho}\frac{\mathrm{d}\rho}{\mathrm{d}t} + \nabla_r \cdot \boldsymbol{u} = 0, \tag{3.26}$$

$$\rho\frac{\mathrm{d}\boldsymbol{u}}{\mathrm{d}t} + \nabla_r \cdot \mathbf{P} = 0, \tag{3.27}$$

$$\rho\frac{\mathrm{d}e}{\mathrm{d}t} + \nabla_r \cdot \boldsymbol{q} + \mathbf{P} : \nabla_r\boldsymbol{u} = 0. \tag{3.28}$$

All details to obtain Eqs. (3.26)–(3.28) are given in Section 4.1 of the book [4]. Note that these equations have been obtained rigorously from the BE without any additional assumptions, that is, they are valid for any gas rarefaction and for any degree of nonequilibrium. However, they are not enough to calculate a gas flow even if they are completed by the state (1.9) and energy (1.14) equations. The fact is that a nonequilibrium state of gas is determined by 16 variables, namely, p, ρ, T, e; 3 components of the vectors $\boldsymbol{u}$ and $\boldsymbol{q}$; and 6 independent components of the tensor **P**, which is symmetric. However, the number of equations available till now is 7, namely, Eqs. (1.9), (1.14), (3.26), (3.28), and tree equations (3.27). Therefore, more 9 equations are needed in order to close the equations system. Such missing equations will be introduced in Section 6.1.

3.4 Entropy Production due to Intermolecular Collisions

To obtain the entropy balance equation, the function $\psi(t, \boldsymbol{r}, \boldsymbol{v})$ given by Eq. (2.34) should be used. Since this function depends on the time t and position $\boldsymbol{r}$, we cannot just write (3.17) for the entropy, but the BE (3.1) must be multiplied by ψ in the form (2.34) and integrated with respect to the velocity. Taking into account Eq. (3.21) for $\psi = 1$, the entropy balance equation takes the form

$$\frac{\partial s}{\partial t} + \nabla_r \cdot \boldsymbol{J}_s = \left(\frac{\partial s}{\partial t}\right)_{\text{coll}}, \tag{3.29}$$

where the notations (2.35) and (2.36) have been used. The entropy variation given by the right-hand-side term is not zero in this case, but it is calculated from (3.15) as

$$\left(\frac{\partial s}{\partial t}\right)_{\text{coll}} = \frac{k_B}{4} \iiiint \ln\left(\frac{f'f'_*}{ff_*}\right)(f'f'_* - ff_*) w \, d^3\boldsymbol{v}' \, d^3\boldsymbol{v}'_* \, d^3\boldsymbol{v}_* \, d^3\boldsymbol{v}. \tag{3.30}$$

Since the expression $\ln\,(x/y)(x-y)$ is never negative, it follows that

$$\left(\frac{\partial s}{\partial t}\right)_{\text{coll}} \geq 0. \tag{3.31}$$

Physically, this inequality expresses the fact that the intermolecular collisions can only increase (or produce) the entropy, but never decrease (or destroy) it. A combination of Eqs. (3.29) and (3.31) leads to the entropy balance equation in the following form:

$$\frac{\partial s}{\partial t} + \nabla_r \cdot \boldsymbol{J}_s \geq 0. \tag{3.32}$$

Historically, the inequality (3.31) was proved by Boltzmann in term of the H-function (H, Greek eta) defined as $H = \int f \ln f \, d^3\boldsymbol{v}$ and is known in the kinetic theory as Boltzmann's H-theorem.

3.5 Intermolecular Collisions Frequency

It would be interesting to know how many collisions happen per unit time and unit volume. First, let us calculate the collision frequency per particle. Let us consider again a target particle and beam of particles moving with a speed g_r similar to those in Section 1.6. When the TCS σ_t is constant, for example, for HS or any other potential with a cut-off impact parameter, the target particle will undergo

$n\sigma_t g_r$ collisions per time unit where n is the beam number density. If the target particle interacts with many beams then the total number of collisions per time unity reads

$$\nu_c = n\sigma_t \overline{g}_r, \tag{3.33}$$

where $\overline{g}_r$ is the average speed over all beams. Since this expression contains the relative speed g_r, it is valid for a moving particle too. In order to calculate the collision frequency for all particles, the average relative speed $\overline{g}_r$ must be calculated for all velocities of the target and beams

$$\overline{g}_r = \frac{1}{n^2} \iint f(\boldsymbol{v}) f(\boldsymbol{v}_*) |\boldsymbol{v}_* - \boldsymbol{v}| \, \mathrm{d}^3\boldsymbol{v}_* \, \mathrm{d}^3\boldsymbol{v}. \tag{3.34}$$

Thus, Eq. (3.33) together with (3.34) provides the frequency of intermolecular collisions per particle.

The average relative speed can be explicitly calculated for a given VDF. Substituting the Maxwellian VDF (2.37) into (3.34), after some mathematical manipulations, we obtain $\overline{g}_r$ in an equilibrium state

$$\overline{g}_r = 4\sqrt{\frac{k_B T}{\pi m}} = \sqrt{2}\langle v \rangle, \tag{3.35}$$

where the average molecular speed $\langle v \rangle$ is given by Eq. (2.44). Then the collision frequency (3.33) takes the form

$$\nu_c = \sqrt{2} n \sigma_t \langle v \rangle. \tag{3.36}$$

For the mathematical details, see Section 2.4 of the book [4].

Exercises

3.1 Show that

$$Q(f^M) = 0. \tag{3.37}$$

Solution: Due to the energy conservation law, we have

$$v^2 + v_*^2 = v'^2 + v_*'^2$$

As a consequence

$$f^M(v) f^M(v_*) = f^M(v'^2) f^M(v_*'^2). \tag{3.38}$$

Substituting it into (3.5), the equability (3.37) is obtained.

3.2 Obtain (3.22).
Hint: Multiply (3.1) by m and integrate it with respect to the velocity $\boldsymbol{v}$. Then use Eqs. (1.6), (2.4), (2.16), and (3.20).

3.3 Obtain (3.27).
Hint: Use Eqs. (2.30), (3.22), (3.23), and (3.25).

3.4 Obtain (3.28).
Hint: Use Eqs. (2.33), (3.22), (3.24), (3.25) and (3.27).

3.5 Show that for $f = f^{\mathrm{M}}$ Eq. (3.31) becomes the equality.
Hint: Use (3.38).
Comments: The equality (3.31) at $f = f^{\mathrm{M}}$ means that in equilibrium the entropy s defined by (2.35) reaches its maximum and it is not produced anymore.

3.6 Obtain (1.31) using (3.33) and (3.35).
Hint: Define the MFP as

$$\mathrm{MFP} = \frac{\langle v \rangle}{\nu}.$$

4
Gas–Surface Interaction

4.1
General form of Boundary Condition for Impermeable Surface

Gas flows are usually bounded by some solid surfaces, which affect the VDF so that the BE (3.1) should be solved taking into account an influence of the gas–surface interaction. Mathematically it means that on a gas–surface boundary, the VDF of incident particles is related to that of reflected molecules. Let us denote the velocity of incident molecules as $\boldsymbol{v}'$, while the velocity of reflected molecules as $\boldsymbol{v}$. Note, if $f(\boldsymbol{v}')$ is the VDF in the gas bulk near a surface, the number of particles having the velocity in $\mathrm{d}^3\boldsymbol{v}'$ near $\boldsymbol{v}'$ and impinging upon the surface per area unit and per time unit is given by

$$-v_n' f(\boldsymbol{v}')\,\mathrm{d}^3\boldsymbol{v}', \qquad v_n' = \boldsymbol{v}' \cdot \boldsymbol{n}, \tag{4.1}$$

where $\boldsymbol{n}$ is the unit vector normal to the surface and directed towards the gas as is shown in Figure 4.1. Analogically, the same surface will reflect the number of particles into $\mathrm{d}^3\boldsymbol{v}$ near $\boldsymbol{v}$ equal to

$$v_n f(\boldsymbol{v})\,\mathrm{d}^3\boldsymbol{v}, \qquad v_n = \boldsymbol{v} \cdot \boldsymbol{n}, \tag{4.2}$$

being $f(\boldsymbol{v})$ the VDF of reflected particles in the gas bulk.

Consider an impermeable surface reflecting all incident particles. Let us write down the probability that a particle impinging to the surface with a velocity $\boldsymbol{v}'$ is reflected into $\mathrm{d}^3\boldsymbol{v}$ near $\boldsymbol{v}$ as

$$R(\boldsymbol{v}', \boldsymbol{v})\,\mathrm{d}^3\boldsymbol{v}, \tag{4.3}$$

where $R(\boldsymbol{v}', \boldsymbol{v})$ is called scattering kernel. Multiplying the expression (4.1) by the probability (4.3) and integrating it over the whole range of incident velocity $\boldsymbol{v}'$, the number of particles reflected into $\mathrm{d}^3\boldsymbol{v}$ near $\boldsymbol{v}$ given by the expression (4.2) is obtained. Omitting $\mathrm{d}^3\boldsymbol{v}$ in both sides, the relation of $f(\boldsymbol{v})$ to $f(\boldsymbol{v}')$ is obtained as

$$v_n f(\boldsymbol{v}) = -\int_{v_n' \le 0} v_n' f(\boldsymbol{v}') R(\boldsymbol{v}', \boldsymbol{v})\,\mathrm{d}^3\boldsymbol{v}'. \tag{4.4}$$

Note that this equation relates the distribution function $f(\boldsymbol{v})$ to $f(\boldsymbol{v}')$ at the same moment of time and in the same point of the surface, that is, we assume that the particles are reflected immediately after their collisions with a surface.

Rarefied Gas Dynamics: Fundamentals for Research and Practice, First Edition. Felix Sharipov.

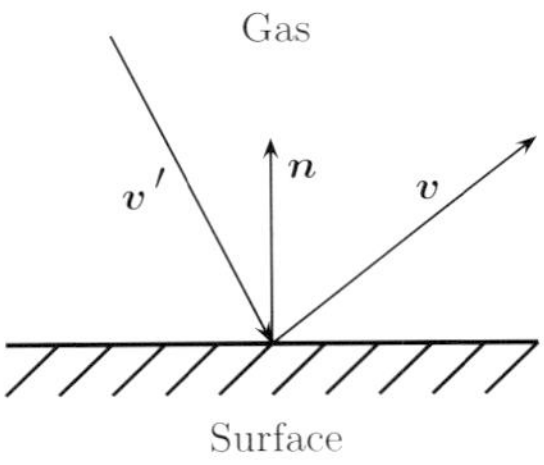

Figure 4.1 Scheme of gas–surface interaction.

The scattering kernel $R(\boldsymbol{v}', \boldsymbol{v})$ is determined by the local temperature T_w of the wall and by its velocity $\boldsymbol{u}_w$. In this chapter, we will consider only a surface being at rest, $\boldsymbol{u}_w = 0$. In case of moving surface, all expressions given below are written in a reference frame related to the surface and then all velocities are shifted by $-\boldsymbol{u}_w$.

The kernel cannot be negative because $R(\boldsymbol{v}', \boldsymbol{v})\, \mathrm{d}^3\boldsymbol{v}$ is the probability that an incident particle will be reflected into the velocity space volume $\mathrm{d}^3\boldsymbol{v}$ near the point $\boldsymbol{v}$, that is,

$$R(\boldsymbol{v}', \boldsymbol{v}) \geq 0. \tag{4.5}$$

Since the particle is always reflected, the sum of all probabilities is equal to unity then the kernel is normalized as

$$\int_{v_n \geq 0} R(\boldsymbol{v}', \boldsymbol{v})\, \mathrm{d}^3\boldsymbol{v} = 1. \tag{4.6}$$

From the macroscopic viewpoint, this condition means the surface impermeability, that is, the normal bulk velocity u_n on the surface is zero

$$u_n = \int v_n f\, \mathrm{d}^3\boldsymbol{v} = 0, \tag{4.7}$$

where the reflected VDF ($v_n > 0$) is related to the incident VDF ($v_n < 0$) by (4.4).

The kernel satisfies one more property that is not so obvious as (4.6), namely, the reciprocity condition

$$v_n' \exp\left(-\frac{v'^2}{v_w^2}\right) R(\boldsymbol{v}', \boldsymbol{v}) = -v_n \exp\left(-\frac{v^2}{v_w^2}\right) R(-\boldsymbol{v}, -\boldsymbol{v}'), \tag{4.8}$$

where $v_n \geq 0$, $v_n' \leq 0$, and v_w is the MPS corresponding to the wall temperature

$$v_w = \sqrt{\frac{2k_\mathrm{B} T_w}{m}}. \tag{4.9}$$

The reciprocity property (4.8) is valid for a surface being in a local equilibrium. It is a consequence of the reversibility of microprocesses of the gas–surface interaction [18, 30] and provides the detailed balance in an equilibrium state.

4.2 Diffuse–Specular Kernel

The explicit expression of the scattering kernel $R(\boldsymbol{v}', \boldsymbol{v})$ depends on the gas–surface interaction law. The well-known diffuse scattering (or cosine) law corresponds to the following kernel

$$R_d(\boldsymbol{v}', \boldsymbol{v}) = \frac{2v_n}{\pi v_w^4} \exp\left(-\frac{v^2}{v_w^2}\right). \tag{4.10}$$

Physically, it means that a particle can be reflected to any direction independent of its velocity before the collision $\boldsymbol{v}'$ with a surface. Such an interaction is called as the complete accommodation. In many practical applications, the diffuse scattering is well justified and provides reliable results. It usually happens for technical surfaces, which are rough and covered by adsorbed gases.

Substituting the kernel (4.10) into Eq. (4.4), it is shown that the reflected VDF is Maxwellian with the wall temperature T_w

$$f(\boldsymbol{v}) = \frac{n_r}{(\sqrt{\pi} v_w)^3} \exp\left(-\frac{v^2}{v_w^2}\right), \quad v_n \geq 0, \tag{4.11}$$

where n_r is the number density of the reflected particles given as

$$n_r = -\frac{2\sqrt{\pi}}{v_w} \int_{v_n' \leq 0} v_n' f(\boldsymbol{v}') \, \mathrm{d}^3\boldsymbol{v}'. \tag{4.12}$$

The opposite limit of the gas–surface interaction is the specular reflection when the tangential velocity component $\boldsymbol{v}_t$ does not change, while the normal component v_n changes it own sign. Mathematically, this gas–surface interaction law takes the form

$$R_s(\boldsymbol{v}', \boldsymbol{v}) = \delta(\boldsymbol{v}_t' - \boldsymbol{v}_t)\delta(v_n' + v_n), \tag{4.13}$$

where $\delta(x)$ is the Dirac delta function (see Appendix (A.4)). The substitution of this kernel into Eq. (4.4) leads to the relation

$$f(v_n, \boldsymbol{v}_t) = f(-v_n', \boldsymbol{v}_t'), \quad v_n \geq 0, \quad v_n' \leq 0. \tag{4.14}$$

In practice, a specular reflection of molecules does not exist.

To take into account a non-complete accommodation, it is assumed that one part α_d of incident particles is scattered diffusely, while the rest of particles $(1 - \alpha_d)$ is reflected specularly so that the scattering kernel is written as a linear combination of (4.10) and (4.13)

$$R_{ds}(\boldsymbol{v}', \boldsymbol{v}) = \alpha_d R_d(\boldsymbol{v}', \boldsymbol{v}) + (1 - \alpha_d) R_s(\boldsymbol{v}', \boldsymbol{v}). \tag{4.15}$$

Such a model of the gas–surface interaction is called *diffuse–specular* model. The VDF of the reflected particles in this case takes the form of linear combination of Eqs. (4.11) and (4.14)

$$f(\boldsymbol{v}) = \alpha_d \frac{n_r}{(\sqrt{\pi} v_w)^3} \exp\left(-\frac{v^2}{v_w^2}\right) + (1-\alpha_d) f(-v'_n, \boldsymbol{v}'_t), \tag{4.16}$$

where $v_n \geq 0$ and $v'_n \leq 0$.

The kernel (4.15) is also widely used to describe a noncomplete accommodation. However, some experimental data contradict theoretical results based on such a kernel. The main reason of the contradiction is that the diffuse–specular kernel contains just one free parameter α_d and cannot describe the complexity of the gas–surface interaction.

4.3 Cercignani–Lampis Kernel

The kernel proposed by Cercignani and Lampis (CL) [31] contains two parameters and reads

$$R_{\mathrm{CL}}(\boldsymbol{v}', \boldsymbol{v}) = \frac{2v_n}{\pi v_w^4 \alpha_n \, \alpha_t(2-\alpha_t)} \exp\left[-\frac{(\boldsymbol{v}_t - (1-\alpha_t)\boldsymbol{v}'_t)^2}{v_w^2 \alpha_t(2-\alpha_t)} - \frac{v_n^2 + (1-\alpha_n) v_n'^2}{v_w^2 \alpha_n}\right] \mathcal{I}_0\left(\frac{2\sqrt{1-\alpha_n} v_n v'_n}{v_w^2 \alpha_n}\right) \tag{4.17}$$

where $\mathcal{I}_0$ denotes the modified Bessel function of first kind and zeroth order defined by (A.27). The coefficient α_n varies from 0 to 1, while α_t varies from 0 to 2. The physical meaning of these coefficients will be given later. In the particular case $\alpha_t = 1$ and $\alpha_n = 1$, the CL kernel becomes the diffuse one (4.10). The other limit $\alpha_n = 0$ and $\alpha_t = 0$ gives the specular scattering (4.13). The limits $\alpha_n = 0$ and $\alpha_t = 2$ correspond to the back-scattering, which cannot be described by the diffuse–specular kernel. In practice, the complete back-scattering does not exist, but a partial back-scattering, when α_t is slightly larger than unity, is possible on rough surfaces.

4.4 Accommodation Coefficients

The knowledge of the scattering kernel is not always necessary, but sometimes it is enough to know only its integral characteristics such as the accommodation coefficients (AC) defined as

$$\alpha(\psi) = \frac{J_n^{(i)}(\psi) + J_n^{(r)}(\psi)}{J_n^{(i)}(\psi) + J_n^{(diff)}(\psi)}, \tag{4.18}$$

where $J_n^{(i)}(\psi)$ and $J_n^{(r)}(\psi)$ are the incident and reflected fluxes given as

$$J_n^{(i)} = \int_{v_n' \leq 0} v_n' \psi(\boldsymbol{v}') f(\boldsymbol{v}') \, \mathrm{d}^3\boldsymbol{v}', \quad J_n^{(r)} = \int_{v_n \geq 0} v_n \psi(\boldsymbol{v}) f(\boldsymbol{v}) \, \mathrm{d}^3\boldsymbol{v}, \tag{4.19}$$

$\psi = \psi(\boldsymbol{v})$ is a function of molecular velocity $\boldsymbol{v}$. Here, the VDF for the incident particles ($v_n' \leq 0$) should be known, while the VDF for the reflected particles ($v_n \geq 0$) is calculated by Eq. (4.4). Then, the quantity $J_n^{(r)}$ is also expressed via the VDF of incident particles as

$$J_n^{(r)} = -\int_{v_n \geq 0} \psi(\boldsymbol{v}) \int_{v_n' \leq 0} v_n' R(\boldsymbol{v}', \boldsymbol{v}) f(\boldsymbol{v}') \, \mathrm{d}^3\boldsymbol{v}' \, \mathrm{d}^3\boldsymbol{v}. \tag{4.20}$$

The quantity $J_n^{(diff)}$ is also the reflected flux calculated by Eq. (4.19), but with the diffuse scattering kernel (4.10)

$$J_n^{(diff)} = \frac{n_r}{(\sqrt{\pi} v_w)^3} \int_{v_n \geq 0} v_n \psi(\boldsymbol{v}) \exp\left(-\frac{v^2}{v_w^2}\right) \, \mathrm{d}^3\boldsymbol{v}, \tag{4.21}$$

where n_r is given by Eq. (4.12) in terms of the incident VDF.

Formally, the function $\psi(\boldsymbol{v})$ can be arbitrary, but only some of its expressions lead to a physically sensible AC. For instance, if $\psi(\boldsymbol{v}) = mv^2/2$, the coefficient α becomes the energy AC. The function describing only the kinetic energy of the normal velocity $\psi = mv_n^2/2$ generates the normal energy AC. The normal momentum AC is calculated with the function $\psi = m|v_n|$. The tangential momentum AC is defined via the function $\psi = mv_t$, where v_t is the tangential velocity component of the incident particle.

Substituting the specular kernel (4.13) into the flux $J_n^{(r)}$ given by (4.20), we obtain

$$J_n^{(r)} = -J_n^{(i)} \quad \text{for specular reflection,} \tag{4.22}$$

for any function $\psi(\boldsymbol{v})$, which is even with respect to v_n. Then, according to (4.18), the AC is zero

$$\alpha = 0 \quad \text{for specular reflection.} \tag{4.23}$$

However, the relation (4.22) is not valid for an odd function $\psi(\boldsymbol{v})$, then the AC is different from zero that does not have any sense for the specular reflection. That is why, only even functions ψ with respect to v_n are considered in the definition of AC (4.18). In the opposite limit of the diffuse scattering, the reflected flux is given by

$$J_n^{(r)} = J_n^{(diff)} \quad \text{for diffuse scattering} \tag{4.24}$$

and according to (4.18), the AC is equal to unity for any function $\psi(\boldsymbol{v})$

$$\alpha = 1 \quad \text{for diffuse scattering.} \tag{4.25}$$

For an arbitrary scattering kernel, the ACs depend on the incident VDF that makes it nonuniversal and nonconvenient for a practical use. Only for particular kernels, the ACs defined by Eq. (4.18) are universal and independent of the VDF. Substituting the diffuse–specular kernel (4.15) into Eq. (4.20), the flux $J_n(\psi)$ takes the form

$$J_n^{(r)} = \alpha_d J_n^{(diff)} - (1 - \alpha_d) J^{(i)} \quad \text{for diffuse–specular scattering,} \tag{4.26}$$

then the AC definition (4.18) yields

$$\alpha(\psi) = \alpha_d \quad \text{for diffuse–specular scattering,} \tag{4.27}$$

for any function $\psi(\boldsymbol{v})$. It means that all ACs are the same and equal to the diffuse part α_d of the scattering. This is a disadvantage of the diffuse–specular kernel because the energy AC and momentum AC are different in practice, but for the diffuse–specular scattering they are the same.

Let us substitute the CL kernel (4.17) into Eq. (4.20) and assume $\psi = mv_t$, where v_t is the tangential velocity of particles. After hard derivations, the reflected flux $J_n^{(r)}$ is related to the incident flux as

$$J_n^{(r)} = -(1 - \alpha_t) J_n^{(i)}. \tag{4.28}$$

Since $J_n^{(diff)} = 0$ for $\psi = mv_t$, then

$$\alpha = \alpha_t \quad \text{for CL kernel and for} \quad \psi = mv_t. \tag{4.29}$$

Now, the parameter α_t of the CL kernel (4.17) takes the physical meaning of the tangential momentum AC. If we do the same operations with the function $\psi = mv_n^2/2$ corresponding to the part of kinetic energy associated with the normal velocity, then the reflected flux will be expressed as

$$J_n^{(r)} = -(1 - \alpha_n) J_n^{(i)} + \alpha_n J_n^{(diff)}. \tag{4.30}$$

Substituting it into the AC definition (4.18), one obtains

$$\alpha = \alpha_n \quad \text{for CL kernel and for} \quad \psi = \frac{1}{2} m v_n^2 \tag{4.31}$$

so that the second parameter of the CL kernel physically is the AC of energy associated with the normal motion of particles. For any other functions, for example, $\psi = mv^2/2$ or $\psi = m|v_n|$, the ACs based on the CL kernel cannot be expressed via the parameters α_t and α_n. Moreover, such ACs depend on the incident VDF.

The ACs are usually extracted from experimental data such as mass flow rate through a capillary, see example, Refs [32–34], or heat transfer through a rarefied gas, see example, Ref. [35]. Numerical values of the coefficient α_d obtained from the experimental results reported in Ref. [32] are given in Table 4.1. The corresponding experimental results were obtained for an atomically clean surface prepared in vacuum conditions by a vapor deposition process. A summary of numerical values of the AC α_t and α_n calculated in the papers [33–35] using the experimental data on the slip coefficients, tube flows [36], and the heat transfer [37] are presented in Table 4.2. The corresponding experimental measurements [36, 37] were carried out for a technical surface without any special treatment.

Table 4.1 Accommodation coefficient α_d extracted from the experimental data [32] applying the diffuse–specular kernel (4.15): a–atomically clean silver, b–atomically clean titanium, c–titanium covered by oxygen.

	α_d		
Gas	**a**	**b**	**c**
He	0.71	0.71	0.96
Ne	0.80		
Ar	0.88	0.87	0.98
Kr	0.92	0.92	1.00

Table 4.2 Accommodation coefficients α_t and α_n extracted in Refs [33–35] from the experimental data [36, 37] applying the CL kernel (4.17): surface is typically technical.

Gas	α_t	α_n
He	0.90	0.10
Ne	0.89	0.75
Ar	0.96	1.00
Kr	1.00	1.00
Xe	1.00	1.00
H_2	0.95	
N_2	0.91	
CO_2	1.00	

These data indicate that for light gases, such as helium and neon, the ACs may differ significantly from unity, while for heavy gases, for example, krypton and xenon, the ACs are close to unity. The experimental results [32] on the ACs for different kinds of surfaces showed that the gas interaction with a contaminated surface is closer to the diffuse scattering than that with an atomically clean surface.

4.5 General form of Boundary Condition for Permeable Surface

If we consider a vapor contacting with its own liquid surface, then incident particles can be partially condensed on the surface and, at the same time, it can evaporate particles independent of the incident flux. In this case, the VDF of particles leaving the surface contains two terms and is written as

$$f(\boldsymbol{v}) = \alpha_c(-\boldsymbol{v}) f_{\mathcal{N}}^{\mathrm{M}}(\boldsymbol{v}) - \frac{1}{v_n} \int_{v_n' \leq 0} v_n' f(\boldsymbol{v}') R(\boldsymbol{v}', \boldsymbol{v})\, \mathrm{d}^3\boldsymbol{v}', \quad v_n \geq 0, \tag{4.32}$$

where the first term describes the evaporation independent of the incident VDF, while the second term in (4.32) corresponds to the reflection of incident particles. The Maxwellian f_w^{M} is determined by the wall temperature T_w and saturated vapor density n_w so that it is given as

$$f_w^{\mathrm{M}}(\boldsymbol{v}) = \frac{n_w}{\left(\sqrt{\pi}v_w\right)^3} \exp\left(-\frac{v^2}{v_w^2}\right). \tag{4.33}$$

In this case, the scattering kernel $R(\boldsymbol{v}', \boldsymbol{v})$ is normalized as

$$\int_{v_n \geq 0} R(\boldsymbol{v}', \boldsymbol{v})\, \mathrm{d}^3\boldsymbol{v} = 1 - \alpha_c(\boldsymbol{v}'), \tag{4.34}$$

where $\alpha_c(\boldsymbol{v}')$ is the probability that an incident particle with a velocity $\boldsymbol{v}'$ will be condensed so that α_c is the condensation coefficient varying from 0 to 1. In the limit $\alpha_c = 1$, all incident particles are condensed, in other words, the complete condensation takes place, but the surface evaporates particles with the wall Maxwellian f_w^{M}. The property (4.34) does not provide the impermeability of the surface for any $\alpha_c > 0$ so that a condensing–evaporating surface can be called permeable. Only in the limit case $\alpha_c = 0$, the surface becomes impermeable and the relation (4.32) takes the form (4.4).

The scattering kernel $R(\boldsymbol{v}', \boldsymbol{v})$ in Eq. (4.32) also satisfies the reciprocal relation (1.8), which is valid for a surface being in a local equilibrium.

4.6
Entropy Production due to Gas–Surface Interaction

The gas–surface interaction modifies the VDF and consequently changes its entropy. In this section, we will show that the collisions of gaseous particles with a surface can only produce the entropy, but cannot destroy it. With the help of Eqs. (4.6) and (4.8), the following inequality can be proved

$$\int v_n f \left(1 - \ln \frac{f}{f_w^{\mathrm{M}}}\right)\, \mathrm{d}^3\boldsymbol{v} \geq 0, \tag{4.35}$$

where the VDF for the reflected particles ($v_n > 0$) is related to the that of incident particles ($v_n < 0$) by Eq. (4.4). The wall Maxwellian f_w^{M} is given by Eq. (4.33) for a permeable surface. For a surface satisfying the impermeability condition (4.7), the density in f_w^{M} does not matter so that any one, for example, its equilibrium value, can be substituted into f_w^{M}.

The proof of the inequality for an impermeable surface is given in Chap.III of the book [18]. In case of condensing–evaporating surface, it is proved in Ref. [38] where it is also shown that the quantity

$$\mathrm{d}\sigma_w = k_{\mathrm{B}} \int v_n f \left(1 - \ln \frac{f}{f_w^{\mathrm{M}}}\right)\, \mathrm{d}^3\boldsymbol{v}\, \mathrm{d}^2\Sigma_w \tag{4.36}$$

represents the entropy production because of the gas interaction with a wall element $d^2\Sigma_w$. To obtain the total entropy production σ_w, it is necessary to integrate (4.36) over the whole wall surface Σ_w

$$\sigma_w = k_B \int_{\Sigma_w} \int v_n f \left(1 - \ln \frac{f}{f_w^M}\right) d^3\boldsymbol{v}\, d^2\Sigma_w. \tag{4.37}$$

The inequality (4.35) leads to the conclusion that the gas–surface interaction can only produce the entropy

$$\sigma_w \geq 0. \tag{4.38}$$

Let us define the chemical potential per particle as

$$\mu_w := k_B T_w \ln \left[\frac{n_w}{(\sqrt{\pi} v_w)^2}\right], \tag{4.39}$$

which corresponds to the saturated vapor at the wall temperature T_w. Using this definition, the entropy production (4.37) takes the form

$$\sigma_w = \int_{\Sigma_w} J_{sn}\, d^2\Sigma_w - \int_{\Sigma_w} \frac{J_{en} - \mu_w J_{Nn}}{T_w}\, d^2\Sigma_w, \tag{4.40}$$

where J_{sn}, J_{en}, and J_{Nn} are the normal components of the entropy flow (2.36), energy flow (2.20), and particle flow (2.18), respectively.

Now we can calculate the entropy variation of a gas confined in a closed container. The entropy S of such a gas is calculated by integrating the specific entropy (2.35) over the whole volume $\mathcal{V}$ of the container

$$S(t) := \int_{\mathcal{V}} s(t, \boldsymbol{r})\, d^3\boldsymbol{r}. \tag{4.41}$$

An integration of (3.29) with respect to the position $\boldsymbol{r}$ over the container volume $\mathcal{V}$ yields

$$\frac{dS}{dt} - \oint_{\Sigma_w} J_{sn}\, d^2\Sigma_w = \sigma_{coll} \tag{4.42}$$

where the integration over the volume $\mathcal{V}$ in the second term has been replaced by the integration over its surface Σ_w using Gauss–Ostrogradsky's theorem (see Eq. (A.37)). The entropy production σ_{coll} due to the intermolecular collisions has been obtained as

$$\sigma_{\text{coll}} = \int_{\mathcal{V}} \left(\frac{\partial s}{\partial t}\right)_{\text{coll}} d^3\boldsymbol{r} \geq 0, \tag{4.43}$$

which is not negative because of (3.31). Extracting the term $\int J_{sn}\, d^2\Sigma_w$ from (4.40) and substituting it into (4.42), the entropy variation takes the form

$$\frac{dS}{dt} = \sigma + \oint_{\Sigma_w} \frac{J_{en} - \mu_w J_{Nn}}{T_w}\, d^2\Sigma_w, \tag{4.44}$$

where σ is the whole entropy production

$$\sigma = \sigma_{\text{coll}} + \sigma_w \geq 0, \tag{4.45}$$

which is not negative because of (4.38) and (4.43). Note that the container wall temperature T_w can be variable, $T_w = T_w(t, \mathbf{r})$. Physically, Eq. (4.44) means that the entropy variation is composed of its production σ, and its inflow is represented by the second term in the right-hand side of (4.44).

Since the total entropy production σ is positive, the equality (4.44) can be written as an inequality

$$\frac{\mathrm{d}S}{\mathrm{d}t} \geq \oint_{\Sigma_w} \frac{J_{en} - \mu_w J_{Nn}}{T_w} \,\mathrm{d}^2\Sigma_w. \tag{4.46}$$

Mathematically, this inequality is a generalization of the H-theorem (3.31) taking into account the gas–surface interaction. Physically, the inequality (4.46) represents the second law of thermodynamics, which is postulated on the macroscopic level, while it is rigorously proved for a dilute gas on the basis of the BE (3.1) and gas–surface interaction law (4.32) written in their quite general forms.

In particular case of a closed system, when no particle exchange is possible $J_{Nn} = 0$, Eq. (4.46) is reduced to

$$\frac{\mathrm{d}S}{\mathrm{d}t} \geq \oint_{\Sigma_w} \frac{J_{en}}{T_w} \,\mathrm{d}^2\Sigma_w. \tag{4.47}$$

When the system is completely insulated and no energy exchange is permitted, then

$$\frac{\mathrm{d}S}{\mathrm{d}t} \geq 0. \tag{4.48}$$

so that the entropy can only increase or be constant.

Exercises

4.1 Consider an impermeable surface being at rest and kept at a temperature T_w. Using Eqs. (4.4), (4.6), and (4.8) show that if the incident VDF is Maxwellian with the temperature T_w

$$f^{\mathrm{M}}(\boldsymbol{v}') = \frac{n}{(\sqrt{\pi} v_w)^3} \exp\left(-\frac{v'^2}{v_w^2}\right), \tag{4.49}$$

then the reflected VDF will be the same Maxwellian $f^{\mathrm{M}}(\boldsymbol{v})$.
Hints: Substitute (4.49) into (4.4). Use (4.8) to replace $R(\boldsymbol{v}', \boldsymbol{v})$ by $R(-\boldsymbol{v}, -\boldsymbol{v}')$. Exchange $-\boldsymbol{v}'$ by $\boldsymbol{v}'$ and use (4.6) in the form $\int_{v_n' \geq 0} R(-\boldsymbol{v}, \boldsymbol{v}')\,\mathrm{d}^3\boldsymbol{v}' = 1$.
Comments: If a gas is staying in equilibrium with a surface, the gas–surface interaction will not change the VDF.

4.2 Do the previous exercise for a condensing–evaporating surface using Eq. (4.32).

4.3 Prove the impermeability condition (4.7) in general case.
Hints: Use the main properties (4.6) and (4.8).

4.4 Check the normalization of the diffuse scattering kernel (4.10).
Hint: Introduce the variable $\boldsymbol{c} = \boldsymbol{v}/v_w$ and then use the integrals (A.18).

4.5 Check the normalization of the CL scattering kernel (4.17).
Hint: Introduce the variables

$$\boldsymbol{c}_t = \frac{\boldsymbol{v}_t - (1-\alpha_t)\boldsymbol{v}'_t}{v_w\sqrt{\alpha_t(2-\alpha_t)}}, \tag{4.50}$$

$$\xi = \frac{v_n \cos\varphi - \sqrt{1-\alpha_n}v'_n}{v_w\sqrt{\alpha_n}}, \quad \eta = \frac{v_n \sin\varphi}{v_w\sqrt{\alpha_n}}. \tag{4.51}$$

Then the normalization condition of the CL kernel is reduced to the integral

$$\frac{1}{\pi^2}\int_{-\infty}^{\infty}\int_{-\infty}^{\infty}\exp(-c_t^2)\,\mathrm{d}^2\boldsymbol{c}_t\int_{-\infty}^{\infty}\int_{-\infty}^{\infty}\exp(-\xi^2-\eta^2)\,\mathrm{d}\xi\,\mathrm{d}\eta$$

which is equal to unity according to (A.18).

4.6 Check the reciprocity (4.8) of both diffuse (4.10) and CL (4.17) kernels.

4.7 Using the normalization (4.6) and reciprocity (4.8), prove the impermeability condition at a surface

$$\int v_n f\,\mathrm{d}^3\boldsymbol{v} = 0.$$

4.8 Obtain (4.26).

4.9 Obtain (4.28) and (4.30).
Hints: Use the variables (4.50), (4.51).

5
Linear Theory

5.1
Small Perturbation of Equilibrium

The VDF corresponding to an equilibrium state is given by Eq. (2.37) in Section 2.4. Now, let us imagine that this state is just slightly perturbed. For instance, one boundary surface can slowly move or it may have a temperature slightly different from its equilibrium value. In this case, the BE and the boundary conditions can be linearized. For such a purpose, we need to choose a small parameter ξ, which must be dimensionless and may have various physical meanings, for example, relative temperature deviation, ratio of surface velocity to some characteristic molecular speed, and so on. Then all expressions are linearized with respect to the small parameter omitting all terms of the second, ξ^2, and higher orders.

The linear form of the BE and boundary conditions may significantly simplify a solution of some problems so that the linearization is desirable. In other situations, gas flows can be calculated only in the linearized form because a perturbation is really small so that the linearization becomes necessary. Following are the general principles of the linearization, which are described in detail. Then, some practical applications are given.

5.2
Linearization Near Global Maxwellian

Let us denote the equilibrium density and temperature as n_0 and T_0, respectively. Then the equilibrium distribution function (2.41) takes the form

$$f_0^{\mathrm{M}} = \frac{n_0}{\left(\sqrt{\pi}v_0\right)^3} e^{-c^2}, \tag{5.1}$$

where the dimensionless velocity $\boldsymbol{c}$ has been defined via the MPS v_0 at the equilibrium temperature T_0

$$\boldsymbol{c} = \frac{\boldsymbol{v}}{v_0}, \quad v_0 = \sqrt{\frac{2k_{\mathrm{B}}T_0}{m}}. \tag{5.2}$$

Rarefied Gas Dynamics: Fundamentals for Research and Practice, First Edition. Felix Sharipov.

Many linearized expressions can be written down in compact forms if the following notation is used

$$(\psi, \phi) = \int \psi(\boldsymbol{v}) f_0^{\mathrm{M}} \phi(\boldsymbol{v})\, \mathrm{d}^3\boldsymbol{v}, \tag{5.3}$$

which can be interpreted as an inner product of two arbitrary functions $\psi(\boldsymbol{v})$ and $\phi(\boldsymbol{v})$. The product (5.3) takes the following form in terms of the dimensionless velocity

$$(\psi, \phi) = \frac{n_0}{\pi^{3/2}} \int \psi(\boldsymbol{c}) e^{-c^2} \phi(\boldsymbol{c})\, \mathrm{d}^3\boldsymbol{c}. \tag{5.4}$$

An integration of the expression (5.3) over the whole volume $\mathcal{V}$ occupied by a gas will be denoted as

$$((\psi, \phi)) = \int_{\mathcal{V}} (\psi, \phi)\, \mathrm{d}^3\boldsymbol{r}, \tag{5.5}$$

which is one more inner product.

If the global equilibrium state is slightly disturbed, the VDF just slightly deviates from the Maxwellian (5.1). The deviation is characterized by the small parameter ξ so that the disturbed VDF can be represented as

$$f(t, \boldsymbol{r}, \boldsymbol{v}) = f_0^{\mathrm{M}}[1 + \xi h(t, \boldsymbol{r}, \boldsymbol{v})], \quad |\xi| \ll 1, \tag{5.6}$$

where the function $h(t, \boldsymbol{r}, \boldsymbol{v})$ is called perturbation function or just perturbation. If we substitute the representation (5.6) into the moment definitions (2.4), (2.16), (2.26), (2.28), and (2.32), we obtain their expressions via the perturbation h and inner product (5.4) as

$$n(t, \boldsymbol{r}) = n_0 + \xi(1, h), \tag{5.7}$$

$$\boldsymbol{u}(t, \boldsymbol{r}) = \frac{\xi}{n_0}(\boldsymbol{v}, h), \tag{5.8}$$

$$T(t, \boldsymbol{r}) = T_0 \left[1 + \frac{\xi}{n_0}\left(\frac{2v^2}{3v_0^2} - 1, h\right)\right], \tag{5.9}$$

$$\mathsf{P}(t, \boldsymbol{r}) = p\,\mathsf{I} + \xi m \left(\boldsymbol{v}\boldsymbol{v} - \frac{1}{3}\mathsf{I}\, v^2, h\right), \tag{5.10}$$

$$\boldsymbol{q}(t, \boldsymbol{r}) = \xi k_{\mathrm{B}} T_0 \left(\left(\frac{v^2}{v_0^2} - \frac{5}{2}\right)\boldsymbol{v}, h\right), \tag{5.11}$$

respectively. The notation I in (5.10) means the unit tensor, see (A.9).

Now, it is convenient to define the dimensionless moments and to write down the dimensional moments as

$$n = n_0(1 + \xi\varrho), \quad \varrho = \frac{1}{n_0}(1, h), \tag{5.12}$$

$$\boldsymbol{u} = \xi v_0 \tilde{\boldsymbol{u}}, \quad \tilde{\boldsymbol{u}} = \frac{1}{n_0}(\boldsymbol{c}, h), \tag{5.13}$$

$$T = T_0(1 + \xi\tau), \quad \tau = \frac{1}{n_0}\left(\frac{2}{3}c^2 - 1, h\right), \tag{5.14}$$

$$\mathsf{P} = p(\mathsf{I} + \xi\Pi), \quad \Pi = \frac{2}{n_0}\left(\boldsymbol{cc} - \frac{1}{3}\mathsf{I}c^2, h\right), \tag{5.15}$$

$$\boldsymbol{q} = \xi p_0 v_0 \tilde{\boldsymbol{q}}, \quad \tilde{\boldsymbol{q}} = \frac{1}{n_0}\left(\left(c^2 - \frac{5}{2}\right)\boldsymbol{c}, h\right). \tag{5.16}$$

The quantities ϱ, τ represent the deviations from equilibrium density and temperature, the quantities $\tilde{\boldsymbol{u}}$ and $\tilde{\boldsymbol{q}}$ are the dimensionless bulk velocity and heat flow vector. The traceless tensor Π, that is, $\Pi_{11} + \Pi_{22} + \Pi_{33} = 0$, represents the deviation of the pressure tensor P from the diagonal tensor $p\mathsf{I}$. The dimensionless moments can be written in the compact form:

$$\begin{bmatrix} \varrho(t, \boldsymbol{r}) \\ \tilde{\boldsymbol{u}}(t, \boldsymbol{r}) \\ \tau(t, \boldsymbol{r}) \\ \Pi(t, \boldsymbol{r}) \\ \tilde{\boldsymbol{q}}(t, \boldsymbol{r}) \end{bmatrix} = \frac{1}{\pi^{3/2}} \int \begin{bmatrix} 1 \\ \boldsymbol{c} \\ \frac{2}{3}c^2 - 1 \\ 2\left(\boldsymbol{cc} - \frac{1}{3}\mathsf{I}c^2\right) \\ \left(c^2 - \frac{5}{2}\right)\boldsymbol{c} \end{bmatrix} e^{-c^2} h(t, \boldsymbol{r}, \boldsymbol{c})\, \mathrm{d}^3\boldsymbol{c}, \tag{5.17}$$

which follow from Eqs. (5.2), (5.4), and (5.7)–(5.11).

Evidently, the global Maxwellian is independent of the time and coordinates

$$\frac{\partial f_0^{\mathrm{M}}}{\partial t} + \boldsymbol{v} \cdot \nabla_r f_0^{\mathrm{M}} = 0. \tag{5.18}$$

Owing to the kinetic energy conservation in each collision, the equality

$$f_0^{\mathrm{M}}(\boldsymbol{v}') f_0^{\mathrm{M}}(\boldsymbol{v}'_*) = f_0^{\mathrm{M}}(\boldsymbol{v}) f_0^{\mathrm{M}}(\boldsymbol{v}_*) \tag{5.19}$$

is fulfilled. Substituting the representation (5.6) into (3.1), taking into account Eqs. (5.18) and (5.19), the linearized BE is derived in the following form:

$$\frac{\partial h}{\partial t} + \boldsymbol{v} \cdot \nabla_r h = \hat{L}h, \tag{5.20}$$

where the linearized collision operator reads

$$\hat{L}h = \iiint f_0^{\mathrm{M}}(\boldsymbol{v}_*)(h'_* + h' - h_* - h) w(\boldsymbol{v}', \boldsymbol{v}'_*; \boldsymbol{v}, \boldsymbol{v}_*)\, \mathrm{d}^3\boldsymbol{v}'\, \mathrm{d}^3\boldsymbol{v}'_*\, \mathrm{d}^3\boldsymbol{v}_* \tag{5.21}$$

or

$$\hat{L}h = \iiint f_0^{\mathrm{M}}(\boldsymbol{v}_*)\,(h'_* + h' - h_* - h) g_r b\, \mathrm{d}b\, \mathrm{d}\epsilon\, \mathrm{d}^3\boldsymbol{v}_*. \tag{5.22}$$

For a cut-off potential, the collision operator $\hat{L}$ can be split as

$$\hat{L}h = \hat{K}h - \nu_0 h, \tag{5.23}$$

where

$$\hat{K}h = \iiint f_0^{\mathrm{M}}(\boldsymbol{v}_*)(h'_* + h' - h_*) g_r b\, \mathrm{d}b\, \mathrm{d}\epsilon\, \mathrm{d}^3\boldsymbol{v}_*, \tag{5.24}$$

$$\nu_0(\boldsymbol{v}) = \sigma_t \int f_0^{\mathrm{M}}(\boldsymbol{v}_*) g_r \, \mathrm{d}^3\boldsymbol{v}_*. \tag{5.25}$$

Doing some hard derivations, the term (5.24) can be presented in the following form:

$$\hat{K}h = \int f_0^{\mathrm{M}} K(\boldsymbol{v}, \boldsymbol{v}_*) h(\boldsymbol{v}_*) \, \mathrm{d}^3\boldsymbol{v}_*, \tag{5.26}$$

where the kernel $K(\boldsymbol{v}, \boldsymbol{v}_*)$ of this operator has a cumbersome expression, which can be found in Refs [4, 18, 21]. The operator $\hat{K}$ written in the form (5.26) significantly simplifies a numerical solving of the linearized BE. The function $\nu_0(\boldsymbol{v})$ can be simplified substituting $f_0^{\mathrm{M}}(\boldsymbol{v}_*)$ given by (5.1) and $g_r = |\boldsymbol{v} - \boldsymbol{v}_*|$ into (5.25)

$$\nu_0(c) = n_0 v_0 \sigma_t \left[\frac{e^{-c^2}}{\sqrt{\pi}} + \left(c + \frac{1}{2c} \right) \mathrm{erf}(c) \right], \tag{5.27}$$

where $c = |\boldsymbol{c}|$, v_0 are given by (5.2) and the error function erf(c) is defined by (A.23).

5.3 Linearization Near Local Maxwellian

In many situations, it is more convenient to linearize near the local Maxwellian (2.46), which should be completely determined, that is, it should not contain unknown quantities. In this case, the Maxwellian becomes a reference distribution function and takes the form

$$f_{\mathrm{R}}^{\mathrm{M}}(t, \boldsymbol{r}, \boldsymbol{v}) = n_{\mathrm{R}} \left(\frac{m}{2\pi k_{\mathrm{B}} T_{\mathrm{R}}} \right)^{3/2} \exp \left[-\frac{m(\boldsymbol{v} - \boldsymbol{u}_{\mathrm{R}})^2}{2k_{\mathrm{B}} T_{\mathrm{R}}} \right], \tag{5.28}$$

where the number density n_{R}, bulk velocity $\boldsymbol{u}_{\mathrm{R}}$, and temperature T_{R} are some given functions of the time t and coordinates $\boldsymbol{r}$ and nominated reference ones. The choice of these functions is arbitrary with the restriction that they should just slightly deviate from their equilibrium values, that is, they can be represented as

$$n_{\mathrm{R}} = n_0(1 + \xi \varrho_{\mathrm{R}}), \quad \boldsymbol{u}_{\mathrm{R}} = \xi v_0 \tilde{\boldsymbol{u}}_{\mathrm{R}}, \quad T_{\mathrm{R}} = T_0(1 + \xi \tau_{\mathrm{R}}), \tag{5.29}$$

where ξ is the small parameter used for the linearization, and ϱ_{R}, $\tilde{\boldsymbol{u}}_{\mathrm{R}}$, and τ_{R} are reference perturbations. Owing to the conditions (5.29), the reference Maxwellian can be represented via the reference perturbation as

$$f_{\mathrm{R}}^{\mathrm{M}} = f_0^{\mathrm{M}}(1 + \xi h_{\mathrm{R}}), \quad h_{\mathrm{R}} = \varrho_{\mathrm{R}} + 2\boldsymbol{c} \cdot \tilde{\boldsymbol{u}}_{\mathrm{R}} + \left(c^2 - \frac{3}{2} \right) \tau_{\mathrm{R}}. \tag{5.30}$$

Now, when the reference Maxwellian is defined, the VDF is represented as

$$f(t, \boldsymbol{r}, \boldsymbol{v}) = f_{\mathrm{R}}^{\mathrm{M}} \, [1 + \xi \, h(t, \boldsymbol{r}, \boldsymbol{v})]. \tag{5.31}$$

Taking into consideration (5.30), the representation (5.31) takes a simple form

$$f = f_0^{\mathrm{M}}[1 + \xi(h + h_{\mathrm{R}})]. \tag{5.32}$$

Substituting it into (3.1), the linearized BE is derived in the following form:

$$\frac{\partial h}{\partial t} + \boldsymbol{v} \cdot \nabla_r h = \hat{L}h + g(t, \boldsymbol{r}, \boldsymbol{v}), \tag{5.33}$$

where the term $g(t, \boldsymbol{r}, \boldsymbol{v})$ is calculated via the reference perturbation as

$$g(t, \boldsymbol{r}, \boldsymbol{v}) = -\left(\frac{\partial h_{\mathrm{R}}}{\partial t} + \boldsymbol{v} \cdot \nabla_r h_{\mathrm{R}} \right). \tag{5.34}$$

Thus, the linearization near the local Maxwellian brings the term $g(t, \boldsymbol{r}, \boldsymbol{v})$ into the BE, which describes a perturbation in the bulk of gas and is called bulk source term.

The linearized moments defined by Eqs. (5.12)–(5.14) will have a slightly different meaning for the representation (5.32). In this case, the deviations of the corresponding reference values must be taken into account

$$n = n_{\mathrm{R}}(1 + \xi\varrho) = n_0[1 + \xi(\varrho_{\mathrm{R}} + \varrho)], \tag{5.35}$$

$$\boldsymbol{u} = \boldsymbol{u}_{\mathrm{R}} + \xi v_0 \tilde{\boldsymbol{u}} = \xi v_0 (\tilde{\boldsymbol{u}}_{\mathrm{R}} + \tilde{\boldsymbol{u}}), \tag{5.36}$$

$$T = T_{\mathrm{R}}(1 + \xi\tau) = T_0[1 + \xi(\tau_{\mathrm{R}} + \tau)]. \tag{5.37}$$

The pressure tensor and heat flux retain the same form (5.15) and (5.16), respectively. The dimensionless moments ϱ, $\tilde{\boldsymbol{u}}$, τ, Π, and $\tilde{\boldsymbol{q}}$ are related to the perturbation function h by the same relations (5.17) as in the previous case.

5.4 Properties of the Linearized Collision Operator

Let us introduce the time reversion operator $\hat{T}$, which inverses a state of particle in the time. For monatomic particle, it changes just the sign of the molecular velocity

$$\hat{T}h(\boldsymbol{r}, \boldsymbol{v}) = h(\boldsymbol{r}, -\boldsymbol{v}). \tag{5.38}$$

It is self-adjoint for the inner product defined by (5.3)

$$(\hat{T}\phi, \psi) = (\hat{T}\psi, \phi). \tag{5.39}$$

Using the relations (3.3) and (3.4), it can be shown that

$$(\hat{T}\hat{L}\phi, \psi) = (\hat{T}\hat{L}\psi, \phi), \tag{5.40}$$

that is, the operator $\hat{T}\hat{L}$ is self-adjoint too.

In case of central intermolecular potential, the operators $\hat{T}$ and $\hat{L}$ are commutative

$$\hat{T}\hat{L}h = \hat{L}\hat{T}h, \tag{5.41}$$

then a combination of Eqs. (5.40), (5.39), and (5.41) leads to the self-adjointness of $\hat{L}$

$$(\hat{L}\phi, \psi) = (\hat{L}\psi, \phi). \tag{5.42}$$

The linearized expression (3.14) for the variation of property ψ takes the form:

$$\left(\frac{\partial \psi}{\partial t}\right)_{coll} = n_0(\psi, \hat{L}h). \tag{5.43}$$

For the collision invariants (3.19), this variation is zero

$$(\psi, \hat{L}h) = 0 \quad \text{for} \quad \psi = m, m\boldsymbol{v}, \frac{1}{2}mv^2. \tag{5.44}$$

A substitution of the invariants into the collision operator (5.21) yields

$$\hat{L}\psi = 0. \tag{5.45}$$

To obtain the linearized entropy production expression, the function $\psi = k_B(1 - \ln f)$ is substituted into (3.14). Then using the representation (5.6) or (5.31), we obtain the compact form of the entropy production

$$\left(\frac{\partial s}{\partial t}\right)_{coll} = -k_B(h, \hat{L}h). \tag{5.46}$$

Since the entropy production because of the intermolecular collisions is positive, the operator $\hat{L}$ satisfies the property

$$(h, \hat{L}h) \leq 0. \tag{5.47}$$

In order to obtain the entropy production in the whole volume $\mathcal{V}$ occupied by a gas, the equality (5.46) must be integrated over the volume as it has been done in Eq. (4.43). Then using the definition (5.5), the entropy production because of intermolecular collisions takes the form:

$$\sigma_{coll} = -k_B\left((h, \hat{L}h)\right). \tag{5.48}$$

An expression of the linearized collision operator $\hat{L}h$ in terms of the DCS section and the proofs of its properties mentioned earlier can be found in the books [4, 5, 18, 20].

5.5
Linearization of Boundary Condition

5.5.1
Impermeable Surface Being at Rest

To illustrate the linearization of the boundary condition (4.4), let us consider the linearization near the global Maxwellian (5.1) corresponding to a temperature T_0. We assume that a solid surface is kept at a temperature T_w just slightly different from T_0 so that

$$T_w = T_0(1 + \xi\tau_w), \tag{5.49}$$

where τ_w is some given value, which can vary from one point of the surface to other. Both incident $f(\boldsymbol{v}')$ and reflected $f(\boldsymbol{v})$ distribution functions related by (4.4) are represented in the linear form (5.6). Then after cumbersome but not difficult derivations, the reflected perturbation function $h(\boldsymbol{v})$ is related to the incident perturbation $h(\boldsymbol{v}')$ as

$$h(\boldsymbol{v}) = \hat{A}h + h_w - \hat{A}h_w, \tag{5.50}$$

where the linearized scattering operator is defined as

$$\hat{A}h = -\frac{1}{v_n f_0^{\mathrm{M}}(\boldsymbol{v})} \int_{v_n' \leq 0} v_n' f_0^{\mathrm{M}}(\boldsymbol{v}') R_0(\boldsymbol{v}', \boldsymbol{v}) h(\boldsymbol{v}') \, \mathrm{d}^3\boldsymbol{v}'. \tag{5.51}$$

The notation $R_0(\boldsymbol{v}', \boldsymbol{v})$ means the scattering kernel corresponding to the equilibrium temperature T_0 so that the operator $\hat{A}$ does not depend on the surface temperature. In contrast to $\hat{A}$, the function h_w is completely determined by the surface temperature and reads

$$h_w = \tau_w \left(c^2 - \frac{3}{2} \right), \tag{5.52}$$

where $\boldsymbol{c}$ is defined by (5.2). This function is called surface source term.

5.5.2
Impermeable Moving Surface

If the surface moves with a velocity $\boldsymbol{u}_w$, which is small in comparison with the MPS

$$\boldsymbol{u}_w = \xi v_0 \tilde{\boldsymbol{u}}_w, \tag{5.53}$$

then all velocities in the boundary condition (4.4) are shifted by $-\boldsymbol{u}_w$. When all nonlinear terms are omitted, the linearized boundary condition will have the same form (5.50) with the surface source term given as

$$h_w = 2\boldsymbol{c} \cdot \tilde{\boldsymbol{u}}_w + \tau_w \left(c^2 - \frac{3}{2} \right). \tag{5.54}$$

This function can be treated as the perturbation of the surface Maxwellian f_w^{M}

$$f_w^{\mathrm{M}} = f_0^{\mathrm{M}}(1 + \xi h_w), \tag{5.55}$$

where

$$f_w^{\mathrm{M}} = n_0 \left(\frac{m}{2\pi k_{\mathrm{B}} T_w} \right)^{3/2} \exp\left[-\frac{m(\boldsymbol{v} - \boldsymbol{u}_w)^2}{2k_{\mathrm{B}} T_w} \right] \tag{5.56}$$

for impermeable moving surface. In this case, the kernel $R_0(\boldsymbol{v}', \boldsymbol{v})$ corresponds to the equilibrium temperature T_0 and to the surface being at rest $\boldsymbol{u}_w = 0$. The details to derive (5.50) with (5.54) are given in Chapter IV of the book [18].

5.5.3
Permeable Surface

Now, let us assume that the surface condensates and evaporates a gas so that the density of saturated vapor n_w slightly differs from the equilibrium one

$$n_w = n_0(1 + \xi\varrho_w). \tag{5.57}$$

If the evaporating surface moves, all velocities in (4.32) are shifted by $-\boldsymbol{u}_w$. For instance, the surface Maxwellian (4.33) takes the form

$$f_w^{\mathrm{M}}(\boldsymbol{v}) = n_w \left(\frac{m}{2\pi k_{\mathrm{B}} T_w} \right)^{3/2} \exp\left[-\frac{m(\boldsymbol{v} - \boldsymbol{u}_w)^2}{2k_{\mathrm{B}} T_w} \right]. \tag{5.58}$$

Then the linearization of the boundary condition (4.32) leads again to the linearized form (5.50) with the following surface source term

$$h_w = \varrho_w + 2\boldsymbol{c} \cdot \tilde{\boldsymbol{u}}_w + \tau_w \left(c^2 - \frac{3}{2} \right). \tag{5.59}$$

In other words, this source term h_w represents the deviation of the surface Maxwellian f_w^{M} given by (5.55).

5.5.4
Linearization Near Reference Maxwellian

If the linearization is performed near the reference Maxwellian (5.28), the linearized boundary conditions are obtained again in the form (5.50) with the surface source term h_w representing the deviation of f_w^{M} from $f_{\mathrm{R}}^{\mathrm{M}}$

$$f_w^{\mathrm{M}} = f_{\mathrm{R}}^{\mathrm{M}}(1 + \xi h_w) = f_0^{\mathrm{M}}[1 + \xi(h_{\mathrm{R}} + h_w)]. \tag{5.60}$$

In this case, the quantities ϱ_w, $\tilde{\boldsymbol{u}}_w$, and τ_w determine the corresponding quantities relative to their reference values

$$n_w = n_{\mathrm{R}}(1 + \xi\varrho_w) = n_0[1 + (\varrho_{\mathrm{R}} + \xi\varrho_w)], \tag{5.61}$$

$$\boldsymbol{u}_w = \xi v_0(\tilde{\boldsymbol{u}}_{\mathrm{R}} + \tilde{\boldsymbol{u}}_w), \tag{5.62}$$

$$T_w = T_{\mathrm{R}}(1 + \xi\tau_w) = T_0[1 + \xi(\tau_{\mathrm{R}} + \tau_w)]. \tag{5.63}$$

5.5.5
Properties of Scattering Operator

Let us introduce the so-called boundary inner produce for two functions $\psi(\boldsymbol{v})$ and $\phi(\boldsymbol{v})$ defined on a surface as

$$(\psi, \phi)_{\mathrm{B}} = \int_{v_n \geq 0} v_n \psi(\boldsymbol{v}) f_0^{\mathrm{M}} \phi(\boldsymbol{v}) \, \mathrm{d}^3\boldsymbol{v}. \tag{5.64}$$

It can be verified that the property (4.8) for the kernel $R_0(\boldsymbol{v}', \boldsymbol{v})$ with the temperature T_0 and the definitions (5.38) and (5.51) leads to the following property of the scattering operator $\hat{A}$

$$(\hat{T}\psi, \hat{A}\phi)_{\rm B} = (\hat{T}\phi, \hat{A}\psi)_{\rm B}. \tag{5.65}$$

To obtain the linearized expression of the entropy production $d\sigma_w$ because of the gas–surface interaction, the representation (5.6) or (5.31) together with (5.55) is substituted into Eq. (4.37). Then the entropy production takes the form:

$$\sigma_w = k_{\rm B} \int_{\Sigma_w} \left[\left(v_n h_w, h - \frac{1}{2} h_w \right) - \frac{1}{2} (v_n h, h) \right] \, d^2\Sigma_w. \tag{5.66}$$

5.5.6 Diffuse Scattering

It is useful to specify the expression (5.50) for the diffuse scattering law. Substituting the kernel (4.10) into (5.51), the scattering operator can be written down in the form:

$$\hat{A}_d h = -\frac{2}{\pi} \int_{c'_n \leq 0} c'_n \exp(-c'^2) h(\boldsymbol{c}') \, d^3\boldsymbol{c}', \tag{5.67}$$

where the dimensionless velocity $\boldsymbol{c}$ is given by (5.2). It is easy to calculate the results of application of the operator $\hat{A}_d$ to various velocity functions, for example,

$$\hat{A}_d 1 = 1, \quad \hat{A}_d \boldsymbol{c}_t = 0, \quad \hat{A}_d c_n = \frac{\sqrt{\pi}}{2}, \quad \hat{A}_d c^2 = 2. \tag{5.68}$$

A linear combination of these expressions leads to the following equality

$$\hat{A}_d h_w = \sqrt{\pi} \tilde{u}_{wn} + \frac{1}{2} \tau_w. \tag{5.69}$$

Substituting this expression into (5.50) with (5.54), the perturbation of the reflected particles can be written as

$$h = \varrho_r + \sqrt{\pi} \tilde{u}_{wn} + 2\boldsymbol{c} \cdot \tilde{\boldsymbol{u}}_w + \tau_w (c^2 - 2), \tag{5.70}$$

where the quantity ϱ_r does not depend on the molecular velocity and expressed via the incident perturbation

$$\varrho_r = -\frac{2}{\pi} \int_{c'_n \leq 0} c'_n \exp(-c'^2) h(\boldsymbol{c}') \, d^3\boldsymbol{c}'. \tag{5.71}$$

5.6 Series Expansion

In some situations, a solution of the full BE can be expressed as a combination of solutions of the linearized BE. Let us expand the VDF in power series with respect to the small parameter ξ

$$f = f_0^{\mathrm{M}}\left(1 + \sum_{n=1}^{\infty} \xi^n h_n\right). \tag{5.72}$$

In the same manner, the scattering kernels used in the boundary condition (4.4) and the surface Maxwellian (5.58) are expanded as

$$R(\boldsymbol{v}', \boldsymbol{v}) = R_0(\boldsymbol{v}', \boldsymbol{v}) + \sum_{n=1}^{\infty} \xi^n R_n(\boldsymbol{v}', \boldsymbol{v}) \tag{5.73}$$

and

$$f_w^{\mathrm{M}} = f_0^{\mathrm{M}}\left(1 + \sum_{n=1}^{\infty} \xi^n h_{wn}\right), \tag{5.74}$$

respectively. The surface Maxwellian f_w^{M} is defined by (5.56) for impermeable surface and by (5.58) for permeable surface.

A substitution of the expansion (5.72) into the BE (3.1) leads to chain of the linearized BEs

$$\frac{\partial h_n}{\partial t} + \boldsymbol{v} \cdot \nabla_r h_n = \hat{L} h_n + g_n, \quad 1 \le n < \infty, \tag{5.75}$$

related recurrently via the bulk source terms given as

$$g_1 = 0, \tag{5.76}$$

$$g_n(t, \boldsymbol{r}, \boldsymbol{v}) = \sum_{k=1}^{n-1} \iiint f_0^{\mathrm{M}}(\boldsymbol{v}_*)[h_k(\boldsymbol{v}')h_{n-k}(\boldsymbol{v}'_*) - h_k(\boldsymbol{v})h_{n-k}(\boldsymbol{v}_*)]$$
$$\times w(\boldsymbol{v}', \boldsymbol{v}'_*; \boldsymbol{v}, \boldsymbol{v}_*)\, \mathrm{d}^3\boldsymbol{v}'\, \mathrm{d}^3\boldsymbol{v}'_*\, \mathrm{d}^3\boldsymbol{v}_*, \quad n \ge 2. \tag{5.77}$$

A derivation of this chain can be found in the book [18].

Substituting the expansion (5.72) into the boundary condition (4.4) or (4.32) produces a chain of the boundary conditions

$$h_n = \hat{A}_0 h_n + h_{wn} - \hat{A}_0 h_{wn} + \eta_n, \tag{5.78}$$

related recurrently via the surface source terms given as

$$\eta_n(t, \boldsymbol{r}, \boldsymbol{v}) = \sum_{k=1}^{n-1} \hat{A}_{n-k}(h_k - h_{wk}). \tag{5.79}$$

Each operator $\hat{A}_n$ is defined via the corresponding term in the kernel expansion (5.73)

$$\hat{A}_n h = -\frac{1}{v_n f_0^{\mathrm{M}}(\boldsymbol{v})} \int_{v'_n \le 0} v'_n f_0^{\mathrm{M}}(\boldsymbol{v}') R_n(\boldsymbol{v}', \boldsymbol{v}) h(\boldsymbol{v}')\, \mathrm{d}^3\boldsymbol{v}'. \tag{5.80}$$

What is the usefulness of the expansion (5.72)? Formally, each equation (5.75) has exactly the same form as Eq. (5.33), while the boundary condition for each term h_n (5.78) is just slightly different from Eq. (5.50). Thus, all methods elaborated to solve the linearized BE can be applied to the chain (5.75) representing the full BE. Moreover, a knowledge of the second term h_2 in the expansion (5.72) allows us to estimate the range of applicability of the linearized equation. Such a range is not universal for all problems, but it depends on the gas rarefaction for any specific problem. The radius of convergence of the expansion (5.72) is also different for each problem and must be analyzed for each specific case.

The details on the power expansion are given in Ref. [39].

5.7 Reciprocal Relations

5.7.1 General Definitions

The reciprocal relations established by Onsager [40, 41] in 1931 are very important results of the nonequilibrium thermodynamics. Casimir [42] generalized the reciprocal relations considering different types of thermodynamic fluxes, namely, odd and even with respect to the time reverse. The paper [43] pointed out that some fluxes are neither odd nor even. Then the reciprocal relations were written down in a more general form.

Why do we need the reciprocal relations? Using these relations, one can couple the kinetic coefficients corresponding to the so-called cross phenomena. As a result, the number of the kinetic coefficients determining irreversible processes is reduced. These relations also can be used as additional criteria of numerical accuracy and experimental uncertainty. In many practical applications, the reciprocal relations allow us to diminish a number of measurements in experiments and to reduce computational efforts in numerical calculations. Nowadays, the reciprocal relations are not just a fundamental property of the nonequilibrium thermodynamics, but they are very useful tools in computational physics and engineering.

The reciprocal relations express the following properties. Consider a system weakly perturbed by N thermodynamic forces described by small parameters ξ_k ($1 \leq k \leq N$). Each force ξ_k causes a thermodynamics flux J_k, for example, a pressure gradient leads to a mass flow, a temperature gradient induces a heat flow, and so on. In a general case, a flux J_k can be induced not only by the corresponding force ξ_k but also by many other forces so that all fluxes linearly depend on all forces

$$J_k = \sum_{n=1}^{N} \Lambda_{kn}\xi_n, \tag{5.81}$$

where Λ_{kn} are the kinetic coefficients. If the set of the thermodynamic fluxes J_k is chosen so that the entropy production in the statistical system is expressed as the sum

$$\sigma = k_B \sum_{k=1}^{N} J_k\xi_k, \tag{5.82}$$

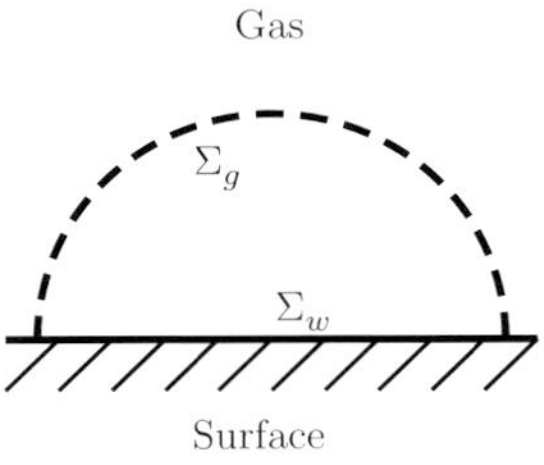

Figure 5.1 Scheme of surfaces Σ_w and Σ_g.

then according to reciprocal theorem in its general form [43], the kinetic coefficients satisfy the following relation:

$$\Lambda_{kn}^t = \Lambda_{nk}^t, \tag{5.83}$$

where the upper index t means the time-reversed kinetic coefficient. Following are the main ideas to obtain the kinetic coefficients.

5.7.2
Kinetic Coefficients

In a general case, a domain occupied by a gas is not always surrounded by a wall surface Σ_w, but it can be infinite. In this case, an imaginary surface Σ_g in the gas volume is introduced so that the surfaces Σ_w and Σ_g compose a closed surface. An example of the surfaces Σ_w and Σ_g is given in Figure 5.1 for a semi-infinite domain.

The expressions of the kinetic coefficients satisfying the reciprocal relations (5.83) were obtained in Ref. [44]. In most general forms, they read

$$\Lambda_{kn} = \left(\left(g^{(k)}, h^{(n)} \right) \right) + \int_{\Sigma_w} \left(v_n h_w^{(k)}, h^{(n)} - \frac{1}{2} h_w^{(n)} \right) \, \mathrm{d}^2\Sigma_w$$
$$+ \frac{1}{2} \int_{\Sigma_g} (v_n h^{(k)}, h^{(n)}) \, \mathrm{d}^2\Sigma_g - \left(\left(\frac{\partial h^{(k)}}{\partial t}, h^{(n)} \right) \right), \tag{5.84}$$

where $g^{(k)}$ and $h_w^{(k)}$ are the source terms corresponding to the force ξ_k, while $h^{(n)}$ is the solution of the linearized BE (5.33) with the boundary condition (5.50) corresponding to the force ξ_n. The time-reversed coefficients take the form

$$\Lambda_{kn}^t = \left(\left(\hat{T} g^{(k)}, h^{(n)} \right) \right) + \int_{\Sigma_w} \left(\hat{T} v_n h_w^{(k)}, h^{(n)} - \frac{1}{2} h_w^{(n)} \right) \mathrm{d}^2\Sigma_w$$
$$+ \frac{1}{2} \int_{\Sigma_g} (\hat{T} v_n h^{(k)}, h^{(n)}) \mathrm{d}^2\Sigma_g - \left(\left(\hat{T} \frac{\partial h^{(k)}}{\partial t}, h^{(n)} \right) \right). \tag{5.85}$$

The proof of the reciprocal relation (5.83) for the coefficients (5.85) is a hard task and out of the scope of this book. Here, we just note that the proof is based only on the property (5.42) of the linearized collision operator $\hat{L}$ and on the property (5.65) of the scattering operator $\hat{A}$. The relation of the kinetic coefficients (5.84) to the entropy production is shown in Ref. [44] and is based on Eqs. (5.48), (5.66), (5.33) and on the divergence Gauss–Ostrogradsky theorem.

The papers [43–46] contain many examples of the application of the reciprocal relations in case of a single gas. When a solution of the full BE is presented as the sum (5.72) of solutions of the linearized BE, the different terms of (5.72) are also coupled by the reciprocal relations [39]. A generalization for gaseous mixtures can be found in Refs [47, 48]. There are some peculiarities in the reciprocal relations for rotating systems [49], for polyatomic gas being in a magnetic field [50], and for gas interacting with a laser radiation [51]. Additional information about the reciprocal relations can be found in Refs [52–55].

In subsequent chapter, the reciprocal relations will be illustrated for some classical problems of fluid mechanics.

Exercises

5.1 Obtain (5.9).
Hint: Substitute (5.6) into (2.26), use (5.7) to express the density together with (A.20). Then omit all terms of the order ξ^2 and higher.

5.2 Obtain (5.11).
Hint: Substitute (5.6) into (2.32). Use (5.8) to express the bulk velocity. Then omit terms of the order ξ^2 and higher.

5.3 Obtain (5.30).
Hint: Substitute (5.29) into (5.28). Use (A.20) and (A.21). Then omit terms of the order ξ^2 and higher.

5.4 Show (5.40).
Hint: Multiplying $\hat{L}\phi$ in the form (5.21) by $\psi(-\boldsymbol{v})$ and integrating it with respect to $\boldsymbol{v}$, the left-hand side of (5.40) is obtained. Then exchange the variables as $(\boldsymbol{v}, \boldsymbol{v}') \to (-\boldsymbol{v}', -\boldsymbol{v})$ and $(\boldsymbol{v}_*, \boldsymbol{v}'_*) \to (-\boldsymbol{v}'_*, -\boldsymbol{v}_*)$. The use of (3.3) leads to the right-hand side of (5.40).

5.5 Obtain (5.46).
Hints: Substitute (5.6) into (3.30). Use (A.22), then omit all terms of the order ξ^2 and higher.

5.6 Prove (5.68).

5.7 Check Eq. (5.50) with (5.54) for a surface reflecting diffusely (4.10) and having a tangential velocity $\boldsymbol{u}_w$.
Hint: Use the kernel (4.10) with the tangential velocity shifted by $-\boldsymbol{u}_w$

$$R_d(\boldsymbol{v}', \boldsymbol{v}) = \frac{m^2\, v_n}{2\pi(k_B T_w)^2} \exp\left\{ -\frac{m\left[v_n^2 + (\boldsymbol{v}_t - \boldsymbol{u}_w)^2\right]}{2k_B T_w} \right\}. \tag{5.86}$$

5.8 Prove (5.65).
Hint: Write (4.8) for $R_0(\boldsymbol{v}', \boldsymbol{v})$ as

$$v'_n f_0^M(\boldsymbol{v}') R_0(\boldsymbol{v}', \boldsymbol{v}) = -v_n f_0^M(\boldsymbol{v}) R_0(-\boldsymbol{v}, -\boldsymbol{v}'). \tag{5.87}$$

Multiply it by $\psi(\boldsymbol{v}')\phi(-\boldsymbol{v})$ and integrate with respect to $\boldsymbol{v}$ ($v_n \geq 0$) and $\boldsymbol{v}'$ ($v'_n \leq 0$). Then change the sign of both velocities $\boldsymbol{v}$ and $\boldsymbol{v}'$ and exchange them, $(\boldsymbol{v}, \boldsymbol{v}') \to (-\boldsymbol{v}', -\boldsymbol{v})$.

6
Transport Coefficients

6.1
Constitutive Equations

In this section, two fundamental laws of mechanics of continuous media, namely, Newton's law and Fourier's law, are given. The first law relates the pressure tensor **P** to a bulk velocity variation, while the second law relates the heat flow vector $\boldsymbol{q}$ to a temperature variation, that is, they describe a response of continuous medium to external stimuli and are called constitutive equations. These laws are valid under the assumption of continuous mechanics discussed in Section 1.1.

Let us define the rate-of-shear tensor **S** as

$$S_{ij} = \frac{1}{2}\left(\frac{\partial u_i}{\partial r_j} + \frac{\partial u_j}{\partial r_i}\right) - \frac{1}{3}\delta_{ij}\nabla_r \cdot \boldsymbol{u}, \tag{6.1}$$

where r_1, r_2, r_3 are components of the position vector $\boldsymbol{r}$ and δ_{ij} is the Kronecker delta (see Eq. (A.14)). According to Newton's law, the pressure tensor is related to the rate-of-shear tensor as

$$\mathbf{P} = p\,\mathbf{I} - 2\mu\mathbf{S}, \tag{6.2}$$

where μ is the shear viscosity. Since both tensors **P** and **S** are symmetric by their definition, the expression (6.2) represents six independent equations. According to Fourier's law, the heat flow vector is proportional to the local temperature gradient

$$\boldsymbol{q} = -\kappa\nabla_r T. \tag{6.3}$$

This expression represents three equations. Thus, the constitutive equations (6.2) and (6.3) together with the conservation laws (3.26)–(3.28), state equation (1.9), and energy equation (1.14) compose a closed system of the 16 equations for the 16 variables listed in the end of Section 3.3. The system represents the so-called Navier–Stokes equations.

According to Chapter 2, the state and energy equations follow from the VDF definition. The conservation laws are general properties of the BE derived in Chapter 3. In other words, these relations are valid for any gas rarefaction and for any nonequilibrium degree. Now one may ask, how to derive the constitutive equations and what are conditions of their validity? According to the

Rarefied Gas Dynamics: Fundamentals for Research and Practice, First Edition. Felix Sharipov.

Chapman–Enskog method [3–5], both Newton's (6.2) and Fourier's (6.3) laws are derived from the BE expanding the VDF in power series with respect to the Knudsen number and retaining just the first-order terms. Physically it means that the method is valid when a characteristic size of gas flow is significantly larger than the molecular mean-free-path. Moreover, this method is based on the assumption that the VDF does not depend explicitly on the time. It is justified when a significant change of macroscopic variable happens during a time interval significantly larger than the mean-free-time. Thus, the derivation of the constitutive equations (6.2) and (6.3) from the BE is based on the two restrictions of the mechanics of the continuous media discussed in Section 1.1. In summary, the Navier–Stokes equations are derived from the BE and valid only for continuous media.

Solutions of some problems on transport phenomena are expressed via the so-called Prandtl number defined via the transport coefficients as

$$\mathrm{Pr} := c_p \frac{\mu}{\kappa}, \quad c_p = \frac{5}{2}\frac{k_B}{m}, \tag{6.4}$$

where c_p is the specific heat per particle at a constant pressure for a monatomic gas. Later it will be shown that this number very weakly depends on the intermolecular potential and is very close to the value 2/3 for monatomic gases.

6.2 Viscosity

To calculate the viscosity coefficient , let us consider a steady gas flow in a boundless region with the bulk velocity having only the second component depending only on the r_1 component

$$u_1 = 0, \quad u_2 = \xi v_0 \frac{r_1}{\ell}, \quad u_3 = 0, \tag{6.5}$$

where ξ is a constant dimensionless gradient of u_2 and ℓ is the EFP. The density n_0 and the temperature T_0 of the gas are assumed to be constant over the whole space. Then the rate-of-shear tensor takes the form:

$$S_{12} = S_{21} = \frac{1}{2}\frac{\mathrm{d}u_2}{\mathrm{d}r_1} = \xi \frac{v_0}{2\ell}, \tag{6.6}$$

while all other components of the tensor are equal to zero. According to Newton's law (6.2), the pressure tensor reads

$$P_{11} = P_{22} = P_{33} = p, \quad P_{23} = P_{32} = P_{13} = P_{31} = 0, \tag{6.7}$$

$$P_{12} = P_{21} = -\mu \frac{\mathrm{d}u_2}{\mathrm{d}r_1}, \tag{6.8}$$

where (6.6) has been used. Let us obtain the last relation from the linearized BE, then the viscosity coefficient μ will be extracted.

The constitutive equations are valid under the assumption that variations of all quantities are small on the distance of the EFP, that is, the bulk velocity gradient

$$\xi = \frac{\ell}{v_0}\frac{du_2}{dr_1}, \quad |\xi| \ll 1, \tag{6.9}$$

must be small and can be used for the linearization representing the VDF by Eq. (5.32). Then the pressure tensor component P_{12} is calculated using the definition (5.10)

$$P_{12} = \xi(\psi_P, h), \quad \psi_P = mv_1v_2. \tag{6.10}$$

If we choose the reference number density and temperature to be equal to their equilibrium values, $n_R = n_0$ and $T_R = T_0$, while the reference bulk velocity u_R to be given by (6.5), the perturbation function will satisfy the spatially homogeneous BE

$$\hat{L}h = \frac{2\psi_P}{mv_0\ell} \tag{6.11}$$

obtained from Eqs. (5.29), (5.30), (5.33), and (5.34). Let us introduce a new perturbation $h^{(\mu)}$ satisfying the equation

$$\hat{L}h^{(\mu)} = \psi_P, \tag{6.12}$$

then the perturbation function of our interest is represented as

$$h = \frac{2}{mv_0\ell}h^{(\mu)}. \tag{6.13}$$

Substituting the expressions (6.13) and (6.9) into (6.10), the shear stress takes the form

$$P_{12} = \frac{(\psi_P, h^{(\mu)})}{k_B T_0}\frac{du_2}{dr_1}. \tag{6.14}$$

Comparing this expression with (6.8), the viscosity is obtained as

$$\mu = -\frac{(\psi_P, h^{(\mu)})}{k_B T_0}. \tag{6.15}$$

Note, f_0^M included into the operator $\hat{L}$, see Eq. (5.21), makes the solution $h^{(\mu)}$ of Eq. (6.12) inversely proportional to the number density n_0, but the product (,) defined by (5.3) is proportional to n_0 so that the viscosity calculated from the BE by (6.15) is independent of the density and hence of the pressure, but it depends only on the temperature and is determined by the intermolecular potential via the collision operator $\hat{L}$.

6.3 Thermal Conductivity

To calculate the thermal conductivity coefficient, we should use the same logic as in the previous section, but with a variable temperature instead of variable bulk velocity. Again, let us consider a gas occupying a boundless region, but being at rest, $\boldsymbol{u} = 0$. In addition, the gas temperature is not constant, but it linearly depends on the r_1 coordinate

$$T = T_0 \left(1 + \xi \frac{r_1}{\ell}\right). \tag{6.16}$$

The Fourier law is valid when the temperature variation is small across the distance ℓ, that is, the temperature gradient ξ should be small

$$\xi = \frac{\ell}{T_0} \frac{\mathrm{d}T}{\mathrm{d}r_1}, \quad |\xi| \ll 1. \tag{6.17}$$

Then the heat flow q_1 is obtained from the definition (5.11) as

$$q_1 = \xi \left(\psi_q, h\right), \quad \psi_q = k_B T_0 \left(\frac{v^2}{v_0^2} - \frac{5}{2}\right) v_1. \tag{6.18}$$

Since the gas is at rest, its pressure $p = nk_B T$ should be constant, and therefore the number density must be distributed as

$$n = n_0 \left(1 - \xi \frac{r_1}{\ell}\right). \tag{6.19}$$

Under such conditions, Fourier's law takes the form:

$$q_1 = -\kappa \frac{\mathrm{d}T}{\mathrm{d}r_1}. \tag{6.20}$$

If we obtain this relation from the linearized BE, the thermal conductivity κ will be extracted.

In the problem under consideration, the reference bulk velocity u_R is zero, while the reference density n_R and temperature T_R are assumed to be equal to (6.19) and (6.16), respectively. Then, the deviations defined by (5.29) are obtained as

$$\varrho_R = -\frac{r_1}{\ell}, \quad \tau_R = \frac{r_1}{\ell}. \tag{6.21}$$

Thus, the perturbation function defined by (5.32) satisfies the spatially homogeneous BE

$$\hat{L}h = \frac{\psi_q}{\ell k_B T_0}. \tag{6.22}$$

Denoting by $h^{(\kappa)}$, the solution of the integral equation

$$\hat{L}h^{(\kappa)} = \psi_q, \tag{6.23}$$

the perturbation function h is represented as

$$h = \frac{h^{(\kappa)}}{\ell k_B T_0}. \tag{6.24}$$

Substituting Eqs. (6.17) and (6.24) into the last relation, the heat flow takes the form:

$$q_1 = \frac{(\psi_q, h^{(\kappa)})}{k_B T_0^2} \frac{\mathrm{d}T}{\mathrm{d}r_1}. \tag{6.25}$$

Comparing this equation with (6.20), the thermal conductivity is extracted as

$$\kappa = -\frac{(\psi_q, h^{(\kappa)})}{k_B T_0^2}. \tag{6.26}$$

The thermal conductivity is independent of the pressure by the same reason as the viscosity.

6.4 Numerical Results

The transport coefficients could be easily calculated if the functions $h^{(\mu)}$ and $h^{(\kappa)}$ would be known. However, it is not so easy to solve the integral equations (6.12) and (6.23). A variational method to such equations using an expansion of the perturbation functions $h^{(\mu)}$ and $h^{(\kappa)}$ with respect to the Sonin polynomials is described in the books [3–5]. Direct numerical solutions of the equations are given in the paper [56] for the HS and in the work [21] for the LJ potential. The results are summarized in the following.

6.4.1 Hard Sphere Potential

The first approximation based on the Sonin polynomials provides the following expressions of the transport coefficients for the HS potential

$$\mu = \frac{5}{16}\frac{\sqrt{\pi m k_B T}}{\sigma_t}, \quad \kappa = \frac{75 k_B}{64 m}\frac{\sqrt{\pi m k_B T}}{\sigma_t}, \quad \sigma_t = \pi d^2, \tag{6.27}$$

where σ_t is the TCS defined by (1.26) and d is the particle diameter. Substituting these expressions into the Prandtl number Pr definition (6.4), one verifies that it is exactly 2/3. The papers [57, 58] reported numerical data for high-order approximations based on the Sonin polynomials. Their results can be written as

$$\mu = 1.01603\frac{5}{16}\frac{\sqrt{\pi m k_B T}}{\sigma_t}, \quad \kappa = 1.02522\frac{75 k_B}{64 m}\frac{\sqrt{\pi m k_B T}}{\sigma_t}, \tag{6.28}$$

that is, the high-order corrections are relatively small. In this case, the Prandtl number is slightly different from 2/3 and equal to $\mathrm{Pr} = 0.660686$.

Numerical solutions of Eqs. (6.23) and (6.25) are reported in Refs [21, 56] where the expressions (6.28) are confirmed.

6.4.2 Lennard-Jones Potential

The variational method described in the books [3–5] can be also applied to any intermolecular potential so that the transport coefficients are expressed via the so-called Ω-integrals calculated numerically.

In the work [21], the coefficients μ and κ were calculated numerically via a solution of Eqs. (6.12) and (6.23) based on the Lennard-Jones potential (2.38). In this

Table 6.1 Dimensionless viscosity $\tilde{\mu}$ and thermal conductivity $\tilde{\kappa}$ based on the LJ potential versus reduced temperature T^*, Ref. [21].

T^*	$\tilde{\mu}$	$\tilde{\kappa}$
29.4	0.1787	0.6740
8.40	0.1480	0.5600
2.42	0.1130	0.4260
1.58	0.09680	0.3645
1.31	0.08919	0.3358

case, it is convenient to present the results in terms of the dimensionless coefficients $\tilde{\mu}$ and $\tilde{\kappa}$ defined as

$$\mu(T) = \tilde{\mu}(T^*)\frac{mv_m}{d^2}, \quad \kappa(T) = \tilde{\kappa}(T^*)\frac{k_B v_m}{d^2}, \tag{6.29}$$

and depending only on the reduced temperature

$$T^* = \frac{k_B T}{\varepsilon}. \tag{6.30}$$

Thus, having data of $\tilde{\mu}$ and $\tilde{\kappa}$ in terms of T^*, the transport coefficients μ and κ can be calculated for any gas at any temperature via the relations (6.29). Some numerical values of $\tilde{\mu}$ and $\tilde{\kappa}$ calculated in Ref. [21] are given in Table 6.1. The values of T^* given in this table correspond to the gases helium, neon, argon, krypton, and xenon, respectively, at $T = 300$ K. The Prandtl number in terms of the dimensionless coefficient is given as $\text{Pr} = 5\tilde{\mu}/2\tilde{\kappa}$. Using the data of Table 6.1, one can verify that the Prandtl number varies in the range $0.6607 \leq Pr \leq 0.6640$, that is, $\text{Pr} \approx 2/3$. More extensive numerical data to calculate the viscosity and thermal conductivity for the LJ potential can be found in the book [3].

6.4.3
Ab Initio Potential

Numerical data on the various properties, including the viscosity and thermal conductivity, of the noble gases calculated on the basis of the AI potential for a wide range of the temperature T can be found in Refs [59–61]. Some of these data are reproduced in Table 6.2 for few values of the temperature. At the moment, the data reported in Refs [59–61] can be considered the most exact theoretical values of the viscosity and thermal conductivity. Empirical data on the transport coefficients reported in Refs [2, 62, 63] are in agreement with these theoretical values within the experimental uncertainty.

Exercises

6.1 Calculate the viscosity μ and thermal conductivity κ of helium, neon, and argon using the data of Tables 1.1 and 6.1. Compare these data with those given in Table 6.2. What is the maximum discrepancy between these data?

Table 6.2 Viscosity μ and thermal conductivity κ based on the AI potential versus temperature T, Refs [59–61].

	μ (μPa s)			κ (mW m^{-1} K^{-1})		
T (K)	He	Ne	Ar	He	Ne	Ar
100	9.5531	14.399	8.1271	74.735	22.277	6.3421
300	19.910	31.860	22.669	155.66	49.410	17.709
500	28.331	44.829	33.921	221.42	69.542	26.534
1000	46.357	71.141	55.450	362.12	110.35	43.430
2000	77.253	113.72	88.604	603.15	176.35	69.416

Solution: 0.6%.

6.2 Extract the diameter d_{He} of helium atom from the numerical data given in Table 6.2 using the viscosity expression (6.28).
Solution:

T (K)	100	300	500	1000	2000
d_{He} (nm)	0.2383	0.2173	0.2070	0.1924	0.1772

Comment: The diameter value depends on the temperature T because the expression (6.28) does not provide the real dependence of viscosity on the temperature.

6.3 Extract the diameter d_{Ar} of argon atom from the numerical data given in Table 6.2 using the viscosity expression (6.28).

T (K)	100	300	500	1000	2000
d_{Ar} (nm)	0.4593	0.3619	0.3362	0.3127	0.2942

6.4 Calculate the ratio of the diameters of helium and argon at various temperatures.
Solution:

T (K)	100	300	500	1000	2000
d_{Ar}/d_{He}	1.927	1.666	1.624	1.625	1.660

Comment: The ratio d_{Ar}/d_{He} is practically independent of the temperature at $T \geq 300$ K. This fact is important for a modeling of gaseous mixture's flows.

6.5 Calculate the Prandtl number using the data of Table 6.2. Check that in all cases $Pr \approx 2/3$.

7
Model Equations

A numerical solution of the BE (3.1) with the exact collision integral (3.5) or (3.10) is still a very hard task even if one uses modern powerful computers. That is why some simplified equations satisfying the main properties of the BE were proposed. Such approximations are called model equations. In this chapter, principal model equations, their advantages, and their shortcomings are described.

7.1
BGK Equation

The first model equation was proposed by Bhatnagar *et al.*, [64] (BGK) and independently by Welander [65]. They assumed that the variation rate of the VDF $f(t, \boldsymbol{r}, \boldsymbol{v})$ is proportional to its deviation from the local Maxwellian f^{M}. Mathematically, it means that the exact collision integral (3.5) is replaced by the following:

$$Q_{\mathrm{B}} = \nu_{\mathrm{B}}(f^{\mathrm{M}} - f)\,, \tag{7.1}$$

where f^{M} contains local values of the number density $n(t, \boldsymbol{r})$, bulk velocity $\boldsymbol{u}(t, \boldsymbol{r})$, and temperature $T(t, \boldsymbol{r})$ calculated *via* the distribution function $f(t, \boldsymbol{r}, \boldsymbol{v})$ in accordance with the definitions (2.4), (2.16), and (2.26), respectively. Thus, the quantities $n(t, \boldsymbol{r})$, $\boldsymbol{u}(t, \boldsymbol{r})$, and $T(t, \boldsymbol{r})$ contained by f^{M} in (7.1) are unknown. The quantity ν_{B} has the order of the intermolecular collision frequency to be obtained below.

One can verify that

$$\int \psi(\boldsymbol{v}) f^{\mathrm{M}}\, \mathrm{d}^3\boldsymbol{v} = \int \psi(\boldsymbol{v}) f\, \mathrm{d}^3\boldsymbol{v}, \quad \text{for} \quad \psi(\boldsymbol{v}) = m, m\boldsymbol{v}, \frac{1}{2} m v^2 \tag{7.2}$$

so that the model collision integral (7.1) satisfies the conservation law properties (3.20). If the expression $\ln f^{\mathrm{M}}$ is represented as a linear combination of the invariants

$$\ln f^{\mathrm{M}} = C_1 m + \boldsymbol{C}_2 \cdot (m\boldsymbol{v}) + C_3 \left(\frac{m v^2}{2} \right), \tag{7.3}$$

where C_1, $\boldsymbol{C}_2$, and C_3 are independent of the velocity $\boldsymbol{v}$, then the conservation law can be written as

$$\int Q_{\mathrm{B}} \ln f^{\mathrm{M}}\, \mathrm{d}^3\boldsymbol{v} = 0. \tag{7.4}$$

Rarefied Gas Dynamics: Fundamentals for Research and Practice, First Edition. Felix Sharipov.

This property allows us to prove the H-theorem doing the following derivation:

$$\int Q_B \ln f \, d^3\boldsymbol{v} = \int Q_B \ln \frac{f}{f^M} \, d^3\boldsymbol{v} = \nu_B \int f^M \left(1 - \frac{f}{f^M}\right) \ln \frac{f}{f^M} \, d^3\boldsymbol{v}. \quad (7.5)$$

It is verified that the expression $(1 - x) \ln x$ is never positive so that adopting $x = f/f^M$ in (7.5), the H-theorem (4.40) is proved for the model collision integral

$$\int Q_B \ln f \, d^3\boldsymbol{v} \leq 0. \quad (7.6)$$

In order to linearize the BKG collision integral Q_B, the representations of the moments (5.35)–(5.37) are substituted into the local Maxwellian f^M, which takes the form

$$f^M = f_0^M \left\{1 + \xi \left[\varrho + 2\boldsymbol{c} \cdot \tilde{\boldsymbol{u}} + \left(c^2 - \frac{3}{2}\right) \tau + h_R\right]\right\}, \quad (7.7)$$

where $\boldsymbol{c}$ is defined by Eq. (5.2), while the moments ϱ, $\tilde{\boldsymbol{u}}$, and τ are calculated via the perturbation function h by Eq. (5.17). Using the expression (7.7) and the representation of the VDF (5.32), the collision integral Q_B is linearized as

$$Q_B = \xi f_0^M \hat{L}_B h, \quad \hat{L}_B h = \nu_B \left[\varrho + 2\boldsymbol{c} \cdot \tilde{\boldsymbol{u}} + \left(c^2 - \frac{3}{2}\right) \tau - h\right]. \quad (7.8)$$

The linearized collision operator $\hat{L}_B$ approximates the exact operator $\hat{L}$ defined by (5.21) or (5.22).

Since the full model Q_B satisfies the H-theorem (7.6), the linearized model (7.8) automatically meets the linearized version (5.47) of the theorem

$$(h, \hat{L}_B h) \leq 0. \quad (7.9)$$

It is verified that the linearized operator (7.8) is self-adjoint

$$\begin{aligned}(\hat{L}_B h_1, h_2) &= (\hat{L}_B h_2, h_1) \\ &= \nu_B n_0 [\varrho_1 \varrho_2 + 2\tilde{\boldsymbol{u}}_1 \cdot \tilde{\boldsymbol{u}}_2 + (3/2)\tau_1 \tau_2] - \nu_B (h_1, h_2),\end{aligned} \quad (7.10)$$

that is, it satisfies (5.42). It is also commutative with the operator $\hat{T}$

$$\hat{T}\hat{L}_B h = \hat{L}_B \hat{T} h = \nu_B \left[\varrho - 2\boldsymbol{c} \cdot \tilde{\boldsymbol{u}} + \left(c^2 - \frac{3}{2}\right) \tau - h(-\boldsymbol{c})\right]. \quad (7.11)$$

Then using Eqs. (7.10), (7.11), and (5.39), the property (5.40) is proved for $\hat{L}_B$

$$(\hat{T}\hat{L}_B h_1, h_2) = (\hat{T}\hat{L}_B h_2, h_1). \quad (7.12)$$

All properties (7.4), (7.6), and (7.12) left the intermolecular collision frequency ν_B undefined. Let us relate it to the transport coefficients μ and κ. If we substitute the linearized operator (7.8) into the integral equations (6.12) and (6.23), which take into account the fact that it has been obtained under the conditions $\varrho = 0$, $\tilde{\boldsymbol{u}} = 0$, and $\tau = 0$, then the perturbation functions $h^{(\mu)}$ and $h^{(\kappa)}$ are obtained analytically as

$$h^{(\mu)} = -\frac{\psi_P}{\nu_B}, \quad h^{(\kappa)} = -\frac{\psi_q}{\nu_B}. \quad (7.13)$$

Substituting them into (6.15) and (6.26), the transport coefficients are obtained as

$$\mu = \frac{p}{\nu_\mathrm{B}}, \quad \kappa = \frac{5}{2}\frac{k_\mathrm{B}p}{\nu_\mathrm{B}m}. \tag{7.14}$$

It is easy to check using (6.4) that the expressions (7.14) lead to Pr = 1 instead of the correct value 2/3 (see Section 1.4.3). Thus, the choice

$$\nu_\mathrm{B} = \frac{p}{\mu} \tag{7.15}$$

provides the compatibility of the BGK model with the viscosity coefficient definition, but not with the thermal conductivity. The other choice

$$\nu_\mathrm{B} = \frac{5}{2}\frac{k_\mathrm{B}p}{\kappa m} \tag{7.16}$$

leads the correct thermal conductivity but not viscosity. However, no choice is compatible with both viscosity and thermal conductivity. This is the main disadvantage of the BGK model.

7.2 S-Model

Several models were proposed to overcome the problem of the wrong Prandtl number provided by the BGK model. The model proposed by Shakhov [66] called as the S-model represents the collision integral as

$$Q_\mathrm{S} = \nu_\mathrm{S}\left\{f^\mathrm{M}\left[1 + \frac{4}{15}\left(\frac{V^2}{v_m^2} - \frac{5}{2}\right)\frac{\boldsymbol{q}\cdot\boldsymbol{V}}{p\, v_m^2}\right] - f(t,\boldsymbol{r},\boldsymbol{v})\right\}, \tag{7.17}$$

where v_m is given by (2.40). It can be verified that

$$\int \psi(\boldsymbol{v}) f^\mathrm{M}\boldsymbol{V}\left(\frac{V^2}{v_m^2} - \frac{5}{2}\right) \mathrm{d}^3\boldsymbol{v} = 0, \quad \psi(\boldsymbol{v}) = 1, m\boldsymbol{v}, \frac{1}{2}mv^2 \tag{7.18}$$

so that the equality (7.2) provides the conservation laws (3.20) also for the S-model. However, the H-theorem (4.40) cannot be proved for the nonlinearized S-model.

Using the same procedure as that for the BGK model, the linearized form of the S-model is derived as

$$\hat{L}_\mathrm{S}h = \nu_\mathrm{S}\left[\varrho + 2\boldsymbol{c}\cdot\tilde{\boldsymbol{u}} + \left(c^2 - \frac{3}{2}\right)\tau + \frac{4}{15}\left(c^2 - \frac{5}{2}\right)\boldsymbol{c}\cdot\tilde{\boldsymbol{q}} - h\right], \tag{7.19}$$

where ϱ, τ, $\tilde{\boldsymbol{u}}$, and $\tilde{\boldsymbol{q}}$ are calculated via the perturbation function by Eq. (5.17). It is easy to verify that the linearized model satisfies the conservation laws in the form (5.45).

The model operator $\hat{L}_\mathrm{S}$ is self-adjoint

$$(\hat{L}_\mathrm{S}h_1, h_2) = (\hat{L}_\mathrm{S}h_2, h_1) = \nu_\mathrm{S}n_0\left(\varrho_1\varrho_2 + 2\tilde{\boldsymbol{u}}_1\cdot\tilde{\boldsymbol{u}}_2 + \frac{2}{3}\tau_1\tau_2 + \frac{4}{15}\tilde{\boldsymbol{q}}_1\cdot\tilde{\boldsymbol{q}}_2\right) - \nu_\mathrm{S}(h_1, h_2). \tag{7.20}$$

The operators $\hat{T}$ and $\hat{L}_S$ are commutative

$$\hat{T}\hat{L}_S h = \hat{L}_S\hat{T}h = \nu_S\left[\varrho - 2\boldsymbol{c}\cdot\tilde{\boldsymbol{u}} + \left(c^2 - \frac{3}{2}\right)\tau - \frac{4}{15}\left(c^2 - \frac{5}{2}\right)\boldsymbol{c}\cdot\tilde{\boldsymbol{q}} - h(-\boldsymbol{c})\right]. \tag{7.21}$$

A combination of Eqs. (7.20) and (7.21) and the self-adjointness of the operator $\hat{T}$ lead to the self-adjointness of the operator $\hat{T}\hat{L}_S$

$$(\hat{T}\hat{L}_S h_1, h_2) = (\hat{T}\hat{L}_S h_2, h_1). \tag{7.22}$$

To prove the linearized H-theorem (5.46), let abbreviate the expression (7.19) as

$$\hat{L}_S h = \nu_S(h_S - h), \tag{7.23}$$

where

$$h_S = \varrho + 2\boldsymbol{c}\cdot\tilde{\boldsymbol{u}} + \left(c^2 - \frac{3}{2}\right)\tau + \frac{4}{15}\left(c^2 - \frac{5}{2}\right)\boldsymbol{c}\cdot\tilde{\boldsymbol{q}}. \tag{7.24}$$

Then

$$(h, \hat{L}_S h) = (h - h_S, \hat{L}_S h) + (h_S, \hat{L}_S h), \tag{7.25}$$

where the first term is negative because it is given as

$$(h - h_S, \hat{L}_S h) = -\nu_S \int f_0^M (h - h_S)^2 \, d^3\boldsymbol{v} \leq 0. \tag{7.26}$$

The second term in (7.25) would be zero if the function h_S could be a linear combination of the collision invariants, but it contains $c^2\boldsymbol{c}$ which is not the invariant. Doing tedious but not complicated derivations, it can be shown that

$$(h_S, \hat{L}_S h) = -\frac{8}{45}\nu_S n_0 \tilde{q}^2 \leq 0. \tag{7.27}$$

Since both terms in the right-hand side of Eq. (7.25) are not positive, the linearized H-theorem (5.47) for the linearized S-model (7.8) is proved.

To determine the frequency ν_S, let us calculate the transport coefficients from the S-model. First, the integral equations (6.12) and (6.23) for the operator $\hat{L}_S$ are solved. The moments ϱ, $\tilde{\boldsymbol{u}}$, and τ are zero because of the choice of the reference Maxwellian f_R^M for the linearization. Representing the solution as $h^{(\mu)} = C_u \psi_P$ and $h^{(\kappa)} = C_T \psi_q$, the constants C_u and C_T are easily found so that the solutions read

$$h^{(\mu)} = -\frac{\psi_P}{\nu_S}, \quad h^{(\kappa)} = -\frac{3}{2}\frac{\psi_q}{\nu_S}, \tag{7.28}$$

where ψ_P and ψ_q are given by Eqs. (6.11) and (6.22), respectively. Then using (6.15) and (6.26), the transport coefficients are derived as

$$\mu = \frac{p}{\nu_S}, \quad \kappa = \frac{15}{4}\frac{k_B p}{\nu_S m}. \tag{7.29}$$

Substituting these expressions into the Prandtl number definition (6.4), we verify that Pr = 2/3, that is, it is very close to the value based on the LJ potential given in Section 1.4.3. Thus, if we extract the frequency ν_S from the viscosity expression (7.29)

$$\nu_S = \frac{p}{\mu}, \tag{7.30}$$

we obtained exactly the same as that obtained for the BGK model (7.15). In contrast to the BGK model, the S-model with the frequency (7.30) provides the correct expressions for both viscosity and thermal conductivity. This is the main advantage of the S-model, but the H-theorem can be proved only for its linearized form.

7.3
Ellipsoidal Model

Another model with the correct Prandtl number has the collision integral in the following form [67]:

$$Q_E = \nu_E \left[\frac{n}{(\sqrt{\pi} v_m)^3 \sqrt{\det(\mathbf{B})}} \exp\left(-\frac{\mathbf{B}^{-1} : \boldsymbol{V}\boldsymbol{V}}{v_m^2} \right) - f \right], \tag{7.31}$$

where the peculiar velocity $\boldsymbol{V}$ is given by (2.22), $\mathbf{B}^{-1}$ is the inverse tensor of $\mathbf{B}$ (see the definition (A.10)), $\det(\mathbf{B})$ means the determinant of $\mathbf{B}$, the double product between two tensors $\mathbf{B}^{-1}$ and $\boldsymbol{V}\boldsymbol{V}$ is defined by (A.8). The tensor $\mathbf{B}$ is related to the pressure tensor $\mathbf{P}$ as

$$\mathbf{B} = \frac{1}{2}\left(3\mathbf{I} - \frac{\mathbf{P}}{p} \right), \tag{7.32}$$

where $\mathbf{I}$ is the unit tensor (A.9). It is possible to show that the model (7.31) satisfies the conservation laws (3.20), but it is impossible to prove the H-theorem for this model.

To linearize this model, Eqs. (5.15) is substituted into (7.32). Then the tensors $\mathbf{B}$, $\mathbf{B}^{-1}$, and determinant $\det(\mathbf{B})$ take the form:

$$\mathbf{B} = \mathbf{I} - \frac{\xi}{2}\Pi, \quad \mathbf{B}^{-1} = \mathbf{I} + \frac{\xi}{2}\Pi, \quad \det(\mathbf{B}) = 1, \tag{7.33}$$

where the terms of the order ξ^2 and higher are omitted. A substitution of (5.12), (5.14), (2.40), and (7.33) into (7.31) leads to the linearized form of the ellipsoidal model

$$\hat{L}_E h = \nu_E \left[\varrho + 2\boldsymbol{c} \cdot \tilde{\boldsymbol{u}} + \left(c^2 - \frac{3}{2} \right) \tau - \frac{1}{2} \boldsymbol{c}\boldsymbol{c} : \Pi - h \right]. \tag{7.34}$$

Note that the expression $\boldsymbol{c}\boldsymbol{c}$ means a tensor with the elements $c_i c_j$. The double product between the tensors $\boldsymbol{c}\boldsymbol{c}$ and Π is given by Eq. (A.6).

The linearized model operator $\hat{L}_E$ is self-adjoint

$$(\hat{L}_E h_1, h_2) = (\hat{L}_E h_2, h_1) = \nu_E n_0 \left(\varrho_1 \varrho_2 + 2\tilde{\boldsymbol{u}}_1 \cdot \tilde{\boldsymbol{u}}_2 + \frac{2}{3}\tau_1 \tau_2 - \frac{1}{4}\Pi_1 : \Pi_2 \right) - \nu_E (h_1, h_2), \tag{7.35}$$

The operators $\hat{T}$ and $\hat{L}_{\rm E}$ are commutative

$$\hat{T}\hat{L}_{\rm E}h = \hat{L}_{\rm E}\hat{T}h = \nu_{\rm E}\left[\varrho - 2\boldsymbol{c}\cdot\tilde{\boldsymbol{u}} + \left(c^2 - \frac{3}{2}\right)\tau - \frac{1}{2}\boldsymbol{cc} : \Pi - h(-\boldsymbol{c})\right]. \tag{7.36}$$

Then with the help of Eqs. (7.35) and (7.36), it is shown that

$$\left(\hat{T}\hat{L}_{\rm E}h_1, h_2\right) = \left(\hat{T}\hat{L}_{\rm E}h_2, h_1\right). \tag{7.37}$$

To prove the linearized H-theorem (5.47), the ellipsoidal operator is expressed as

$$\hat{L}_{\rm E} = \hat{L}_{\rm B} - \frac{\nu_{\rm E}}{2}\boldsymbol{cc} : \Pi. \tag{7.38}$$

Simple derivations show that

$$(h, \hat{L}_{\rm E}h) = (h, \hat{L}_{\rm B}h) - \frac{\nu_{\rm E}}{2}(h, \boldsymbol{cc}) : \Pi = (h, \hat{L}_{\rm B}h) - \frac{\nu_{\rm E}n_0}{4}\Pi : \Pi \le 0, \tag{7.39}$$

where the first term is nonpositive because of (7.9) and the second term is obtained from the definition (5.15).

To obtain the quantity $\nu_{\rm E}$, the transport coefficients are obtained solving Eqs. (6.12) and (6.23) for the operator $\hat{L}_{\rm E}$. The solutions can be expressed via the functions $\psi_{\rm P}$ and ψ_q (see Eqs. (6.11) and (6.22)) as

$$h^{(\mu)} = -\frac{2}{3}\frac{\psi_{\rm P}}{\nu_{\rm E}}, \quad h^{(\kappa)} = -\frac{\psi_q}{\nu_{\rm E}}, \tag{7.40}$$

Their substitutions into (6.15) and (6.26) give

$$\mu = \frac{2}{3}\frac{p}{\nu_{\rm E}}, \quad \kappa = \frac{5}{2}\frac{k_{\rm B}p}{\nu_{\rm E}m}. \tag{7.41}$$

It is checked that the Prandtl number (6.4) based on the expressions (7.41) is equal to the correct value 2/3. The frequency extracted from the thermal conductivity (7.41) coincides with the second choice (7.16) used for the BGK model. If the frequency is extracted from the viscosity (7.41), it takes the form

$$\nu_{\rm E} = \frac{2}{3}\frac{p}{\mu}, \tag{7.42}$$

which is different from the expressions (7.15) and (7.30).

7.4 Dimensionless Form of Model Equations

Since most of problems of rarefied gas dynamics are solved in a dimensionless form, it is better to write down the model equations in their dimensionless form. In any problem, it is possible to adopt a characteristic size a so that the dimensionless coordinates can be defined as

$$\boldsymbol{x} = \frac{\boldsymbol{r}}{a}. \tag{7.43}$$

Table 7.1 Constants A_1, A_2, and A_3 of model collision operator.

Model	A_1	A_2	A_3
BGK	1 or 2/3	0	0
S	1	0	1
Ellipsoidal	2/3	1	0

In addition, the dimensionless velocity $\boldsymbol{c}$ defined by (5.2) will be used. Then all three model equations can be written in a unique form as

$$\frac{a}{v_0}\frac{\partial f}{\partial t} + \boldsymbol{c}\cdot\nabla_x f = \delta A_1 \frac{p\mu_0}{p_0\mu}(F - f), \tag{7.44}$$

where

$$F = f^{\mathrm{M}} \exp\left[A_2\left(\frac{(\mathsf{I}-\mathsf{B}^{-1}) : \boldsymbol{V}\boldsymbol{V}}{v_m^2} - \ln\sqrt{\det(\mathsf{B})}\right)\right] \times\left[1 + A_3\frac{4}{15}\left(\frac{V^2}{v_m^2} - \frac{5}{2}\right)\frac{\boldsymbol{q}\cdot\boldsymbol{V}}{p\,v_m^2}\right]. \tag{7.45}$$

The nabla operator ∇_x is defined by (A.11) with the vector $\boldsymbol{x}$ instead of $\boldsymbol{r}$. The rarefaction parameter defined by (1.34) is expressed in terms of the equilibrium pressure p_0 and viscosity μ_0 at the equilibrium temperature T_0

$$\delta = \frac{ap_0}{\mu_0 v_0}, \tag{7.46}$$

The constants A_1, A_2, and A_3 are different for each model and are given in Table 7.1.

The linearized model equations (7.8), (7.19), and (7.34) take the following general form

$$\frac{a}{v_0}\frac{\partial h}{\partial t} + \boldsymbol{c}\cdot\nabla_x h = A_1\delta(H - h) + \tilde{g}. \tag{7.47}$$

where

$$H = \varrho + 2\boldsymbol{c}\cdot\tilde{\boldsymbol{u}} + \left(c^2 - \frac{3}{2}\right)\tau - \frac{1}{2}A_2\boldsymbol{c}\boldsymbol{c} : \Pi + \frac{4}{15}A_3\left(c^2 - \frac{5}{2}\right)\boldsymbol{c}\cdot\tilde{\boldsymbol{q}}. \tag{7.48}$$

and $\tilde{g}$ is the dimensionless source term related to the dimensional one (5.34) as

$$\tilde{g}(t, \boldsymbol{x}, \boldsymbol{c}) = \frac{a}{v_0}g(t, \boldsymbol{r}, \boldsymbol{v}) = -\frac{a}{v_0}\frac{\partial h_{\mathrm{R}}}{\partial t} - \boldsymbol{c}\cdot\nabla_x h_{\mathrm{R}}. \tag{7.49}$$

Exercises

7.1 Prove (7.2).
Hint: Replace the integration variable $\boldsymbol{v}$ by $\boldsymbol{c} = \sqrt{m/2k_BT}(\boldsymbol{v} - \boldsymbol{u})$ in the left-hand sides and then use Eqs. (2.4), (2.16), and (2.25) to calculate the right-hand sides.

7.2 Obtain the constant C_1, $\boldsymbol{C}_2$, and C_3 in Eq. (7.3)
Solution:

$$C_1 = \frac{1}{m} \ln \left[n \left(\frac{m}{2\pi k_B T} \right)^{3/2} \right] - \frac{u^2}{2k_B T}, \quad \boldsymbol{C}_2 = \frac{\boldsymbol{u}}{k_B T}, \quad C_3 = -\frac{1}{k_B T}.$$

7.3 Show that $Q_S = 0$ and $Q_E = 0$ in equilibrium.
Hint: Use Eq. (2.47).

7.4 Show that $\hat{L}\psi = 0$ ($\psi = 1, \boldsymbol{v}, v^2$) for all model collision operators, $\hat{L}_B$, $\hat{L}_S$, and $\hat{L}_E$.

7.5 Calculate $\hat{L}\psi_1$ and $\hat{L}\psi_2$ for all model operators, where $\psi_1 = c_1 c_2$ and $\psi_2 = c_1 c^2$.
Solution:
$\hat{L}_B\psi_1 = -\nu c_1 c_2$; $\hat{L}_S\psi_1 = -\nu c_1 c_2$; $\hat{L}_E\psi_1 = -\nu(3/2)c_1 c_2$
$\hat{L}_B\psi_2 = -\nu c_1 c^2$; $\hat{L}_S\psi_2 = -(2/3)\nu c_1(c^2 - 5/2)$; $\hat{L}_E\psi_2 = -\nu c_1 c^2$

7.6 Check the inequalities $(\psi_1, \hat{L}\psi_1) \leq 0$ and $(\psi_2, \hat{L}\psi_2) \leq 0$ for all three model collision operators. The functions ψ_1 and ψ_2 are defined in the previous exercise.

8
Direct Simulation Monte Carlo Method

8.1
Main Ideas

Nowadays, the direct simulation Monte Carlo (DSMC) method became a powerful tool in modeling of gas flows. The excellent books by Bird [16, 17, 68] supplied by numerical codes motivated an application of this method to many technological fields related to gas dynamics. To use the DSMC method, neither grids in the velocity space nor finite difference scheme are necessary. The physical cells can be easily adapted to any geometrical configuration. It is not difficult to model nonelastic collisions occurring in polyatomic gases. Even more complicated phenomena such as dissociation, ionization, and so on are considered without a great effort. Surely, this method is most used among all of the rarefied gases because of these advantages. However, the DSMC method has its own shortcomings that restrict its application. The main difficulty of the method is the statistical scattering (or statistical noise), which is commented in the end of present chapter.

The DSMC method consists of numerical simulations of molecule motion, interaction between them, and their interaction with a solid surface. Following this method, the region of gas flow is divided into a network of cells with dimensions such that the change in flow properties across each cell is small. Then a huge number (about 10^7 or 10^8) of molecules are distributed over the gas flow region, that is, their positions $\boldsymbol{r}_i$ and velocities $\boldsymbol{v}_i$ are stored in a computer memory. The time is advanced in discrete steps of magnitude Δt. The particle motion and intermolecular collisions are uncoupled over the time increment Δt by the repetition of the following procedures:

Step I: The particles are moved through the distance determined by their velocities $\boldsymbol{v}_i$ and Δt and new positions are calculated as

$$\boldsymbol{r}_{i,\text{new}} = \boldsymbol{r}_{i,\text{old}} + \boldsymbol{v}_i \Delta t. \tag{8.1}$$

If the straight trajectory crosses a solid surface, a simulation of the gas–surface interaction is performed according to a given law, then a new velocity $\boldsymbol{v}_i$ is generated and the particle continues to move with the new velocity. If the new position $\boldsymbol{r}_{i,\text{new}}$ is out of the computational region, then the information about the corresponding particle is removed. It happens if the gas flow region is not closed, but some surfaces allow in-flux and out-flux of the gas. To simulate an in-flux, new particles are generated according to a boundary condition. Usually, a local Maxwellian (2.46) with the given values of the density n, bulk velocity $\boldsymbol{u}$, and temperature T is generated.

Step II: A representative number of intermolecular collisions in each cell are simulated.

Step III: The macroscopic characteristics are calculated.

These steps are repeated many times in order to establish a stationary flow. Then the simulations must be continued in order to calculate the average values of macroscopic quantities over many iterations. More details about these steps are given in the following.

8.2 Generation of Specific Distribution Function

The Monte Carlo method is based on random numbers. Let us denote a random fraction (or random number) between 0 and 1 as R_f. Each random number generated during simulation can be used only once so that if the notation R_f appears several times it is understood that the value of R_f is different every time. In calculations, one needs to generate various distributions departing from the set of random numbers. There are two main methods of the generation, namely, cumulative distribution function and acceptance–rejection method.

Let us consider a variable x distributed according to a function $f(x)$ in the range from a to b. The distribution function is normalized as

$$\int_a^b f(x)\,\mathrm{d}x = 1. \tag{8.2}$$

The cumulative function F related to the distribution function f is defined as

$$F(x) = \int_a^x f(y)\,\mathrm{d}y. \tag{8.3}$$

If it is possible to invert the cumulative function analytically, then the set of values x generated as

$$x = F^{-1}(R_f) \tag{8.4}$$

satisfies the distribution $f(x)$. Here, the notation F^{-1} means the inverse function of F.

If the function $F(x)$ cannot be inverted, then the variable x is generated uniformly in the range $[a, b]$

$$x = a + (b - a)R_f, \tag{8.5}$$

but only those values satisfying the condition

$$\frac{f(x)}{f_{\max}} > R_f \tag{8.6}$$

are accepted, while all others are rejected. Here, $f_{\max}$ is the maximum of the distribution function f.

8.3
Simulation of Gas–Surface Interaction

8.3.1
Kernel Decomposition

According to the definition of scattering kernel $R(\boldsymbol{v}', \boldsymbol{v})$ in Section 4.1, it represents the distribution of velocity $\boldsymbol{v}$ of scattered particle having its incident velocity equal to $\boldsymbol{v}'$. Every component of the velocity $\boldsymbol{v}$ can be generated independent of each other if the kernel $R(\boldsymbol{v}', \boldsymbol{v})$ is decomposed as

$$R(\boldsymbol{v}', \boldsymbol{v}) = R_n(v_n', v_n) R_{t1}(v_{t1}', v_{t1}) R(v_{t2}', v_{t2}), \tag{8.7}$$

where v_n is the normal component, while v_{t1} and v_{t2} are two tangential components. All kernel components are normalized as

$$\int_0^\infty R_n(v_n', v_n)\, \mathrm{d}v_n = 1, \tag{8.8}$$

$$\int_{-\infty}^\infty R_i(v_{ti}', v_{ti})\, \mathrm{d}v_{ti} = 1, \quad i = 1, 2, \tag{8.9}$$

providing the normalization (4.6). Thus, each component v_n, v_{t1}, and v_{t2} can be generated applying the procedure described in the previous section using the corresponding kernel component as the distribution function.

8.3.2
Diffuse Scattering

The diffuse scattering kernel (4.10) has the following components:

$$R_n(v_n', v_n) = \frac{2v_n}{v_w^2} \exp\left(-\frac{v_n^2}{v_w^2}\right), \tag{8.10}$$

$$R_{ti}(v_{ti}', v_{ti}) = \frac{1}{\sqrt{\pi} v_w} \exp\left(-\frac{v_{ti}^2}{v_w^2}\right), \quad i = 1, 2. \tag{8.11}$$

The cumulative function for the normal component is obtained as

$$F(v_n) = \int_0^{v_n} R_n(v_n', x)\, \mathrm{d}x = 1 - \exp\left(-\frac{v_n^2}{v_w^2}\right). \tag{8.12}$$

Equating it to a random number R_f, the normal velocity is expressed as

$$v_n = v_w \sqrt{-\ln R_f}, \tag{8.13}$$

where the expression $(1 - R_f)$ has been replaced by R_f.

The tangential velocity components v_{t1} and v_{t2} can be generated independently by the acceptance–rejection method. The procedure described in Appendix C of the book [17] suggests the generation of pair v_{t1} and v_{t2} using the polar coordinate

$$v_{t1} = v_t \cos\theta, \quad v_{t2} = v_t \sin\theta. \tag{8.14}$$

To derive the distributions of v_t and θ, the product $R_{t1}R_{t2}$ is transformed with help of (8.14) into

$$R_{t1}(v'_{t1}, v_{t1})R_{t2}(v'_{t2}, v_{t2})\,\mathrm{d}v_{t1}\,\mathrm{d}v_{t2} = \frac{1}{\pi v_m^2}\exp\left(-\frac{v_t^2}{v_w^2}\right) v_t\,\mathrm{d}v_t\,\mathrm{d}\theta. \tag{8.15}$$

Then the distribution of v_t takes the form

$$R_t(v'_t, v_t) = \frac{2v_t}{v_w^2}\exp\left(-\frac{v_t^2}{v_w^2}\right), \tag{8.16}$$

that is, the magnitude v_t has the same cumulative function as that for the normal velocity (8.12). The polar angle θ is uniformly distributed between 0 and 2π. Once the variables v_t and θ are generated as

$$v_t = v_w\sqrt{-\ln R_f}, \quad \theta = 2\pi R_f, \tag{8.17}$$

the tangential velocity components are calculated by (8.14).

To simulate the diffuse–specular scattering (4.15), the α_d fraction of incident particles is reflected according the above-mentioned procedure, while the rest of the particles are reflected specularly, that is, the tangential components remain the same, while the normal component v_n change its own sign.

8.3.3
Cercignani–Lampis Scattering

The CL kernel (4.17) is decomposed on the following components:

$$R_n(v'_n, v_n) = \frac{2v_n}{v_w^2\alpha_n}\exp\left[-\frac{v_n^2 + (1-\alpha_n)v_n'^2}{v_w^2\alpha_n}\right] I_0\left(\frac{2\sqrt{1-\alpha_n}v_n v'_n}{v_w^2\alpha_n}\right), \tag{8.18}$$

$$R_{ti}(v'_{ti}, v_{ti}) = \frac{1}{\sqrt{\pi\alpha_t(2-\alpha_t)}v_w}\exp\left[-\frac{(v_{ti} - (1-\alpha_t)v'_{ti})^2}{\alpha_t(2-\alpha_t)v_w^2}\right], \tag{8.19}$$

where $i = 1, 2$ mean two components of tangential velocity.

Since the kernel component $R_n(v'_n, v_n)$ does not have a cumulative function that can be inverted, the normal velocity v_n could be generated by the acceptance–rejection method. However, the Bessel function should be calculated several times and many random numbers must be generated to obtain one value of v_n by this method. To reduce the computational effort, Lord [69] proposed an interesting trick that allows to obtain v_n generating two intermediate variables v_t and θ according to (8.17) using only two random numbers. Then, the normal component of reflected particle is calculated by

$$v_n = \left[\alpha_n v_t^2 + (1-\alpha_n)v_n'^2 + 2\sqrt{\alpha_n(1-\alpha_n)}v_t v'_n\cos\theta\right]^{1/2}. \tag{8.20}$$

The paper [69] contains a proof that the distribution of the normal velocity v_n obtained by this procedure corresponds to Eq. (8.18). It is verified that in the limit

$\alpha_n = 0$ the normal component of reflected particle is given by $v_n = |v'_n|$, while it is generated according to the diffuse scattering law at $\alpha_n = 1$.

To generate the tangential velocity, the variables v_t and θ are related to v_{t1} and v_{t2} as

$$\frac{v_{t1} - (1-\alpha_t)v'_{t1}}{\sqrt{\alpha_t(2-\alpha_t)}} = v_t \cos\theta, \quad \frac{v_{t2} - (1-\alpha_t)v'_{t2}}{\sqrt{\alpha_t(2-\alpha_t)}} = v_t \sin\theta. \tag{8.21}$$

The product $R_{t1}R_{t2}$ calculated from (8.19) using these relations takes the form (8.15), that is, the quantities v_t and θ are distributed exactly by the same way as those in the case of the diffuse scattering. Once the variables v_t and θ are generated by (8.17), the tangential components of the reflected particle are calculated from (8.21) via the corresponding components of the incident velocity as

$$v_{t1} = \sqrt{\alpha_t(2-\alpha_t)}v_t \cos\theta + (1-\alpha_t)v'_{t1}, \tag{8.22}$$

$$v_{t2} = \sqrt{\alpha_t(2-\alpha_t)}v_t \sin\theta + (1-\alpha_t)v'_{t2}. \tag{8.23}$$

It is checked that at $\alpha_t = 0$ the tangential velocity remains the same after the scattering, while it corresponds to the diffuse scattering at $\alpha_t = 1$.

8.4 Intermolecular Interaction

The intermolecular collisions are simulated for each cell separately. Let us consider a cell of volume $\mathcal{V}_C$ having N_m number of model particles. In order to simulate a correct number of collision between these particles during the time interval Δt, we should know the real number of intermolecular collisions. If each real particle undergoes ν_c number of collisions per unit time and per unit volume, then the total number of collisions during the time interval Δt inside a volume $\mathcal{V}_C$ reads

$$N_{\text{coll}} = \frac{1}{2} n \nu_c \mathcal{V}_C \, \Delta t, \tag{8.24}$$

where the factor of half has appeared because each collision involves two particles. Using the number density definition (1.5) and the collision frequency ν_c (3.33), the number of real collisions is expressed as

$$N_{\text{coll}} = \frac{N^2}{2\mathcal{V}_C} \, \sigma_t \, \bar{g}_r \, \Delta t. \tag{8.25}$$

However, in the DSMC, one deals with model particles representing real ones. Let us introduce the representation of model particle F_N defined as

$$F_N = \frac{N}{N_m}. \tag{8.26}$$

Thus, in calculations, we have F_N times less particles than in reality. The number of model collisions N_{mc} is also F_N times smaller than its real number $N_{\rm coll}$

$$N_{mc} = \frac{N_{\rm coll}}{F_N} = \frac{N_m^2 F_N}{2\mathcal{V}_C}\,\sigma_t\,\overline{g}_r\,\Delta t, \tag{8.27}$$

so that if a cell contains N_m model particles, N_{mc} collisions must be simulated. Note that the calculation of the average relative speed $\overline{g}_r$ over all N_m particles requires significant computational effort. In addition, we should take into account the fact that the collisions with higher relative speed happen more often. To overcome these difficulties, it is suggested to calculate the maximum number of possible collisions as

$$N_{mc}^{(\max)} = \frac{N_m^2 F_N}{2\mathcal{V}_C}\,\sigma_t\,g_{r,\max}\,\Delta t, \tag{8.28}$$

where $g_{r,\max}$ is a maximum relative speed. Then $N_{mc}^{(\max)}$ pairs are chosen randomly independent of their relative speed. Among these pairs, only those with the relative speed g satisfying the condition

$$\frac{g_r}{g_{r,\max}} > R_f \tag{8.29}$$

are accepted for collision.

This procedure is called no time counter method and described in details in Section 11.1 of the book by Bird [17]. There are some variations of the expression (8.28) where the term N_m^2 is replaced by the product $N_m\overline{N}_m$ or by $N_m(N_m - 1)$. Here, $\overline{N}_m$ is the average value of N_m over previous time intervals Δt. The substitution of N_m by $(N_m - 1)$ is justified by the more exact calculation of the number of all pairs composed by N_m particles. However, when the number N_m is sufficiently large, all these expressions lead to the same results.

Another procedure to calculate the number of pairs to be tested for acceptance and rejection is called majorant frequency method [70]. For sufficiently large number of particles per cell N_m, this scheme leads to the same number of pairs to be tested and as a consequence it provides the same results. A comparative analysis can be found in Refs [71, 72].

8.5 Calculation of Post-Collision Velocities

Once a pair of particles for collision is accepted, their pre-collision velocities $\boldsymbol{v}$ and $\boldsymbol{v}_*$ must be replaced by post-collision velocities $\boldsymbol{v}'$ and $\boldsymbol{v}'_*$ according to the following steps:

(i) The relative $\boldsymbol{g}_r$ and center mass $\boldsymbol{G}$ velocities before collisions are calculated

$$\boldsymbol{g}_r = \boldsymbol{v} - \boldsymbol{v}_*, \quad \boldsymbol{G} = \frac{\boldsymbol{v} + \boldsymbol{v}_*}{2}. \tag{8.30}$$

(ii) The distance b and azimuthal ϵ impact parameters are generated as

$$b = b_M \sqrt{R_f}, \quad \epsilon = 2\pi R_f, \tag{8.31}$$

where b_M is the potential cut-off distance. Physically, the expressions (8.31) mean that the parameter b is distributed in the range $[0, b_M]$ in accordance with the function $f(b) = 2b/b_M^2$, and the parameter ϵ is distributed uniformly in the range $[0, 2\pi]$.

(iii) The deflection angle χ is calculated as a function of b and g_r. The references explaining such calculations are given in Section 1.5.

(iv) The vector of relative velocity $\boldsymbol{g}'_r = (g'_1, g'_2, g'_3)$ after collision is calculated via the vector of pre-collision relative velocity $\boldsymbol{g}_r = (g_1, g_2, g_3)$

$$g'_1 = g_1 \cos\chi + \sqrt{g_2^2 + g_3^2} \,\sin\epsilon\, \sin\chi, \tag{8.32}$$

$$g'_2 = g_2 \cos\chi + \frac{g_r g_3 \cos\epsilon - g_1 g_2 \sin\epsilon}{\sqrt{g_2^2 + g_3^2}} \sin\chi, \tag{8.33}$$

$$g'_3 = g_3 \cos\chi - \frac{g_r g_2 \cos\epsilon + g_1 g_3 \sin\epsilon}{\sqrt{g_2^2 + g_3^2}} \sin\chi. \tag{8.34}$$

(v) The post-collision velocities are calculated

$$\boldsymbol{v}' = \boldsymbol{G} + \frac{1}{2}\boldsymbol{g}'_r, \quad \boldsymbol{v}'_* = \boldsymbol{G} - \frac{1}{2}\boldsymbol{g}'_r. \tag{8.35}$$

In case of the HS potential, the deflection angle is easily calculated by (1.21). Taking into account that this potential is already cut-off, $b_M = d$, it is shown that the function $\cos\chi$ is uniformly distributed over the interval $[-1, 1]$. Then, instead of the impact parameter b generation, one may generate directly the deflection angle via its function as

$$\cos\chi = 2R_f - 1, \tag{8.36}$$

while the parameter ϵ is generated by (8.31). The uniformity of $\cos\chi$ means that all directions of post-collision relative velocity $\boldsymbol{g}'_r$ are equally probable. Then Eqs. (8.32)–(8.34) are also replaced by simpler ones

$$g'_1 = g_r \cos\chi, \quad g'_2 = g_r \sin\chi \sin\epsilon, \quad g'_3 = g_r \sin\chi \cos\epsilon, \tag{8.37}$$

that is, the direction of the vector $\boldsymbol{g}'_r$ does not depend on the direction of $\boldsymbol{g}_r$. Thus, the HS potential requires very few operations to calculate the post-collision velocities.

For an arbitrary potential, the step (iii) requires a significant computational effort that can increase the total computational time for several orders. To overcome such a difficulty, it was suggested [73] to calculate a look-up table for the deflection angle χ and to store it. The regularly distributed values of the collision energy E_r defined by Eq. (1.22) can be introduced as

$$E_{rj} = \frac{j-0.5}{N_e} E_{rm}, \quad 1 \le j \le N_e, \tag{8.38}$$

where E_{rm} is the maximum energy and N_e is an integer. The impact parameter values b_i are distributed so that they would be equally probable

$$b_i = b_{\mathrm{M}} \sqrt{\frac{i-0.5}{N_b}}, \quad 1 \le i \le N_b, \tag{8.39}$$

where N_b is an integer. Thus, the matrix χ_{ij} of dimension $N_e \times N_b$ is calculated, stored, and can be used in calculations of any kind of gas flows. To choose the deflection angle χ_{ij} from the precalculated matrix for a specific collision, the following rules are used

$$i = N_b R_f + 1, \quad j = \frac{E_r}{E_{rm}} N_e + 1, \tag{8.40}$$

so that the index i is randomly chosen from the range $[1, N_b]$, while the index j corresponds to the discrete value of energy E_{rj} closest to the energy E_r in the specific collision calculated by (1.22). Typical parameter values of the matrix χ_{ij} are given in Refs [73–76]. It is recommended to adopt the values about 600 for N_e and N_b. The reasonable cut-off for the parameter b_{M} is $3d$, where d is the zero point of the potential $U(d) = 0$. The value of the maximum energy E_{rm} depends on typical gas temperature T and should be about $20k_{\mathrm{B}}T/\epsilon$, where ϵ is the potential well depth given in Table 1.1.

8.6 Calculation of Macroscopic Quantities

In practice, we are interested in macroscopic quantities that are calculated according to their definitions given in Section 2.2. Let us consider again a cell having the volume $\mathcal{V}_C$ and containing N_m model particles with velocities $\boldsymbol{v}_i$ ($1 \le i \le N_m$). The number density defined by (1.5) is calculated via the number of real particles $N_m F_N$ divided by the cell volume

$$n = \frac{N_m F_N}{\mathcal{V}_C}. \tag{8.41}$$

The bulk velocity defined by (2.16) is calculated as the average velocity over all particles

$$\boldsymbol{u} = \frac{1}{N_m} \sum_{i=1}^{N_m} \boldsymbol{v}_i. \tag{8.42}$$

The temperature defined by (2.26) is calculated as

$$T = \frac{m}{3N_m k_B} \sum_{i=1}^{N_m} (\boldsymbol{v}_i - \boldsymbol{u})^2 \tag{8.43}$$

so that the average kinetic energy is equal to $(3/2)k_B T$. The pressure tensor **P** and heat flux vector $\boldsymbol{q}$ are calculated as

$$\mathbf{P} = m \sum_{i=1}^{N_m} (\boldsymbol{v}_i - \boldsymbol{u})(\boldsymbol{v}_i - \boldsymbol{u}), \tag{8.44}$$

$$\boldsymbol{q} = \frac{m}{2} \sum_{i=1}^{N_m} (\boldsymbol{v}_i - \boldsymbol{u})^2 (\boldsymbol{v}_i - \boldsymbol{u}), \tag{8.45}$$

according to (2.28) and (2.32), respectively.

In many applications, flow rates of mass and energy through some surfaces are calculated directly. At each time step Δt, a number of particles crossing a surface Σ in both positive N^+ and negative N^- directions are counted. Then the mass flow rate through this surface is given as

$$\dot{M} = mF_N \frac{N^+ - N^-}{\Delta t}. \tag{8.46}$$

The same mass flow rate can be calculated via the bulk velocity as

$$\dot{M} = \int_\Sigma \rho u_n \, \mathrm{d}^2\Sigma, \tag{8.47}$$

where u_n is the component of the bulk velocity $\boldsymbol{u}$ normal to the surface Σ.

The energy flux through a surface can be calculated by the analogous procedure. At each time step Δt, the total energy, in case of monatomic gases it is just the kinetic energy $mv^2/2$ of particles, is counted in positive E^+ and negative E^- directions. Then the energy flow rate through the surface Σ reads

$$\dot{E} = F_N \frac{E^+ - E^-}{\Delta t}. \tag{8.48}$$

The same quantity is calculated using the energy flow vector $\boldsymbol{J}_e$ defined by (2.20)

$$\dot{E} = \int_\Sigma J_{ne} \, \mathrm{d}^2\Sigma, \tag{8.49}$$

where J_{ne} is the component of $\boldsymbol{J}_{ne}$ normal to the surface Σ.

8.7 Statistical Scatter

The statistical scatter (or statistical noise) of calculated quantities is "Achilles' heel" of the DSMC method. Each numerical result obtained by this method must be analyzed in order to estimate the statistical scatter that depends on many factors. The numerical scheme parameters such as the number of model particles and time increment Δt affect the statistical scatter error. In addition, physical input parameters such pressure drop and temperature difference also determine this error. It is difficult to estimate the error a priori especially in complex flows. It is recommended that each physical situation would be calculated several times using different sequences of random numbers. Then the scatter of these results will give an idea about the statistical numerical error. Using the rule that the relative statistical error is inversely proportional to the square root of the number of samples, one can reduce this error up to desirable value. In other words, to reduce a statistical scatter twice, the number of samples should be increased four times.

Initially, the DSMC method was elaborated for aerothermodynamic problems, where the Mach number is extremely high. Under such conditions, the statistical noise is very low and a small number of samples provide reliable results. When the Mach number is small, the number of samples needed to reduce the noise up to a reasonable value is so huge that the computational time becomes inadmissibly long. In this case, an approach based on a numerical solution of the kinetic equation is more advantageous. One of such methods is described in the consequent chapter.

Exercises

8.1 Check that the relations (8.32)–(8.34) provide the equality $g_r' = g_r$.

8.2 Check that the angle between vectors $\boldsymbol{g}_r$ and $\boldsymbol{g}_r'$ is equal to χ.
Hint: Calculate the inner product $\boldsymbol{g}_r \cdot \boldsymbol{g}_r'$ and verify that it is $g_r^2 \cos \chi$.

8.3 Consider a surface at the temperature $T_w = 300$ K with the ACs $\alpha_n = 1$ and $\alpha_t = 0.8$. An incident atom of helium has the following velocity components: $v_n' = -1200$ m/s, $v_{t1} = 900$ m/s, and $v_{t2} = 0$ m/s. Calculate its velocity after the interaction with the surface using the following random numbers: $R_{f1} = 0.2678$, $R_{f2} = 0.5673$, $R_{f3} = 0.9456$.
Solution: Since $\alpha_n = 1$, v_n is calculated by (8.13) using R_{f1}. The quantities v_t and θ are calculated by (8.17) using R_{f2} and R_{f3}, respectively. The quantities v_{t1} and v_{t2} are calculated by (8.22) and (8.23). Then $v_n = 1281$ m/s, $v_{t1} = 956$ m/s, $v_{t2} = -276$ m/s.

8.4 Calculate the velocity $[v_n, v_{t1}, v_{t2}]$ under the conditions of the previous exercise assuming the complete accommodation, $\alpha_t = 1$.
Solution: $v_n = 1281$ m/s, $v_{t1} = 792$ m/s, $v_{t2} = -282$ m/s.

9
Discrete Velocity Method

9.1
Main Ideas

The discrete velocity method (DVM) is the main deterministic method applied to solve the BE and its models. Its applications to the model equations is easier than to the BE, therefore, the details will be given only for the models. To simplify the explanations, we assume that the VDF does not depend on the time. In addition, all explanations will be given in terms of the dimensionless coordinates $\boldsymbol{x}$ defined by (7.43) and dimensionless velocity $\boldsymbol{c}$ defined by (5.2).

The idea of the DVM is to choose a set of values (or nodes) of the molecular velocity $\boldsymbol{c}_i$ and to substitute the VDF $f(\boldsymbol{x},\boldsymbol{c})$ by a set of functions $f_i(\boldsymbol{x})$

$$f_i(\boldsymbol{x}) = f(\boldsymbol{x}, \boldsymbol{c}_i). \tag{9.1}$$

Then all integrals with respect to the molecular velocity become quadratures, in other words, the sum with respect to the velocity nodes. For instance, the integral of a function ψ representing a moment (2.11) is approximated as

$$\int \psi(\boldsymbol{c}) f(\boldsymbol{x},\boldsymbol{c})\, \mathrm{d}^3\boldsymbol{c} \approx \sum_i \psi(\boldsymbol{c}_i)\, f_i(\boldsymbol{x}) W_i, \tag{9.2}$$

where W_i is the weight of the velocity node $\boldsymbol{c}_i$. The weight values depend on the velocity node distribution, which should be carefully chosen for each specific problem.

Thus, the integro-differential kinetic equation (7.44) is replaced by a system of differential equations for the functions $f_i(\boldsymbol{x})$ written as

$$\boldsymbol{c}_i \cdot \nabla_x f_i = \delta A_1 \frac{p/p_0}{\mu/\mu_0}(F_i - f_i), \tag{9.3}$$

where $F_i(\boldsymbol{x}) = F(\boldsymbol{c}_i, \boldsymbol{x})$ being $F(\boldsymbol{c}_i, \boldsymbol{x})$ given by (7.45). To solve each differential equation (9.3) by some finite difference method, it is necessary to know the terms F_i that contain the moments n, $\boldsymbol{u}$, T, P, and $\boldsymbol{q}$ calculated by some quadrature (9.2) via all functions f_i, which are not known. Thus, an iterative procedure should be organized.

Rarefied Gas Dynamics: Fundamentals for Research and Practice, First Edition. Felix Sharipov.

The boundary condition (4.4) presented in terms of the dimensionless velocity $\boldsymbol{c}$

$$c_n f(\boldsymbol{c}) = -\int_{c'_n \leq 0} c'_n f(\boldsymbol{c}') R(\boldsymbol{c}', \boldsymbol{c}) \, \mathrm{d}^3\boldsymbol{c}' \tag{9.4}$$

is also calculated via a quadrature as

$$c_{nj} f_j = -\sum_i c_{ni} f_i R_{ij} W_i, \quad R_{ij} = R(\boldsymbol{c}_i, \boldsymbol{c}_j), \tag{9.5}$$

where the left-hand side contains only the nodes with $c_{nj} \geq 0$, while the right-hand side considers only the nodes of incident particles $c_{ni} \leq 0$.

The same ideas of the discretization is applied to the linearized model equations (7.48). In this case, the perturbation function $h(\boldsymbol{x}, \boldsymbol{c})$ is substituted by a set of the function $h_i(\boldsymbol{x})$ so that

$$h_i(\boldsymbol{x}) = h(\boldsymbol{x}, \boldsymbol{c}_i). \tag{9.6}$$

The moments defined by (5.17) are approximated as

$$\frac{1}{\pi^{3/2}} \int e^{-c^2} \psi(\boldsymbol{c}) h(\boldsymbol{x}, \boldsymbol{c}) \, \mathrm{d}^3\boldsymbol{c} \approx \sum_i \psi(\boldsymbol{c}_i) \, h_i(\boldsymbol{x}) W_i, \tag{9.7}$$

so that the expression $e^{-c_i^2}/\pi^{3/2}$ is included into the weights W_i. The kinetic equation (7.47) is replaced by the set of differential equations

$$\boldsymbol{c}_i \cdot \nabla_x h_i = A_1 \delta (H_i - h_i) + \tilde{g}_i, \tag{9.8}$$

where $H_i(\boldsymbol{x}) = H(\boldsymbol{c}_i, \boldsymbol{x})$ and $\tilde{g}_i(\boldsymbol{x}) = \tilde{g}(\boldsymbol{c}_i, \boldsymbol{x})$. The linearized boundary condition (5.51) are discretized as

$$h_j = \sum_i A_{ij} (h_i - h_{wi}) W_i + h_{wj}, \tag{9.9}$$

where A_{ij} follows from Eq. (5.51), while h_{wj} is the discretization of h_w given by (5.54) or (5.59)

$$A_{ij} = -\frac{c_{ni} f_0^{\mathrm{M}}(\boldsymbol{c}_i)}{c_{nj} f_0^{\mathrm{M}}(\boldsymbol{c}_j)} R_0(\boldsymbol{c}_i, \boldsymbol{c}_j), \quad h_{wj} = h_w(\boldsymbol{c}_j). \tag{9.10}$$

In summary, to solve the system (9.3) or (9.8), three main issues must be settled:

(i) How to discretize the molecular velocity and what quadrature formula (9.2) is used to calculate the moments of the VDF.
(ii) How to organize of the iterative procedure and what is the criterion of its convergence.
(iii) How to approximate the derivative in the left-hand side of Eq. (9.3) or (9.8), that is, what finite difference scheme must be used to solve the differential equations.

Later in the book, each issue is discussed in detail.

9.2 Velocity Discretization

9.2.1 Onefold Integral

Initially, let us consider quadrature formulas to calculate the moments (9.2) of the VDF for a function f of one variable c, which could be a component of the dimensionless molecular velocity $\boldsymbol{c}$. In this particular case, the quadrature (9.2) reads

$$\int_{-\infty}^{\infty} \psi(c) f(c)\, \mathrm{d}c \approx \sum_{i=1}^{N} \psi(c_i) f(c_i) W_i. \tag{9.11}$$

where c_i are nodes, W_i are weights, and N is the number of the nodes.

Simple quadratures are based on a regularly distributed nodes in a fixed range $c_{\min} \leq c \leq c_{\max}$. The velocity range $[c_{\min}, c_{\max}]$ depends on a specific problem to be solved and should be sufficiently large to reach a desirable accuracy. According to Exercise 2.3, a very small fraction (about 5×10^{-7}) of molecules have a dimensionless speed c higher than 4 so that the values $c_{\min} = -4$ and $c_{\max} = 4$ provide a good accuracy in most of the problems.

Let us consider Simpson's quadrature, which combines the simplicity and reasonable accuracy. It is based on regularly distributed nodes

$$c_i = c_{\min} + i\Delta c, \ \Delta c = \frac{c_{\max} - c_{\min}}{N_c}, \quad 0 \leq i \leq N_c, \tag{9.12}$$

where N_c is an even integer. The nodes have the following weights

$$W_i = W_i^S \Delta c, \tag{9.13}$$

where the following notation is introduced

$$W_i^S = \begin{cases} 1/3 & \text{for} \quad i = 0 \text{ and } N, \\ 2/3 & \text{for} \quad \text{all even } i \text{ except } 0 \text{ and } N, \\ 4/3 & \text{for} \quad \text{all odd } i. \end{cases} \tag{9.14}$$

The accuracy of quadrature can be significantly improved if a nonuniform distribution of the nodes is used. The Gauss rule is one such quadrature and provides the highest accuracy for a given number N_c of nodes. The numerical values of the nodes and weights and the technique to calculate them are well described in the book [77]. This rule is appropriate to solve the linearized kinetic equations where the integrals of the type (9.7) are very often. If the nodes are distributed symmetrically, then the onefold integral of this type can be approximated as

$$\frac{1}{\sqrt{\pi}} \int_{-\infty}^{\infty} e^{-c^2} \psi(c) h(c)\, \mathrm{d}c = \frac{1}{\sqrt{\pi}} \int_{0}^{\infty} e^{-c^2} [\psi(-c)h(-c) + \psi(c)h(c)]\, \mathrm{d}c$$
$$\approx \sum_{i=1}^{N_c} [\psi(-c_i)h(-c_i) + \psi(c_i)h(c_i)] W_i. \tag{9.15}$$

Following the procedure described in the book [77] for arbitrary weighting function, the nodes c_i and weights W_i can be calculated numerically for the weighting function

$$W(c) = \frac{1}{\sqrt{\pi}} e^{-c^2}. \tag{9.16}$$

Their values for $N = 4$, 6, and 8 are given in Appendix, Section B.1.1.

9.2.2 Twofold Integral

Quadrature formulas for twofold integrals are similar to those for onefold integrals. Let f be a function of two velocity components, $f = f(c_1, c_2)$. Both components c_1 and c_2 are discretized using the nodes defined by Eq. (9.12). Then, a moment $\psi(c_1, c_2)$ is calculated as

$$\int_{-\infty}^{\infty} \int_{-\infty}^{\infty} \psi(c_1, c_2) f(c_1, c_2)\, \mathrm{d}c_1\, \mathrm{d}c_2 \approx \sum_{i=1}^{N_c} \sum_{j=1}^{N_c} \psi(c_i, c_j) f(c_i, c_j) W_i W_j. \tag{9.17}$$

This expression is easily generalized when the discretization for each velocity component is different from each other.

In many situations, the polar coordinates are used for the velocity when the two-dimensional vector $\boldsymbol{c}_p = (c_1, c_2)$ is presented in terms of its magnitude c_p and direction θ as

$$c_1 = c_p \cos\theta, \quad c_2 = c_p \sin\theta. \tag{9.18}$$

A twofold integral for the perturbation function of the type (9.7) can be written in these variables as

$$\begin{aligned} &\frac{1}{\pi} \int_{-\infty}^{\infty} \int_{-\infty}^{\infty} e^{-c_1^2 - c_2^2} \psi(c_1, c_2) h(c_1, c_2)\, \mathrm{d}c_1\, \mathrm{d}c_2 \\ &\quad = \frac{1}{\pi} \int_0^{2\pi} \int_0^{\infty} e^{-c_p^2} \psi(c_p, \theta) h(c_p, \theta) c_p\, \mathrm{d}c_p\, \mathrm{d}\theta. \end{aligned} \tag{9.19}$$

In some cases, instead of the whole range $0 \le \theta \le 2\pi$, it is enough to consider a smaller range $0 \le \theta \le \pi$ or $0 \le \theta \le \pi/2$ so that the regularly distributed nodes can be introduced as

$$\theta_j = \frac{\theta_{\max}}{N_\theta} j, \quad 0 \le j \le N_\theta, \tag{9.20}$$

where $\theta_{\max}$ is equal to π or $\pi/2$ and N_θ is an integer. If we adopt Simpson's rule, the integer N_θ must be even and each node has the weight

$$W_j^\theta = \frac{2\pi}{N_\theta} W_j^S, \tag{9.21}$$

where the notation W_j^S is defined by (9.14). If one wishes to use the Gauss quadrature for the velocity magnitude c_p, the weighting function

$$W(c) = \frac{1}{\pi} e^{-c_p^2}\, c_p \tag{9.22}$$

is recommended. Then the integral (9.19) is approximated as

$$\frac{1}{\pi}\int_0^{2\pi}\int_0^{\infty} e^{-c_p^2}\psi(c_p,\theta)h(c_p,\theta)c_p\,\mathrm{d}c_p\,\mathrm{d}\theta$$
$$\approx \sum_{i=1}^{N_c}\sum_{j=0}^{N_\theta}\psi(c_i,\theta_j)h(c_i,\theta_j)W_i^c W_j^\theta. \tag{9.23}$$

The quantities of c_i and W_i^c are calculated numerically following the technique described in the book [77]. Their values for $N_c = 4$, 6, and 8 are given in Section B.1.2.

9.3 Iterative Procedure

To organize the iterative procedure, the kinetic equation (9.3) is written down as

$$\boldsymbol{c}_i\cdot\nabla_x f_i + \delta' f_i = \delta' F_i, \quad \delta' = A_1\delta\frac{p/p_0}{\mu/\mu_0}. \tag{9.24}$$

To begin the iterations, an initial ($n = 0$) approximation of the functions $f_i^{(0)}$ is assumed. It can be the global Maxwellian $f_0^{\mathrm{M}}(\boldsymbol{c}_i)$ defined by (5.1). Then the following steps are realized:

1) Once the nth approximations of the functions $f_i^{(n)}$ are known, the corresponding moments $n^{(n)}$, $\boldsymbol{u}^{(n)}$, $T^{(n)}$, $\mathbf{P}^{(n)}$, and $\boldsymbol{q}^{(n)}$ are calculated by a quadrature formula (9.11) using their definitions (2.12), (2.16), (2.26), (2.28), and (2.32), respectively. The corresponding approximations of the pressure $p^{(n)} = n^{(n)}k_{\mathrm{B}}T^{(n)}$ and viscosity $\mu^{(n)}$ at the temperature $T^{(n)}$ are calculated. Note that none of the model equations considered here provides the function $\mu = \mu(T)$, but such a function should be assumed. The quantities $p^{(n)}$ and $\mu^{(n)}$ are substituted into $\delta'^{(n)}$ (see Eq. (9.24)).
2) The nth approximations of $F_i^{(n)}$ for each value of the molecular velocity $\boldsymbol{c}_i$ are calculated according to Eqs. (7.45).
3) The $(n+1)$th approximations of the functions $f_i^{(n+1)}$ at boundaries are calculated from (9.5) as

$$c_{nj}f_j^{(n+1)} = -\sum_i c_{ni}f_i^{(n)}R_{ij}W_i, \tag{9.25}$$

4) The $(n+1)$th approximations of the functions $f_i^{(n+1)}$ are calculated solving the differential equation

$$\boldsymbol{c}_i\cdot\nabla_x f_i^{(n+1)} + \delta'^{(n)}f_i^{(n+1)} = \delta'^{(n)}F_i^{(n)} \tag{9.26}$$

with the boundary condition (9.25).

All steps are repeated until a convergence criterion is satisfied.

The same procedure with small modifications is applied to the linearized equation (7.48). First, Eq. (7.47) is written down as

$$\boldsymbol{c}_i\cdot\nabla_x h_i + \delta' h_i = \delta' H_i + \tilde{g}_i(\boldsymbol{x}), \quad \delta' = A_1\delta. \tag{9.27}$$

In the beginning ($n = 0$), it can be assumed that $h^{(0)} = 0$. Then the following steps are realized:

1) Once the nth approximations of the perturbation function $h_i^{(n)}$ is known, the corresponding moments of ϱ, $\tilde{\boldsymbol{u}}$, τ, Π, and $\tilde{\boldsymbol{q}}$ are calculated by the quadrature formula (9.15) using the definitions (5.17).
2) The nth approximations of $H_i^{(n)}$ for each value of the molecular velocity $\boldsymbol{c}_i$ are calculated according to Eqs. (7.48).
3) The $(n+1)$th approximations of the functions $h_i^{(n+1)}$ at boundaries are calculated from (9.9) as

$$h_j^{(n+1)} = \sum_j A_{ij}(h_i^{(n)} - h_{wi})W_i + h_{wj}. \tag{9.28}$$

4) The $(n+1)$th approximation of the functions $h_i^{(n+1)}$ are calculated solving the differential equation

$$\boldsymbol{c}_i \cdot \nabla_x h_i^{(n+1)} + \delta' h_i^{(n+1)} = \delta' H_i^{(n)} + \tilde{g}_i(\boldsymbol{x}). \tag{9.29}$$

with the boundary condition (9.28).

Again, all steps are repeated until a convergence criterion is satisfied.

9.4 Finite Difference Schemes

9.4.1 Main Principles

The left-hand side of the kinetic equation to be approximated by finite differences has the same structure for both full (9.3) and linearized (9.8) forms. Therefore, only its linearized form (9.8) will be considered here, but some modifications of numerical schemes needed for the nonlinearized equation will be pointed out in the end of chapter. To solve a differential equation such as (9.8), two main issues should be settled: (i) how to approximate derivatives and (ii) how to sweep the nodes of position $\boldsymbol{x}$.

Let us consider a function of one variable $h = h(x)$. Its derivative in a node x_k can be calculated by the one-sided approximation based on the nearest nodes

$$\left.\frac{\mathrm{d}h}{\mathrm{d}x}\right|_{x=x_k} \approx \frac{h(x_k) - h(x_{k-l})}{x_k - x_{k-l}}, \quad l = \pm 1, \tag{9.30}$$

where the index l determines the sweeping direction, that is, when $l = 1$ the nodes are swept from left to right, while $l = -1$ means the opposite direction. The approximation (9.30) is rather coarse though a more exact derivative calculation can be done by the centered approximation as

$$\left.\frac{\mathrm{d}h}{\mathrm{d}x}\right|_{x=x_{k-l/2}} \approx \frac{h(x_k) - h(x_{k-l})}{x_k - x_{k-l}}, \quad x_{k-l/2} = \frac{x_k + x_{k-l}}{2}, \quad l = \pm 1 \tag{9.31}$$

so that the derivative is referred to the point being in the middle between the nodes x_k and x_{k-l}. In order to choose the correct sweeping direction, one should follow the simple rule: the sweeping direction must coincide with the molecular velocity direction.

9.4.2
One-Dimensional Planar Flows

Let us begin from the simplest one-dimensional flow when the perturbation function depends only on one coordinate x_1. Since x_1 is the unique coordinate, the notation x_1 will be replaced by x in this section. Note that the dependence on the unique coordinate does not mean that the perturbation function depends only on the corresponding velocity component c_1. In a general case, the perturbation function depends on all three velocity components even in case of one-dimensional flow. Thus, we assume that the perturbation function depends on the four variables

$$h = h(x, \boldsymbol{c}) \tag{9.32}$$

and satisfies the following kinetic equation

$$c_1 \frac{\partial h}{\partial x} + \delta' h(x, \boldsymbol{c}) = \delta' H(x, \boldsymbol{c}) + \tilde{g}(x, \boldsymbol{c}), \tag{9.33}$$

obtained from (9.27). Since the discretization of the position coordinate and that of the velocity are independent of each other, the subscript i at the velocity $\boldsymbol{c}_i$ is also omitted in this section, while c_1 means the velocity component along the x-axis.

Consider $N_x + 1$ nodes x_k $(0 \leq k \leq N_x)$ distributed along the x-axis by some arbitrary way. Then the following notations are introduced

$$h_k = h(x_k, \boldsymbol{c}), \quad H_k = H(x_k, \boldsymbol{c}), \quad \tilde{g}_k = \tilde{g}(x_k, \boldsymbol{c}), \tag{9.34}$$

The one-sided approximation (9.30) for the kinetic equation (9.33) reads

$$c_1 \frac{h_k - h_{k-l}}{x_k - x_{k-l}} + \delta' h_k = \delta' H_k + \tilde{g}_k, \quad l = \mathrm{sgn}(c_1), \tag{9.35}$$

where the derivative is referred to the kth node. The function sgn(x) defined by (A.26) determines the sign of the index l to provide the correct direction for the sweeping, that is, if $c_1 > 0$, then the nodes are swept from left to right, while they are swept in the opposite direction for $c_1 < 0$. Now, the perturbation function h is calculated in the kth node as

$$h_k = \frac{C_k h_{k-l} + \delta' H_k + \tilde{g}_k}{C_k + \delta'}, \tag{9.36}$$

where

$$C_k = \frac{c_1}{x_k - x_{k-l}}. \tag{9.37}$$

The index k varies from 1 to N_x for $c_1 > 0$. The value of h_0 is calculated from the boundary conditions. In case of $c_1 < 0$, the index k varies in the inverse order from $N_x - 1$ to 0. The point h_{N_x} is calculated from the boundary conditions. Thus, in any case, the coefficient C_k is positive.

The one-sided scheme is very simple but rather coarse. Moreover, the node number in the physical space increases by increasing the rarefaction parameter δ that requires even significant computational effort using the optimized version [78] of the scheme. However, it is useful if one needs to make quickly some estimative calculations. More exact results for the same number of nodes N_x are obtained for the centered approximation (9.31) of the derivative, which is referred to the middle point between the node k and the nearest one. Then, the average value of h_k and h_{k-l}, and the average value of H_k and H_{k-l} are used for the middle point, while the source term $\tilde{g}_{k-l/2}$ is calculated directly in this middle point. Finally, Eq. (9.33) is approximated as

$$c_1 \frac{h_k - h_{k-l}}{x_k - x_{k-l}} + \frac{1}{2}\delta'(h_k + h_{k-l}) = \frac{1}{2}\delta'(H_k + H_{k-l}) + \tilde{g}_{k-l/2}. \tag{9.38}$$

The index l is calculated again via the velocity component c_1 by (9.35). For this scheme, it is more convenient to define the coefficient C_k as

$$C_k = \frac{2c_1}{x_k - x_{k-l}}. \tag{9.39}$$

Then the value of the perturbation function in the kth node is extracted from (9.38) as

$$h_k = \frac{(C_k - \delta')h_{k-l} + \delta'(H_k + H_{k-l}) + 2\tilde{g}_{k-l/2}}{C_k + \delta'}. \tag{9.40}$$

9.4.3
Two-Dimensional Planar Flows

Let us consider a two-dimensional planar flow when the perturbation function depends on two position variables, but it can be dependent on the all three velocity components

$$h = h(x_1, x_2, \boldsymbol{c}). \tag{9.41}$$

Then the kinetic equation (9.8) is reduced to

$$c_1 \frac{\partial h}{\partial x_1} + c_2 \frac{\partial h}{\partial x_2} + \delta' h(x_1, x_2, \boldsymbol{c}) = \delta' H(x_1, x_2, \boldsymbol{c}) + \tilde{g}(x_1, x_2, \boldsymbol{c}). \tag{9.42}$$

The nodes x_{k_1} $(0 \le k_1 \le N_{x1})$ and x_{k_2} $(0 \le k_2 \le N_{x2})$ and the following notations

$$\begin{aligned} h_{k_1,k_2} &= h(x_{k_1}, x_{k_2}, \boldsymbol{c}), \\ H_{k_1,k_2} &= H(x_{k_1}, x_{k_2}, \boldsymbol{c}), \quad \tilde{g}_{k_1,k_2} = \tilde{g}(x_{k_1}, x_{k_2}, \boldsymbol{c}) \end{aligned} \tag{9.43}$$

are introduced.

The simplest scheme in this case is based on the one-sided approximation of Eq. (9.42) as

$$c_1 \frac{h_{k_1,k_2} - h_{k_1-l_1,k_2}}{x_{k_1} - x_{k_1-l_1}} + c_2 \frac{h_{k_1,k_2} - h_{k_1,k_2-l_2}}{x_{k_2} - x_{k_2-l_2}} + \delta' h_{k_1,k_2}$$
$$= \delta' H_{k_1,k_2} + \tilde{g}_{k_1,k_2}, \tag{9.44}$$

with the indexes

$$l_1 = \text{sng}(c_1), \quad l_2 = \text{sng}(c_2) \tag{9.45}$$

providing the correct direction of the sweeping. In other words, the derivatives are referred to the node (k_1, k_2). The perturbation function h in this node is expressed from (9.44) as

$$h_{k_1,k_2} = \frac{C_{k_1} h_{k_1-l_1,k_2} + C_{k_2} h_{k_1,k_2-l_2} + \delta' H_{k_1,k_2} + \tilde{g}_{k_1,k_2}}{\delta' + C_{k_1} + C_{k_2}}, \tag{9.46}$$

where C_{k_1} and C_{k_2} are defined as

$$C_{k_1} = \frac{c_1}{x_{k_1} - x_{k_1-l_1}}, \quad C_{k_2} = \frac{c_2}{x_{k_2} - x_{k_2-l_2}}. \tag{9.47}$$

As in case of one-dimensional flows, the derivative can be referred to the central point between the nodes with the coordinates

$$x_{k_1-l_1/2} = \frac{x_{k_1} + x_{k_1-l_1}}{2}, \quad x_{k_2-l_2/2} = \frac{x_{k_2} + x_{k_2-l_2}}{2}. \tag{9.48}$$

Then Eq. (9.42) is approximated as

$$\frac{c_1}{2\left(x_{k_1} - x_{k_1-l_1}\right)} (h_{k_1,k_2} + h_{k_1,k_2-l_2} - h_{k_1-l_1,k_2} - h_{k_1-l_1,k_2-l_2})$$
$$+ \frac{c_2}{2(x_{k_2} - x_{k_2-l_2})} \left(h_{k_1,k_2} + h_{k_1-l_1,k_2} - h_{k_1,k_2-l_2} - h_{k_1-l_1,k_2-l_2}\right)$$
$$+ \frac{1}{4}\delta' \left(h_{k_1,k_2} + h_{k_1-l_1,k_2} + h_{k_1,k_2-l_2} + h_{k_1-l_1,k_2-l_2}\right)$$
$$= \frac{1}{4}\delta' \left(H_{k_1,k_2} + H_{k_1-l_1,k_2} + H_{k_1,k_2-l_2} + H_{k_1-l_1,k_2-l_2}\right)$$
$$+ \tilde{g}_{k_1-l_1/2,k_2-l_2/2}, \tag{9.49}$$

where h and H are calculated as their average over the four nodes, while g is calculated in the middle point with the coordinates (9.48). The perturbation function h in the node (x_{k_1}, x_{k_2}) is obtained from (9.49) as

$$h_{k_1,k_2} = \left[\left(C_{k_1} - C_{k_2} - \delta'\right) h_{k_1-l_1,k_2} + \left(C_{k_2} - C_{k_1} - \delta'\right) h_{k_1,k_2-l_2}\right.$$
$$+ \left(C_{k_1} + C_{k_2} - \delta'\right) h_{k_1-l_1,k_2-l_2}$$
$$+ \delta' \left(H_{k_1,k_2} + H_{k_1-l_1,k_2} + H_{k_1,k_2-l_2} + H_{k_1-l_1,k_2-l_2}\right)$$
$$\left. + 4\tilde{g}_{k_1-l_1/2,k_2-l_2/2}\right] \left(C_{k_1} + C_{k_2} + \delta'\right)^{-1}, \tag{9.50}$$

where

$$C_{k_1} = \frac{2c_1}{x_{k_1} - x_{k_1-l_1}}, \quad C_{k_2} = \frac{2c_2}{x_{k_2} - x_{k_2-l_2}}. \tag{9.51}$$

This scheme is rather cumbersome, but it provides more exact results than the one-sided scheme for the same number of the nodes N_{x1} and N_{x2}.

Later in this book, it will be shown that some rarefied gas flows are characterized by discontinuous perturbation function. As a consequence, a numerical solution based on both one-sided (9.46) and centered (9.50) schemes becomes nonphysical, namely, some macroscopic quantities have a saw-type behavior. Such a numerical phenomenon can be reduced if the derivative is calculated along the characteristic line having a direction coinciding with that of the molecular velocity. Let us adopt the polar coordinates for the two-dimensional velocity $\boldsymbol{c}_p = (c_1, c_2)$ defined by (9.18). Then the quadrature (9.23) can be used for full and linearized kinetic equations, respectively. The physical coordinates x_1 and x_2 can be related to the variables s and θ as

$$x_1 = x_{k_1} - s\ \cos\theta, \quad x_2 = x_{k_2} - s\ \sin\theta. \tag{9.52}$$

The sketch given in Figure 9.1 shows the s-axis. Then the two derivatives are replaced by one with respect to the variable s

$$c_1\frac{\partial h}{\partial x_1} + c_2\frac{\partial h}{\partial x_2} = -c_p\frac{\partial h}{\partial s}. \tag{9.53}$$

In terms of these variables, the kinetic equation (9.42) reads

$$-c_p\frac{\partial h}{\partial s} + \delta' h = \delta' H + \tilde{g}. \tag{9.54}$$

The advantage of this form is that the perturbation function is never discontinuous with respect to the variable s. The one-sided approximation of the derivatives leads to the expression

$$c_p\frac{h_{k_1,k_2} - h_{\text{int}}}{\Delta s} + \delta' h_{k_1,k_2} = \delta' H_{k_1,k_2} + \tilde{g}_{k_1,k_2}, \tag{9.55}$$

where Δs is the distance from the node (x_{k_1}, x_{k_2}) to the horizontal line at $x_{k_2-l_2}$ or to the vertical line at $x_{k_1-l_2}$ depending on the velocity direction

$$\Delta s = \begin{cases} c_p/C_{k_1} & \text{if} \quad C_{k_1} > C_{k_2}, \\ c_p/C_{k_2} & \text{if} \quad C_{k_1} < C_{k_2}, \end{cases} \tag{9.56}$$

where C_{k_1} and C_{k_2} are given by (9.47). The quantity h_{int} is calculated as an interpolation of h in the nearest nodes

$$h_{\text{int}} = \begin{cases} h_{k_1-l_1,k_2}(1 - \tan\theta) + h_{k_1-l_1,k_2-l_2}\tan\theta, & \text{if} \quad C_{k_1} > C_{k_2}, \\ h_{k_1,k_2-l_2}(1 - \cot\theta) + h_{k_1-l_1,k_2-l_2}\cot\theta, & \text{if} \quad C_{k_1} < C_{k_2}. \end{cases} \tag{9.57}$$

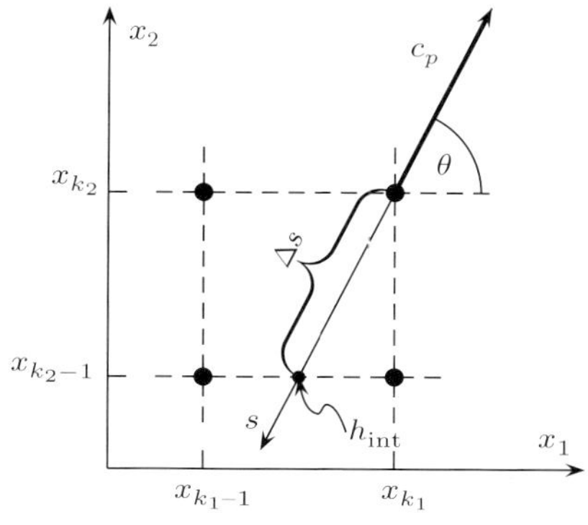

Figure 9.1 One-sided scheme along characteristic.

The position of h_{int} is indicated in Figure 9.1. The perturbation in the node (x_{k_1}, x_{k_2}) is calculated as

$$h_{k_1,k_2} = \frac{(c_p/\Delta s)h_{\text{int}} + \delta' H_{k_1,k_2} + \tilde{g}_{k_1,k_2}}{\delta' + c_p/\Delta s}. \tag{9.58}$$

9.4.4 One-Dimensional Axisymmetric Flows

One-dimensional axisymmetric flows are usually calculated in the polar coordinates replacing the physical coordinates x_1 and x_2 by x and ϕ

$$x_1 = x\cos\phi, \quad x_2 = x\sin\phi. \tag{9.59}$$

In contrast to the previous section, here x is defined differently, $x = \sqrt{x_1^2 + x_2^2}$. Then, the two components c_1 and c_2 of the two-dimensional velocity vector $\boldsymbol{c}_p$ are replaced by its radial component c_x and its azimuthal component c_ϕ shown in Figure 9.2. In terms of these variables, the derivatives read 20,

$$c_1\frac{\partial h}{\partial x_1} + c_2\frac{\partial h}{\partial x_2} = c_x\frac{\partial h}{\partial x} + \frac{c_\phi}{x}\frac{\partial h}{\partial \phi} + \frac{c_\phi^2}{x}\frac{\partial h}{\partial c_x} - \frac{c_x c_\phi}{x}\frac{\partial h}{\partial c_\phi}. \tag{9.60}$$

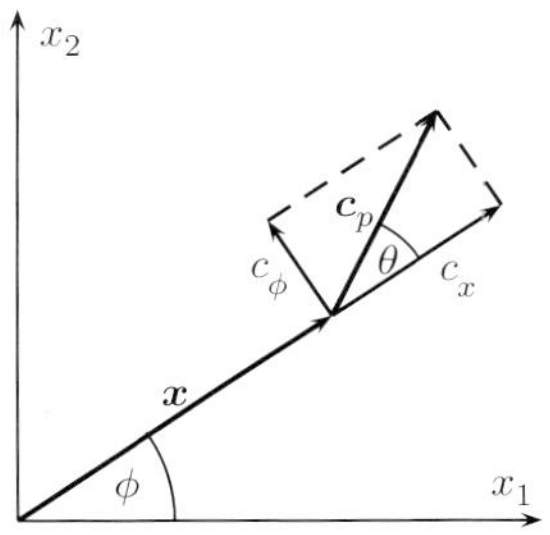

Figure 9.2 Polar coordinates.

An axial symmetry in terms of the polar coordinates means that

$$\frac{\partial h}{\partial \phi} = 0, \tag{9.61}$$

that is, the perturbation h depends only on the three coordinates x, c_x, c_ϕ. The velocity components c_x and c_ϕ can be expressed via the magnitude c_p and angle θ as

$$c_x = c_p \cos\theta, \quad c_\phi = c_p \sin\theta \tag{9.62}$$

so that θ is the angle between the position vector $\boldsymbol{x} = (x_1, x_2)$ and velocity vector $\boldsymbol{c} = (c_1, c_2)$ (see Figure 9.2). Thus, the axial symmetry reduces the number of variables from four x_1, x_2, c_1, c_2 to three x, θ, c_p, $h = h(x, \theta, c_p)$. In terms of these three variables, the last two terms of Eq. (9.60) are simplified and the left-hand side of the kinetic equation is transformed to

$$c_1 \frac{\partial h}{\partial x_1} + c_2 \frac{\partial h}{\partial x_2} = c_p \cos\theta \frac{\partial h}{\partial x} - \frac{c_p \sin\theta}{x} \frac{\partial h}{\partial \theta}. \tag{9.63}$$

Then the kinetic equation takes the form

$$c_p \cos\theta \frac{\partial h}{\partial x} - \frac{c_p \sin\theta}{x} \frac{\partial h}{\partial \theta} + \delta' h(x, \theta, c_p) = \delta' H(x, \theta, c_p) + \tilde{g}(x, \theta, c_p). \tag{9.64}$$

To approximate the derivatives of this equation, some nodes x_k of the radial coordinate x and those θ_j of the angle θ are introduced. The range and distribution of the variable x depend on specific problem, while the angle nodes θ_j are usually distributed uniformly in the internal $0 \le \theta \le \pi$

$$\theta_j = j\frac{\pi}{N_\theta}, \quad 0 \le j \le N_\theta, \tag{9.65}$$

where N_θ is an integer. Once the nodes x_k and θ_j are determined, the discrete functions are defined as

$$h_{kj} = h\left(x_k, \theta_j, c_p\right), \quad H_{kj} = H\left(x_k, \theta_j, c_p\right), \quad \tilde{g}_{kj} = \tilde{g}\left(x_k, \theta_j, c_p\right). \tag{9.66}$$

For the first and last angle nodes, $j = 0$ and $j = N_\theta$, the derivative $\partial h/\partial\theta$ in (9.64) disappears and then the kinetic equation is approximated by the scheme similar to that for the one-dimensional one (9.38)

$$lc_p \frac{h_{kj} - h_{k-l,j}}{x_k - x_{k-l}} + \frac{1}{2}\delta'(h_{kj} + h_{k-l,j}) = \frac{1}{2}\delta'(H_{kj} + H_{k-l,j}) + \tilde{g}_{k-l/2,j}, \tag{9.67}$$

where

$$l = \begin{cases} 1 & \text{for} \quad j = 0, \\ -1 & \text{for} \quad j = N_\theta, \end{cases} \tag{9.68}$$

and the semi-integer subscript $k - l/2$ means that the middle point between x_k and x_{k-l} is used

$$\tilde{g}_{k-l/2,j} = \tilde{g}(x_{k-l/2}, \theta_j, c_p), \quad x_{k-l/2} = \frac{x_k + x_{k-l}}{2}. \tag{9.69}$$

Then the perturbation in the point x_k is extracted as

$$h_{kj} = \frac{(C_k^x - \delta')h_{k-l,j} + \delta'(H_{kj} + H_{k-l,j}) + 2\tilde{g}_{k-l/2,j}}{C_k^x + \delta'}, \tag{9.70}$$

where the coefficient

$$C_k^x = \frac{2lc_p}{x_k - x_{k-l}} \tag{9.71}$$

has been introduced. The nodes x_k always must be swept so that this coefficient is positive according to the subscript l defined by (9.68).

For the other values of the angle θ_j $(1 \leq j \leq N_\theta - 1)$, except $\theta_j = \pi/2$, the kinetic equation (9.63) is approximated as

$$\begin{aligned} & c_p \cos\theta_{j+1/2} \frac{h_{k,j} + h_{k,j+1} - h_{k-l,j} - h_{k-l,j+1}}{2(x_k - x_{k-l})} \\ & - \frac{c_p \sin\theta_{j+1/2}}{x_{k-l/2}} \frac{h_{k,j+1} + h_{k-l,j+1} - h_{k,j} - h_{k-l,j}}{2\pi/N_\theta} \\ & + \frac{\delta'}{4}(h_{k,j+1} + h_{k-l,j+1} + h_{kj} + h_{k-l,j}) \\ & = \frac{\delta'}{2}(H_{k,j+1/2} + H_{k-l,j+1/2}) + \tilde{g}_{k-l/2,j+1/2}, \end{aligned} \tag{9.72}$$

where the centered approximations are used for both derivatives $\partial h/\partial x$ and $\partial h/\partial\theta$. The regular distribution of the angle nodes θ_j has allowed to substitute $\theta_{j+1} - \theta_j$ by π/N_θ. The subscript l depends on the angle θ_j as

$$l = \text{sgn}\left(\cos\theta_j\right). \tag{9.73}$$

Moreover, the index $j + 1/2$ means the average angle between θ_j and θ_{j+1}

$$\theta_{j+1/2} = \frac{\theta_j + \theta_{j+1}}{2}. \tag{9.74}$$

Then the perturbation h_{kj} is obtained from (9.72) as

$$\begin{aligned} h_{kj} = & \left[\left(C_{kj}^x - \frac{C_j^\theta}{x_{k-l/2}} - \delta' \right) h_{k-l,j} + \left(-C_{kj}^x + \frac{C_j^\theta}{x_{k-l/2}} - \delta' \right) h_{k,j+1} \right. \\ & + \left(C_{kj}^x + \frac{C_j^\theta}{x_{k-l/2}} - \delta' \right) h_{k-l,j+1} + 2\delta' \left(H_{k,j+1/2} + H_{k-l,j+1/2}\right) \\ & \left. + \, 4\tilde{g}_{k-l/2,j+1/2} \right] \left(C_{kj}^x + \frac{C_j^\theta}{x_{k-l/2}} + \delta' \right)^{-1}, \end{aligned} \tag{9.75}$$

where the coefficients

$$C_{kj}^x = 2c_p \frac{\cos\theta_{j+1/2}}{x_k - x_{k-l}}, \quad C_j^\theta = 2c_p \frac{N_\theta \sin\theta_{j+1/2}}{\pi} \tag{9.76}$$

has been defined. The nodes θ_j are usually swept from $N_\theta - 1$ to 1, while the sweeping direction of the nodes x_k is determined by the index l defined by (9.73).

9.4.5 Full Kinetic Equation

The finite difference scheme for the nonlinearized kinetic equation has just two differences: (i) the bulk source term is absent; (ii) the coefficient δ' defined by (9.24) is a function of the position $\boldsymbol{x}$ via the pressure p and viscosity μ. Thus, all schemes shown earlier should be slightly modified. The modification will be shown only for the one-dimensional scheme with the centered approximation (9.46).

In case of one-dimensional flow, the kinetic equation (9.24) is reduced to

$$c_1 \frac{\partial f}{\partial x} + \delta' f = \delta' F, \tag{9.77}$$

The discretized functions are defined as

$$f_k = f(x_k, \boldsymbol{c}), \quad F_k = F(x_k, \boldsymbol{c}), \quad \delta'_k = \delta'(x_k) \tag{9.78}$$

Then Eq. (9.46) is replaced by

$$f_k = \frac{(C_k - \delta'_{k-l}) f_{k-l} + \delta'_k F_k + \delta'_{k-l} F_{k-l}}{C_k + \delta'_k}. \tag{9.79}$$

All other notations are the same.

Exercises

9.1 Check that the quadrature based on the nodes and weights given in Section B.1.1 provides exact values (within the roundoff error) of the following integral calculated as

$$\frac{1}{\sqrt{\pi}} \int_0^\infty e^{-c^2} c^n \, \mathrm{d}c = \sum_{i=1}^{N_c} c_i^n W_i \tag{9.80}$$

for any power n varying from 0 to $2N_c - 1$.

9.2 Do the previous exercise with the nodes and weights given in Section B.1.2 and the following integral

$$\frac{1}{\pi} \int_0^\infty e^{-c^2} c^{n+1} \, \mathrm{d}c = \sum_{i=1}^{N} c_i^n W_i. \tag{9.81}$$

9.3 Calculate the integral

$$\frac{1}{\sqrt{\pi}} \int_0^4 e^{-c^2} c^n \, \mathrm{d}c \tag{9.82}$$

using Simpson's rule and compare the result with that obtained in Exercise 9.1. for the same number N_c.

9.4 Check (9.53) using (9.52).

10
Velocity Slip and Temperature Jump Phenomena

10.1
General Remarks

To calculate gas flows in the hydrodynamic regime, usually the nonslip velocity and temperature continuity are assumed on a solid surface. Since numerical methods to solve equations on continuous mechanics are well elaborated and widely implemented into commercial codes, it is attractive to extend these methods to rarefied gas flows. If the rarefaction is moderate, it can be taken into account by the so-called velocity slip and temperature jump boundary conditions, which are widely used for modeling of microfluidics, gas flows in vacuum systems, and so on. Thus, the velocity slip and temperature jump coefficients have the same importance in gas dynamics as the transport coefficients such as viscosity and thermal conductivity.

A lot of effort was done to calculate the velocity slip and temperature jump coefficients applying the model equations and the BE itself with different potentials of the intermolecular interaction and for different laws of the gas–surface interaction. A comparative analysis of these numerical data is given in the review paper [79] where the reader can find an extensive list of papers about this topic. In this chapter, just the main results on the velocity slip and temperature jump coefficients are given. The description is restricted by a monatomic single gas, but most of the results described here are applicable to polyatomic gases and mixture.

To define the velocity slip and temperature jump coefficients, consider a plane solid surface restricting a flow of gas occupying a semi-infinite space. The Cartesian coordinates (r_1, r_2) with the origin at the solid surface are introduced so that the axis r_1 is normal to the surface and directed toward the gas, while the axis r_2 is tangential to the surface as is shown in Figure 10.1. In such flows, no characteristic size is available; therefore, the EFP ℓ defined by (1.32) can be used to define the dimensionless coordinates as

$$x_1 = \frac{r_1}{\ell}, \quad x_2 = \frac{r_2}{\ell}. \tag{10.1}$$

Each coefficient is calculated considering different kinds of driving forces. In spite of the fact that the definitions and calculations correspond to a plane surface, the

Rarefied Gas Dynamics: Fundamentals for Research and Practice, First Edition. Felix Sharipov.

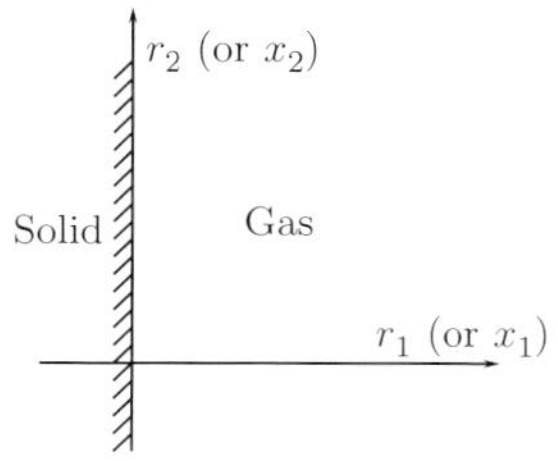

Figure 10.1 Coordinates for the calculation of slip and jump coefficients.

obtained results can be used for a curve surface too if the curvature radius is significantly larger than the EFP ℓ.

10.2 Viscous Velocity Slip

10.2.1 Definition and Input Equation

The viscous slip occurs when a gas flowing along a solid surface in the r_2 direction has only the u_2 component depending only on the normal coordinate r_1. Under these conditions, the slip boundary condition reads

$$u_2 = \sigma_\mathrm{P}\, \ell\, \frac{\mathrm{d}u_2}{\mathrm{d}r_1} \quad \text{at} \quad r_1 = 0, \tag{10.2}$$

where σ_P is the viscous slip coefficient (VSC), the EFP ℓ is calculated by Eq. (1.32) using the local pressure p and local temperature T.

To calculate the VSC, a thin layer adjacent to the solid surface having the thickness of few EFPs ℓ, or Knudsen layer, is considered. It is assumed that the pressure p and temperature T of the gas are constant in this layer and equal to their equilibrium values p_0 and T_0, respectively. The wall temperature T_w is also equal to that of the gas T_0. The calculation of the VCS is based on the linearized kinetic equation (5.33). For the sake of simplicity, the explanations will be given only for the S-model equations (see (7.47) and Table 7.1).

The dimensionless velocity gradient

$$\xi_u = \frac{\ell}{v_0}\frac{\mathrm{d}u_2}{\mathrm{d}r_1}, \quad |\xi_u| \ll 1 \tag{10.3}$$

is used as the small parameter of the linearization. Here, v_0 is the MPS (5.2) corresponding to the equilibrium temperature T_0. The linearization of the kinetic equation is performed near a reference Maxwellian defined by Eq. (5.31) with (5.28) where the reference pressure p_R and temperature T_R are constant and equal to p_0 and T_0, respectively. However, the reference velocity $\boldsymbol{u}_\mathrm{R}$ is not zero, but it is convenient to use its linear function of the coordinate r_1 with the gradient ξ_u given by (10.3). It means that the reference perturbations defined by (5.29) are expressed as

$$\varrho_R = 0, \ \tilde{\boldsymbol{u}}_R = [0, x_1, 0], \ \tau_R = 0. \tag{10.4}$$

Under such conditions, the reference perturbation function (5.30) and bulk source term (7.49) take the form

$$h_R = 2c_2 x_1, \tilde{g}^{(u)} = -c_1 \frac{\partial h_R}{\partial x_1} = -2c_1 c_2. \tag{10.5}$$

Then the linearized BE (5.33) reads

$$c_1 \frac{\partial h(x_1, \boldsymbol{c})}{\partial x_1} = \frac{\ell}{v_0} \hat{L} h - 2c_1 c_2. \tag{10.6}$$

The linearized S-model, see (7.47) and Table 7.1, takes the following particular form for the flow in question

$$c_1 \frac{\partial h(x_1, \boldsymbol{c})}{\partial x_1} = 2c_2 \tilde{u}_2 + \frac{4}{15}\left(c^2 - \frac{5}{2}\right) c_2 \tilde{q}_2 - h(x_1, \boldsymbol{c}) - 2c_1 c_2. \tag{10.7}$$

The dimensionless moments $\tilde{u}_2$ and $\tilde{q}_2$ are related to the perturbation function as

$$\tilde{u}_2 = \frac{1}{\pi^{3/2}} \int e^{-c^2} c_2 h \, \mathrm{d}^3\boldsymbol{c}, \ \tilde{q}_2 = \frac{1}{\pi^{3/2}} \int e^{-c^2} \left(c^2 - \frac{5}{2}\right) c_2 h \, \mathrm{d}^3\boldsymbol{c}, \tag{10.8}$$

following from (5.17). The velocity u_2 and heat flow q_2 are expressed as

$$u_2 = \xi_u v_0 (x_1 + \tilde{u}_2), \ q_2 = \xi_u p_0 v_0 \tilde{q}_2, \tag{10.9}$$

which are consequences of (5.36) and (5.16), respectively. The BGK model has the same form (10.7) but without the term containing the heat flow $\tilde{q}_2$.

The boundary condition on the solid surface ($x_1 = 0$) is expressed by (5.50). Since we assumed the wall temperature T_w to be equal to its equilibrium temperature T_0, the surface perturbation h_w given by (5.52) is zero, $h_w = 0$. Far from the surface, $x_1 \to \infty$, the flow tends to that considered to calculate the viscosity coefficient (see Section 6.2). In other words, the perturbation tends to that expressed by Eq. (6.13) shifted by the velocity $\tilde{u}_2$

$$\lim_{x_1 \to \infty} h = 2\left(c_2 \tilde{u}_2 + \frac{h^{(\mu)}}{m v_0 \ell}\right), \tag{10.10}$$

where $h^{(\mu)}$ satisfies Eq. (6.12). In particular case of the S-model, the solution $h^{(\mu)}$ is given by (7.28) together with (1.32), (5.2), and (7.30) as

$$h^{(\mu)} = -m v_0 \ell c_1 c_2. \tag{10.11}$$

Then the condition (10.10) for the S-model takes the form

$$\lim_{x_1 \to \infty} h = 2c_2(\tilde{u}_2 - c_1), \tag{10.12}$$

which can be obtained directly from (10.7) assuming that

$$\lim_{x_1 \to \infty} \frac{\partial h}{\partial x_1} = 0, \quad \text{and} \quad \lim_{x_1 \to \infty} \tilde{q}_2 = 0. \tag{10.13}$$

Since the condition (10.10) or (10.12) contains the unknown velocity $\tilde{u}_2$, it must be corrected in each iteration solving numerically Eq. (10.6) or (10.7).

To satisfy the condition (10.2), the bulk velocity calculated in the Knudsen layer must have the following asymptotic behavior outside of the Knudsen layer

$$\lim_{x_1 \to \infty} \tilde{u}_2 = \sigma_{\mathrm{P}}, \tag{10.14}$$

which follows from (10.1) and (10.9). Once Eq. (10.6) or (10.7) is solved, then the VSC σ_{P} is calculated from the asymptotic values of the velocity deviation $\tilde{u}_2$.

10.2.2
Velocity and Heat Flow Profiles

The typical velocity deviations $\tilde{u}_2$ based on the numerical solution of the S-model (10.7) are depicted on Figure 10.2(a). One curve represents the diffuse scattering ($\alpha_t = 1$ and $\alpha_n = 1$) on the solid surface, while the second one is calculated using the CL scattering kernel with the ACs smaller than unity, namely, $\alpha_t = 0.75$ and $\alpha_n = 0.25$. From this figure we see that both curves have the same shape tending to constant values, which is the VSC according to (10.14), far from the surface. As is expected, the noncomplete accommodation leads to a larger value of the VSC. To calculate the velocity profile u_2, Eq. (10.9) is used. Its dimensionless part $u_2/(v_0\xi_u) = x_1 + \tilde{u}_2$ is plotted on Figure 10.2(b). From this plot, we see that the real velocity on the surface is not that calculated by the slip boundary condition (10.2), but it is slightly smaller. However, such a velocity defect just near the surface has the second order with respect to the Knudsen number and usually is neglected when the slip condition (10.2) is applied.

The gas flow under question is isothermal, but nevertheless, the heat flow in the Knudsen layer does exist [80]. Typical heat flow profiles of $\tilde{q}_2$ are depicted in Figure 10.3, which shows that the heat flow is positive, that is, the heat flows in the same direction as the gas flows, and it vanishes far from the solid surface,

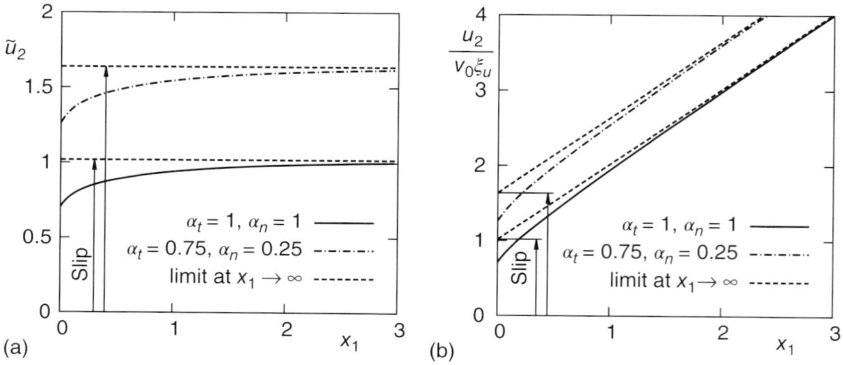

Figure 10.2 Velocity deviation $\tilde{u}_2$ (a) and velocity itself u_2 (b) versus coordinate x_1 in viscous slip problem.

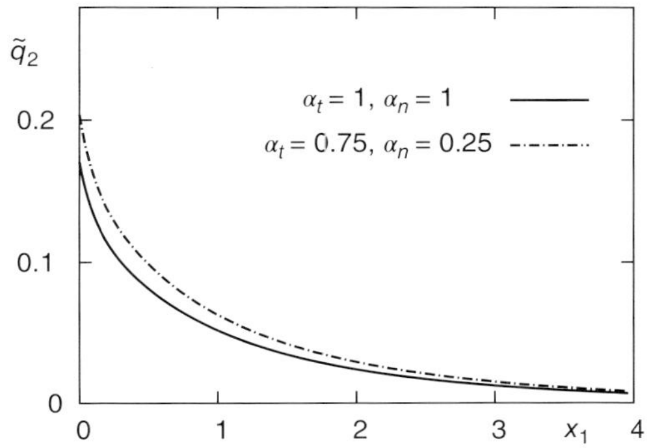

Figure 10.3 Dimensionless heat flow $\tilde{q}_2$ versus coordinate x_1 in viscous slip problem.

Table 10.1 Heat flow rate $Q^{(u)}$ based on the S-model and CL kernel, Ref. [80].

α_t	α_n	$Q^{(u)}$
0.5	0.25	0.2327
0.75	0.25	0.1967
1	1	0.1626

at $x_1 \to \infty$. It is affected by the boundary conditions though the influence is not strong. Two quite different boundary conditions, namely, complete accommodation ($\alpha_t = 1$, $\alpha_n = 1$) and small ACs ($\alpha_t = 0.75$, $\alpha_n = 0.25$), yielded the heat flow profiles similar to each other.

Let us introduce the whole heat flow rate in the Knudsen layer because of the velocity gradient as

$$Q^{(u)} = \int_0^\infty \tilde{q}_2(x_1)\, \mathrm{d}x_1. \tag{10.15}$$

Some values of this quantity based on Eq. (10.7) and the CL scattering law were reported in Ref. [80], which partially are reproduced in Table 10.1. We see that the heat flow rate $Q^{(u)}$ is smaller for the diffuse scattering than that for the noncomplete accommodation.

10.2.3
Numerical and Experimental Data

First, let us fix the complete accommodation and compare results obtained from different kinetic equations. Such a comparison is performed in Table 10.2. The BGK model was used by many authors to calculate the VSC (see example, Refs [81–87]). All of them obtained the value $\sigma_\mathrm{P} = 1.016$. An application of the S-model just slightly changed this value. The BE with the HS molecules was solved numerically by the authors of Refs [88–92] who reported that the value $\sigma_\mathrm{P} = 0.987$

Table 10.2 Viscous slip coefficient σ_P based on different equations and potentials. Diffuse gas–surface interaction.

σ_P	Equat.	Potential	References
1.016	BGK		[81–87]
1.018	S-model		[33, 85]
0.987	BE	HS	[88–92]
0.996	BE	LJ ($T^* = 30$)	[93]
0.997	BE	LJ ($T^* = 10$)	[93]
1.01	BE	LJ ($T^* = 2.40$)	[93]
1.02	BE	LJ ($T^* = 1.58$)	[94]
1.03	BE	LJ ($T^* = 1.31$)	[94]
1.02	BE	LJ ($T^* = 1.10$)	[93]

is slightly lower than that obtained from the model equations. Finally, the paper [93] contains the data of VSC based on the BE with the LJ potential for different values of the reduced temperature T^* (see the definition (6.30)). As was shown in the review [79], the same values of σ_P can be extracted from the Poiseuille flow data [94]. In summary, if we compare the values of the VSC σ_P based on the model equations and BE with different kinds of potential, we find that the values of σ_P is about 1.00 ± 0.03. Thus, the value $\sigma_P = 1.00$ can be successfully used in practical calculations.

Now, let us fix kinetic equation and analyze the VSC for different gas–surface interaction laws. The diffuse–specular kernel was applied by many authors, see example, Refs [82, 84, 86, 87, 90, 95, 96]. A comparative analyze of these data is performed in the review [79]. As is expected, the VSC is strongly affected by the diffuse part of scattering α_d. However, this scattering law does not allow to distinguish the momentum AC and energy AC, which are quite different, while the CL kernel containing two coefficients α_t and α_n allows such a distinction.

Numerical data of VSC based on the CL scattering law reported in Ref. [92], where the BE was solved, and in Refs [33, 85], where the S-model was applied. Those data based on the S-model are partially reproduced in Table 10.3, which shows that the slip viscous coefficient σ_P is weakly affected by the accommodation coefficient α_n, but it depends significantly on the momentum accommodation coefficient α_t. The formula containing only the momentum AC α_t

$$\sigma_P = \frac{1.772}{\alpha_t} - 0.754 \tag{10.16}$$

perfectly interpolates the numerical results of Refs [33, 85] (see the sixth column of Table 10.3) and, at present, it is the best for practical calculations.

It is not so easy to measure the VSC. Indeed, it cannot be measured directly, but some quantity dependent on this coefficient is measured and then the VSC is extracted. Many examples of such quantities, methods of extraction, and experimental values of VSC are given in Ref. [79] where experimental data reported in

Table 10.3 Viscous slip coefficient σ_P based on the S-model and CL kernel, Refs [33, 85].

	σ_P				
α_t	$\alpha_n = 0.25$	0.5	0.75	1	Eq. (10.16)
0.5	2.845	2.825	2.807	2.791	2.790
0.75	1.636	1.626	1.617	1.609	1.609
1	1.018	1.018	1.018	1.018	1.018
1.25	0.638	0.647	0.656	0.664	0.664

Table 10.4 Experimental values of viscous slip coefficient σ_P and the corresponding values of tangential momentum accommodation coefficient α_t.

Gas	Surface	σ_P	α_t	References
He	Glass	1.255	0.882	[36]
He	Steel	1.364	0.837	[101]
He	Silica	1.252	0.883	[103]
Ne	Glass	1.333	0.849	[36]
Ne	Fluoroplastic	1.52	0.779	[36]
Ar	Glass	1.180	0.916	[36]
Ar	Fluoroplastic	1.20	0.907	[36]
Ar	Silica	1.558	0.766	[103]
Ar	Steel	1.472	0.796	[101]
Kr	Glass	1.023	0.997	[36]
Xe	Glass	0.993	1.014	[36]
CO_2	Glass	1.030	0.993	[36]

Refs [36, 97–103] were used. Some of these data are presented in Table 10.4 which shows that the results have a significant dispersion, even those measured by the same authors. In spite of the significant uncertainty of the experimental results, all of them are in the reasonable range and close to the theoretical value.

The values of α_t calculated by (10.16) using the corresponding VSC σ_P are also given in Table 10.4. Since the uncertainty of the VSC is large, the uncertainty of the accommodation coefficient α_t is also large. However, it is possible to outlook some tendencies. First, the accommodation coefficient α_t varies in a small range. The smallest value is 0.766. Second, the heavy gases, such as xenon or carbon dioxide, have the accommodation coefficient close to unity, while for the light gases, such as helium, it is close to 0.9.

The influence of the gas chemical composition on the VSC was studied in the paper [104] where it was pointed out that for a mixture with a large ratio of atomic weights, for example, helium–xenon, the VSC can be 40% larger than that for a single gas. When the atomic weights ratio is not so large, for example, neon–argon, the VSC differs from that for a single gas just for 4%.

10.3 Thermal Velocity Slip

10.3.1 Definition and Input Equation

In this section, another type of velocity slip will be considered, namely, the slip due to a longitudinal temperature gradient. The fact is that, near a solid surface with a nonuniform temperature distribution, a gas begins to flow from a colder region to a hotter region. This phenomenon is called *thermal creep*. It is very weak and usually neglected in the hydrodynamic regime. However, if a gas is maintained at a constant pressure, the thermal creep becomes a main phenomenon disturbing the equilibrium state.

To define the slip condition because of a longitudinal temperature gradient, we also consider a gas occupying the semi-infinite space $r_1 \geq 0$, as is shown in Figure 10.1. The difference from the previous situation is that the wall temperature T_w is not uniform but it is a function of the r_2 coordinate, $T_w = T_w(r_2)$. Then, the slip boundary condition reads

$$u_2 = \sigma_\mathrm{T} \frac{\mu}{\rho T_w} \frac{\mathrm{d}T_w}{\mathrm{d}r_2} \quad \text{at} \quad r_1 = 0, \tag{10.17}$$

where the quantity σ_T is called *the thermal slip coefficient* (*TSC*).

To calculate the TSC, the dimensionless temperature gradient

$$\xi_\mathrm{T} = \frac{\ell}{T_w} \frac{\mathrm{d}T_w}{\mathrm{d}r_2}, \quad |\xi_\mathrm{T}| \ll 1 \tag{10.18}$$

is used as the small parameter for the linearization. Moreover, we assume that the temperatures of the surface T_w and that of the gas T linearly depend on the x_2 coordinate

$$T(x_2) = T_w(x_2) = T_0(1 + x_2\xi_\mathrm{T}), \tag{10.19}$$

where T_0 is an equilibrium temperature. The pressure of the gas is assumed to be constant and equal to p_0 over the whole semi-infinite space $x_1 \geq 0$. Consequently, the gas density must be given as

$$n(x_2) = n_0(1 - x_2\xi_\mathrm{T}) \tag{10.20}$$

in order to maintain the product nT to be constant within the linear approximation.

To linearize the kinetic equation, see Eq. (5.31), the reference velocity is assumed to be zero, $\boldsymbol{u}_\mathrm{R} = 0$, while the reference density n_R and temperature T_R are given by Eqs. (10.20) and (10.19), respectively. Then the reference perturbations for the flow in question take the form

$$\varrho_\mathrm{R} = -x_2, \quad \tilde{\boldsymbol{u}}_\mathrm{R} = 0, \quad \tau_\mathrm{R} = x_2. \tag{10.21}$$

Substituting these expressions into (5.30), the reference perturbation and bulk source term are obtained as

$$h_{\mathrm{R}} = \left(c^2 - \frac{5}{2}\right) x_2, \quad \tilde{g}^{(\mathrm{T})} = -c_2 \frac{\partial h_{\mathrm{R}}}{\partial x_2} = -c_2 \left(c^2 - \frac{5}{2}\right). \tag{10.22}$$

Then the linearized BE (5.33) is reduced to

$$c_1 \frac{\partial h(x_1, \mathbf{c})}{\partial x_1} = \frac{\ell}{v_0} \hat{L} h - c_2 \left(c^2 - \frac{5}{2}\right). \tag{10.23}$$

The linearized S-model follows from Eq. (7.47), considering that the density and temperature deviations from their reference values are zero, $\varrho = 0$ and $\tau = 0$,

$$c_1 \frac{\partial h(x_1, \mathbf{c})}{\partial x_1} = 2c_2 \tilde{u}_2 + \frac{4}{15}\left(c^2 - \frac{5}{2}\right) c_2 \tilde{q}_2 - h(x_1, \mathbf{c}) - c_2 \left(c^2 - \frac{5}{2}\right), \tag{10.24}$$

where the dimensionless velocity $\tilde{u}_2$ and heat flux $\tilde{q}_2$ are given by Eq. (10.8). They are related to the dimensional quantities as

$$u_2 = \xi_{\mathrm{T}} v_0 \tilde{u}_2, \quad q_2 = \xi_{\mathrm{T}} p_0 v_0 \tilde{q}_2. \tag{10.25}$$

The kinetic equation (10.23) or (10.24) is solved together with the boundary condition (5.50) at the surface $x_1 = 0$. From Eq. (5.63), we conclude that $\tau_w = 0$ so that the surface source term (5.52) is zero too. Far from the surface, at $x_1 \to \infty$, the solution tends to that obtained to calculate the thermal conductivity (6.22) with the shift by the velocity

$$\lim_{x_1 \to \infty} h = 2c_2 \tilde{u}_2 + \frac{h^{(\kappa)}}{k_{\mathrm{B}} T_0 \ell}, \tag{10.26}$$

where $h^{(\kappa)}$ satisfies (6.23) with a small modification of ψ_q, namely, the component v_1 is replaced by v_2

$$\psi_q = k_{\mathrm{B}} T_0 \left(\frac{v^2}{v_0^2} - \frac{5}{2}\right) v_2 \tag{10.27}$$

instead of the expression (6.18). Particularly for the S-model, the function $h^{(\kappa)}$ is given by (7.28) together with (1.32), (5.2), and (7.30) as

$$h^{(\kappa)} = -\frac{3}{2} k_{\mathrm{B}} T_0 \ell \; c_2 \left(c^2 - \frac{5}{2}\right). \tag{10.28}$$

Then the expression (10.26) becomes

$$\lim_{x_1 \to \infty} h = c_2 \left[2\tilde{u}_2 - \frac{3}{2}\left(c^2 - \frac{5}{2}\right)\right]. \tag{10.29}$$

To obtain the same expression from (10.24), the asymptotic values of the heat flow $\tilde{q}_2$ are needed. It can be obtained from Fourier's law (6.3) , which is valid far from the surface. Thus, combining Eqs. (6.3) (1.32), (10.25), and (6.4), the dimensionless heat flow is obtained as

$$\lim_{x_1 \to \infty} \tilde{q}_2 = -\frac{5}{4\,(\text{Pr})} = -\frac{15}{8}, \tag{10.30}$$

where the value of the Prandtl number Pr = 2/3 has been used. Thus, assuming that $\partial h/\partial x_1 \to 0$ and regarding (10.30) at $x_1 \to \infty$, the asymptotic (10.29) is derived. In numerical calculations, the unknown values of the velocity $\tilde{u}_2$ is taken from a previous iteration.

Once the dimensionless velocity $\tilde{u}_2$ is known, the TSC is calculated as

$$\sigma_\text{T} = 2 \lim_{x_1 \to \infty} \tilde{u}_2, \tag{10.31}$$

which is a consequence of Eqs. (1.9), (10.1), (10.25), (10.17), and (10.18).

10.3.2
Velocity and Heat Flow Profiles

Typical velocity profiles $\tilde{u}_2$ obtained from the S-model (10.24) are shown in Figure 10.4. One curve was obtained for the complete accommodation, $\alpha_t = 1$ and $\alpha_n = 1$, while the second curve corresponds to the CL scattering law with the ACs equal to $\alpha_t = 0.75$ and $\alpha_n = 0.25$. From this figure we see that the velocity multiplied by the factor 2 tends to be a constant value at $x_1 \to \infty$, which is equal to the TSC σ_T. The limit values of $2\tilde{u}_2$ are also depicted in Figure 10.4. It is curious that the curves are only slightly different in spite of the different boundary conditions.

The profiles of heat flow with the minus sign, $-\tilde{q}_2$, are depicted on Figure 10.5 for the diffuse scattering ($\alpha_t = 1$ and $\alpha_n = 1$) and for the CL scattering law with $\alpha_t = 0.75$ and $\alpha_n = 0.25$. Note that the quantity $\tilde{q}_2$ is naturally negative because the heat flows against the temperature gradient. Surprisedly, the profiles are practically the same meaning that the heat flow is not sensitive to the boundary condition. As is expected, it tends to be the value 1.875, far from the surface according to Eq. (10.30).

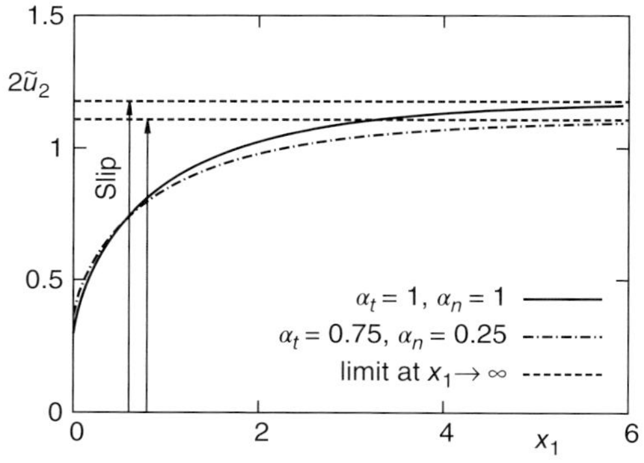

Figure 10.4 Dimensionless velocity $\tilde{u}_2$ versus coordinate x_1 in thermal slip problem.

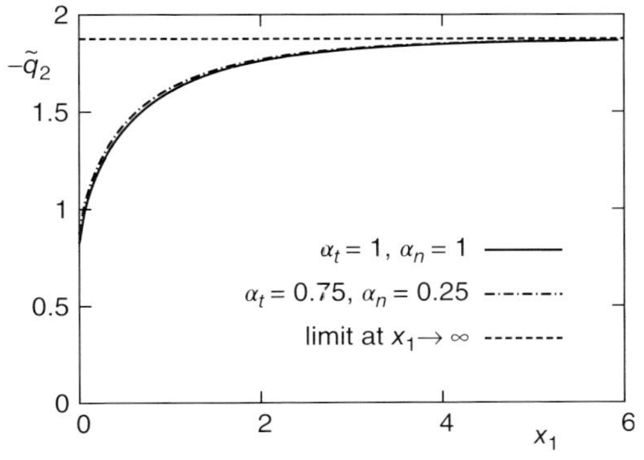

Figure 10.5 Dimensionless heat flow $\tilde{q}_2$ versus coordinate x_1 in thermal slip problem.

10.3.3
Numerical and Experimental Data

The main results corresponding to the diffuse gas–surface interaction, namely, Refs [33, 85, 89, 91–94, 105], are given in Table 10.5. The TSC obtained in Refs [33, 85] from the S-model is equal to $\sigma_T = 1.175$, while the value obtained from the BE is smaller, namely, it is 1.018 for the HS potential and it varies from 1.06 to 1.16 for the LJ potential. Summarizing these data, it can be said that the value of σ_T is about 1.10 ± 0.07. If the uncertainty of 7% is acceptable, the value $\sigma_T = 1.1$ can be used in practical calculations. Otherwise, the value of σ_T should be calculated for each specific gas.

The TSC based on the CL scattering law was calculated from the BE in Ref. [92] and from the S-model in Refs [33, 85]. The data of the last sources are partially presented in Table 10.6. From these data, we see that the TSC is sensitive to both ACs α_t and α_n. The numerical data can be interpolated by simple analytical formula as

$$\sigma_T = 1.175 - 0.37\left(1 - \alpha_t\right)\left(1 - \alpha_n\right). \tag{10.32}$$

Table 10.5 Thermal slip coefficient σ_T based on different equations and potentials. Diffuse gas–surface interaction.

σ_T	Equat.	Potential	References
1.175	S-model		[33, 85]
1.018	BE	HS	[89, 91–93, 105]
1.06	BE	LJ ($T^* = 30$)	[93]
1.06	BE	LJ ($T^* = 10$)	[93]
1.10	BE	LJ ($T^* = 2.40$)	[93]
1.16	BE	LJ ($T^* = 1.58$)	[94]
1.18	BE	LJ ($T^* = 1.31$)	[94]
1.15	BE	LJ ($T^* = 1.10$)	[93]

Table 10.6 Thermal slip coefficient σ_T based on the S-model and CL kernel, Refs [33, 85].

	σ_T			
α_t	$\alpha_n = 0.25$	0.5	0.75	1
0.5	1.034	1.081	1.127	1.172
0.75	1.107	1.129	1.152	1.174
1	1.175	1.175	1.175	1.175
1.25	1.240	1.219	1.197	1.175

Table 10.7 Thermal slip coefficient σ_T extracted from experimental data.

Gas	Surface	σ_T	Ref.
He	Pyrex	0.97	[106]
He	Glass	1.004	[107]
Ne	Pyrex	0.92	[106]
Ne	Glass	0.988	[107]
Ar	Pyrex	0.97	[106]
Ar	Glass	1.030	[107]
Kr	Pyrex	1.06	[106]
Xe	Glass	1.116	[107]

An experimental measurement of TSC is more difficult than that of the VSC. Some experimental data for the noble gases are presented in Table 10.7, which shows that these data of σ_T are close to the theoretical values for the diffuse scattering presented in Table 10.5. The light gases, such as neon, represent a smaller value, namely, 0.92, than that given in Table 10.5, but to get this value using the interpolation formula (10.32) both coefficients α_t and α_n must be about 0.17 that is not compatible with the data given in Table 10.4. Such an incompatibility can be related to both large uncertainty of the experiments and imperfection of the theory.

As is pointed out in the review [79], the TSC of polyatomic gases is slightly smaller than that of a monatomic gas. The influence of the intermolecular structure on the TSC is about 20%. According to the paper [108], the influence of the chemical composition on the TSC depends on its atomic weights ratio. For instance, the TSC of a helium–xenon mixture can be larger than that of a single gas for 35%, but for a neon–argon mixture, this difference is about 1%.

10.4
Reciprocal Relation

The problems of calculation of the VSC and TSC are coupled via the reciprocal relation (5.83). In this case, the pair of thermodynamic forces, namely, ξ_u defined by (10.3) and ξ_{T} defined by (10.18), is considered. Explicit expressions of the corresponding kinetic coefficients $\Lambda_{u\mathrm{T}}$ and $\Lambda_{\mathrm{T}u}$ were obtained in Ref. [43]. This is the case when one of these coefficients, $\Lambda_{\mathrm{T}u}$, is neither odd nor even with respect to the time reversion.

The time-reversed kinetic coefficients are obtained by using the source terms $g^{(u)}$ and $g^{(\mathrm{T})}$ obtained from their dimensionless expressions (10.5) and (10.22) with the help of (7.49)

$$g^{(u)} = -2\frac{v_0}{\ell}c_1c_2, \quad g^{(\mathrm{T})} = -\frac{v_0}{\ell}c_2\left(c^2 - \frac{5}{2}\right). \tag{10.33}$$

Moreover, the asymptotic perturbations (10.10) and (10.26) are substituted into the definition (5.85) adopting the notation $h^{(u)}$ for the solution of Eq. (10.6) or (10.7) and $h^{(\mathrm{T})}$ for that of Eq. (10.23) or (10.24). Note that for both solutions, the surface source terms h_w are zero.

Since the solutions $h^{(u)}$ and $h^{(\mathrm{T})}$ are one-dimensional depending only on the r_1 (or x_1) coordinate, the region of integration in (5.85) is the one-dimensional interval $[0, \infty]$, while the integration on the surface Σ_g is substituted by the integrant at $r_1 \to \infty$. Then the expression (5.85) takes the simplified form:

$$\Lambda^t_{kn} = \int_0^\infty \left(\hat{T}g^{(k)}, h^{(n)}\right) \mathrm{d}r_1 - \frac{1}{2}\lim_{r_1\to\infty}\left(\hat{T}v_1h^{(k)}, h^{(n)}\right), \tag{10.34}$$

where $k, n = u$, T. Substituting Eqs. (10.33), (10.10), and (10.26) into (10.34), the kinetic coefficients take the form

$$\Lambda^t_{u\mathrm{T}} = -\frac{\sigma_{\mathrm{T}}p_0}{2mv_0} + \frac{2(c_1h^{(\mu)}, h^{(\kappa)})}{(mv_0\ell)^2}, \tag{10.35}$$

$$\Lambda^t_{\mathrm{T}u} = \frac{2p_0}{mv_0}Q^{(u)} + \frac{\sigma_{\mathrm{T}}p_0}{2mv_0} - \frac{2(c_1h^{(\mu)}, h^{(\kappa)})}{(mv_0\ell)^2}, \tag{10.36}$$

where the quantity $Q^{(u)}$ is defined by (10.15). According to (5.83), these coefficients are equal to each other so that

$$\sigma_{\mathrm{T}} = \frac{4\left(c_1h^{(\mu)}, h^{(\kappa)}\right)}{mv_0p_0\ell^2} - 2Q^{(u)}. \tag{10.37}$$

In particular case of the S-model, the functions $h^{(\mu)}$ and $h^{(\kappa)}$ are given by Eqs. (10.11) and (10.28), which lead to the following simple relation of the TSC to the quantity $Q^{(u)}$

$$\sigma_{\mathrm{T}} = \frac{3}{2} - 2Q^{(u)}. \tag{10.38}$$

This equality can be checked using the numerical values of $Q^{(u)}$ given in Table 10.1 and values of σ_{T} given in Table 10.6.

Thus, the quantity σ_{T} obtained from Eq. (10.23) (or (10.24)) is coupled by the relation (10.37) (or (10.38)) with the quantity $Q^{(u)}$ obtained from Eq. (10.6) (or (10.7)). What is its utility? Using this coupling, the TSC σ_{T} can be obtained together with the VSC σ_{P} from the unique equation (10.6), reducing the computational effort to calculate both slip coefficients. If both equations (10.6) and

(10.24) are solved, the reciprocal relation (10.37) can be used as an additional criterion of the numerical error.

10.5 Temperature Jump

10.5.1 Definition and Input Equation

The temperature jump phenomenon is similar to the viscous velocity slip. The difference is that a temperature of gas is considered near a solid surface instead of its velocity. To define the temperature jump coefficient, again consider a gas occupying a semi-infinite space shown in Figure 10.1. Let us denote T_w as the temperature of the solid wall. If the gas temperature T varies along the r_1 coordinate, then it is not equal to the wall temperature T_w on the solid surface but it is given by the temperature jump condition at $r_1 = 0$

$$T = T_w + \zeta_T \ell \frac{\mathrm{d}T}{\mathrm{d}r_1}, \quad \text{at} \quad r_1 = 0, \tag{10.39}$$

where ζ_T is the temperature jump coefficient (TJC).

To calculate the TJC, we assume that the wall temperature is equal to the equilibrium one, $T_w = T_0$. The dimensionless temperature gradient

$$\xi_T = \frac{\ell}{T_0} \frac{\mathrm{d}T}{\mathrm{d}r_1}, \quad |\xi_T| \ll 1 \tag{10.40}$$

is used as the small parameter for the linearization. It is assumed that the pressure p of the gas is constant over the semi-infinite space and is equal to its equilibrium value p_0. The reference perturbations for the problem in question are similar to those (10.21) adopted to calculate the TSC, replacing x_2 by x_1

$$\varrho_R = -x_1, \quad \tilde{u}_R = 0, \quad \tau_R = x_1. \tag{10.41}$$

Then the linearized BE reads

$$c_1 \frac{\partial h(x_1, \boldsymbol{c})}{\partial x_1} = \frac{\ell}{v_0} \hat{L}h - c_1 \left(c^2 - \frac{5}{2} \right). \tag{10.42}$$

To write down the S-model for this situation, we should consider that the heat flow $\tilde{q}_1$ is constant over the whole space and equal to the same value as that obtained far from the surface in the previous problem (10.30)

$$\tilde{q}_1 = -\frac{15}{8}. \tag{10.43}$$

Moreover, we take into account that the gas is at rest so that its velocity is zero. Thus, among all moments, only the density and temperature deviations are variable and unknown. Then the S-model reads

$$c_1 \frac{\partial h(x_1, \boldsymbol{c})}{\partial x_1} = \varrho + \tau \left(c^2 - \frac{3}{2} \right) - h(x_1, \boldsymbol{c}) - \frac{3}{2} c_1 \left(c^2 - \frac{5}{2} \right). \tag{10.44}$$

The unknown moments are calculated via the perturbation as

$$\varrho = \frac{1}{\pi^{3/2}} \int e^{-c^2} h \, \mathrm{d}^3\boldsymbol{c}, \quad \tau = \frac{1}{\pi^{3/2}} \int e^{-c^2} \left(\frac{2}{3}c^2 - 1\right) h \, \mathrm{d}^3\boldsymbol{c}, \tag{10.45}$$

that is extracted from the definitions (5.17). The dimensional density and temperature are expressed by (5.35) and (5.37) and take the following form:

$$n = n_0[1 + \xi_{\mathrm{T}}(-x_1 + \varrho)], \quad T = T_0[1 + \xi_{\mathrm{T}}(x_1 + \tau)] \tag{10.46}$$

in this particular case.

The kinetic equation (10.42) or (10.44) is solved together with the boundary condition (5.50). Since we assumed that $T_w = T_0$, the surface perturbation h_w given by (5.52) is zero. Far from the surface ($x_1 \to \infty$), the perturbation function tends to that obtained to calculate the thermal conductivity with a shift of the density and temperature

$$\lim_{x_1 \to \infty} h = \varrho + \left(c^2 - \frac{3}{2}\right)\tau + \frac{h^{(\kappa)}}{k_{\mathrm{B}} T_0 \ell}, \tag{10.47}$$

where $h^{(\kappa)}$ is the solution of Eq. (6.23) with ψ_q given by (6.18). Since the pressure far from the surface is equal to that of equilibrium, then $\varrho = -\tau$ so that Eq. (10.47) is simplified to

$$\lim_{x_1 \to \infty} h = \left(c^2 - \frac{5}{2}\right)\tau + \frac{h^{(\kappa)}}{k_{\mathrm{B}} T_0 \ell}. \tag{10.48}$$

For the S-model, the function $h^{(\kappa)}$ is similar to (10.28), but with c_1 instead of c_2

$$h^{(\kappa)} = -\frac{3}{2} k_{\mathrm{B}} T_0 \ell \left(c^2 - \frac{5}{2}\right) c_1. \tag{10.49}$$

Then Eq. (10.48) is reduced to

$$\lim_{x_1 \to \infty} h = \left(c^2 - \frac{5}{2}\right)\left(\tau - \frac{3}{2}c_1\right). \tag{10.50}$$

The same expression is obtained directly from the S-model (10.44) assuming that

$$\lim_{x_1 \to \infty} \frac{\partial h}{\partial x_1} = 0, \quad \lim_{x_1 \to \infty} (\varrho + \tau) = 0. \tag{10.51}$$

In numerical calculations, the unknown value of the temperature deviation τ in (10.48) (or (10.50)) is substituted by that obtained in a previous iteration.

Once the temperature deviation τ is known, the jump coefficient is calculated as the asymptotic value of τ far from the surface

$$\zeta_{\mathrm{T}} = \lim_{x_1 \to \infty} \tau \tag{10.52}$$

that is a consequence of Eqs. (10.1), (10.39), (10.46), and (10.40).

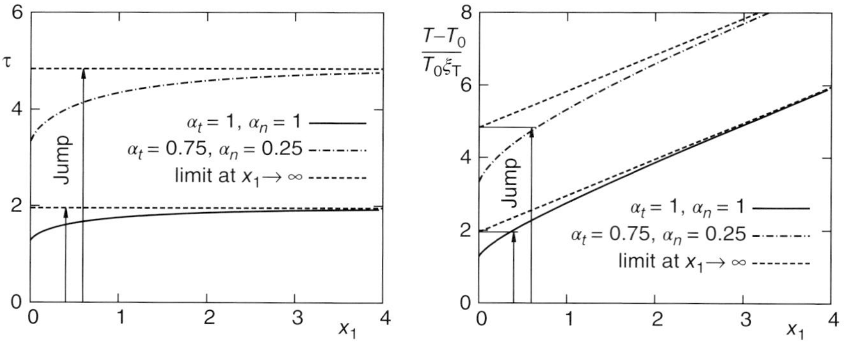

Figure 10.6 Deviations of gas temperature from reference temperature T_R (a) and from surface temperature T_0 (b) versus coordinate x_1 in temperature jump problem.

10.5.2
Temperature Profile

The typical deviations τ from the reference temperature T_R are shown in Figure 10.6(a) based on the solution of the S-model (10.44) and CL scattering law. The lower curve represents a solution for the complete accommodation ($\alpha_t = 1$, $\alpha_n = 1$) and the upper curve corresponds to the following ACs $\alpha_t = 0.75$ and $\alpha_n = 0.25$. Qualitatively, these curves are very similar to those representing the velocity deviation shown in Figure 10.2(a), namely, they have the same shape and tend to have constant values equal to the TJC depicted by the dashed line. As is expected, the TJC is smaller for the complete accommodation. The deviations from the wall temperature T_0 are plotted on Figure 10.6(b), which points out that the real temperature on the surface is not that calculated by the jump boundary condition (10.39), but it is slightly smaller. Again, such a behavior of the temperature is similar to that of the velocity profile shown in Figure 10.2 (b). As in case of the velocity, the temperature defect just near the surface has the second order with respect to the Knudsen number and usually is neglected when the jump condition (10.39) is applied.

10.5.3
Numerical Data

The numerical values of TJC obtained from different equations and for the complete accommodation on the solid surface are given Table 10.8. The earliest paper reporting the TJC based on the kinetic equation was published by Welander [65]. Without acquaintance with the paper proposing the BGK model [64], Welander approximated the BE by this model. One should note that applying the BGK model to the temperature jump problem, the expression (7.16) for the parameter ν_B should be used. Under this condition, the value reported in Ref. [65] in terms of our notations is given by $\zeta_T = 1.949$. The more exact solution of the BGK model

Table 10.8 Temperature jump coefficients ζ_T based on different equations and potentials. Diffuse gas–surface interaction.

ζ_T	Equat.	Potential	References
1.949	BGK		[65]
1.954	BGK		[105, 109, 110]
1.954	S-model		[33]
1.890	BE	HS	[105, 110, 111]
1.91	BE	LJ ($T^* = 30$)	[93]
1.91	BE	LJ ($T^* = 10$)	[93]
1.92	BE	LJ ($T^* = 2.40$)	[93]
1.95	BE	LJ ($T^* = 1.10$)	[93]

Table 10.9 Temperature jump coefficients ζ_T based on the CL kernel (4.17), Ref. [33].

	ζ_T			
α_t	$\alpha_n = 0.25$	0.5	0.75	1
0.5	5.828	4.170	3.088	2.335
0.75	4.833	3.549	2.670	2.040
1	4.567	3.376	2.551	1.954

in Refs [105, 109, 110] leads to a slightly different value of $\zeta_T = 1.954$. The same value is obtained from the S-model in Ref. [33]. The BE with the HS potential, see Refs [105, 110, 111], yields the smaller value $\zeta_T = 1.890$, while the same equation for the LJ potential, Ref. [93], provides the TJC varying from 1.91 to 1.95. Thus, any value from this small range can be used in practical calculations.

The S-model subject to the CL scattering condition was solved in Ref. [33]. The results are compared in Table 10.9. Analyzing the numerical data of ζ_T for the values of α_t and α_n smaller than unity, we see that the TJC depends significantly on both ACs α_t and α_n. The dependence of ζ_T on the energy AC α_n is understandable, while its dependence on the momentum AC α_t is rather unexpected.

The review paper [79] provides an estimation of the TJC of polyatomic gases showing that it is about 20% smaller than that of a monatomic gas. The data on TJC reported in the paper [112] can be used to estimate the influence of the gas chemical composition. According to these data, the TJC of a neon–argon mixture differs from that of a single gas just for 2%, while the TJC of a helium–xenon mixture exceeds its value of a single gas for 24%.

Exercises

10.1 Obtain (10.30).

10.2 Obtain (10.31).

10.3 Check Eq. (10.38) using the data given in Tables 10.1 and 10.6.

10.4 Consider a helium gas flowing near a solid surface at the temperature $T = 300$ K. The normal velocity gradient is given as $\mathrm{d}u_2/\mathrm{d}r_1 = 100\ \mathrm{s}^{-1}$. Calculate the slip velocity just near the surface at (a) the standard pressure (see Table A.1); (b) $p = 1$ Pa.
Solution: (a) $u_2 = 22\ \mu$m/s; (b) $u_2 = 2.3$ m/s.

10.5 Consider a longitudinal temperature gradient $\mathrm{d}T/\mathrm{d}r_2 = 100$ K/m instead of the velocity gradient in the previous exercise. Calculate the thermal slip velocity.
Solution: (a) $u_2 = 49\ \mu$m/s; (b) $u_2 = 4.9$ m/s.

10.6 Consider a normal temperature gradient $\mathrm{d}T/\mathrm{d}r_1 = 100$ K/m instead of the longitudinal one in the previous exercise. Calculate the temperature jump near the surface.
Solution: (a) $\Delta T = 43\ \mu$K; (b) $\Delta T = 4.3$ K.
Comments: Under standard conditions, the velocity slip and temperature jump are negligible, while they are significant in a rarefied gas.

11
One-Dimensional Planar Flows

11.1
Planar Couette Flow

The planar Couette flow is the simplest problem of fluid mechanics that is why it is the first to be considered in this chapter.

11.1.1
Definitions

Consider two parallel plates placed at $r_1 = \pm a/2$ confining a gas as is shown in Figure 11.1. The left plate ($r_1 = -a/2$) is moving up with a speed $u_w/2$, while the right plate ($r_1 = a/2$) is moving down with the same speed so that the relative speed between the plates is equal to u_w. The plates are considered so large that the gas flow is invariant in both r_2 and r_3 directions. Moreover, we consider a steady, or time independent, flow. Mathematically, it means that the VDF depends only on the r_1 coordinate and on the three components of the molecular velocity $\boldsymbol{v}$

$$f = f(r_1, \boldsymbol{v}). \tag{11.1}$$

We are going to calculate the shear stress P_{12} and velocity profile u_2 between the plates. Since they are calculated via the VDF, they depend only on the coordinate r_1 because of (11.1).

The distance a between the plates can be considered as the characteristic size of the problem and it is used to define the rarefaction parameter δ by (7.46), assuming that the equilibrium pressure p_0 and temperature T_0 correspond to the situation when the plates are at rest, $u_w = 0$. Since the gas flow depends only on one dimensionless coordinate x_1, the subscript "1" will be omitted in this section so that the dimensionless coordinate x is defined as

$$x = \frac{r_1}{a} \tag{11.2}$$

according to (7.43). The ratio

$$\xi = \frac{u_w}{v_0}, \quad |\xi| \ll 1 \tag{11.3}$$

Rarefied Gas Dynamics: Fundamentals for Research and Practice, First Edition. Felix Sharipov.

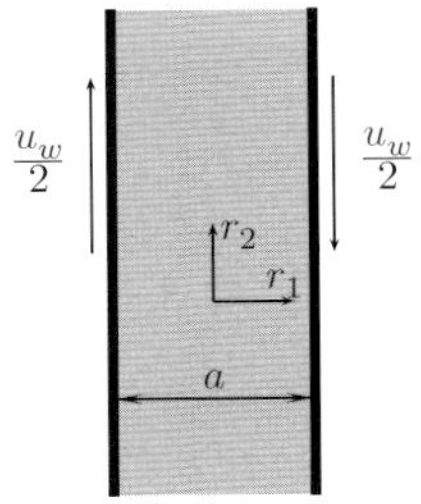

Figure 11.1 Scheme and coordinates of planar Couette flow.

is used as the small parameter to linearize the kinetic equation so that the moments of our interest are linearized as

$$u_2 = \xi v_0 \tilde{u}, \quad P_{12} = \xi p_0 \Pi, \tag{11.4}$$

according to Eqs. (5.13) and (5.15), respectively. Since $\tilde{u}_2$ is the unique component of the bulk velocity $\tilde{\boldsymbol{u}}$ and Π_{12} is the unique element of the tensor Π, the subscripts "2" and "12" are omitted in this section.

11.1.2
Free-Molecular Regime

In the free-molecular regime ($\delta \to 0$), the collision integral vanishes and then the BE (3.1) is reduced to

$$v_1 \frac{\partial f}{\partial x} = 0. \tag{11.5}$$

In other words, the VDF does not depend on the coordinate x, but it depends only on the velocity, $f = f(\boldsymbol{v})$. The distribution of those particles moving from the left surface to the right surface is determined only by the boundary conditions on the left surface. The particles moving to the left surface have the VDF determined by the right surface. Thus, the VDF can be split into two parts

$$f(\boldsymbol{v}) = \begin{cases} f_+ & \text{for} \quad v_1 > 0, \\ f_- & \text{for} \quad v_1 < 0, \end{cases} \tag{11.6}$$

each of them defined in the half of the velocity space.

Under assumption of the diffuse scattering, the VDF is Maxwellian with the equilibrium number density n_0 and temperature T_0, but with the bulk velocity equal to that of the corresponding wall. The particles reflected by the left wall and moving to right ($v_1 > 0$) have the bulk velocity $[0, u_w/2, 0]$ and vice versa, then the VDF reads

$$f_{\pm} = n_0 \left(\frac{m}{2\pi k_B T_0} \right)^{3/2} \exp \left\{ -\frac{m[v_1^2 + (v_2 \mp u_w/2)^2 + v_3^2]}{2k_B T_0} \right\}. \tag{11.7}$$

Substituting this expression into Eqs. (2.12), (2.16), (2.26), and (2.28), it is concluded that the total density is equal to its equilibrium value n_0, the bulk velocity is equal to zero, $u_2 = 0$, the temperature is equal to

$$T = T_0 \left[1 + \frac{1}{6} \left(\frac{u_w}{v_0} \right)^2 \right] = T_0 \left(1 + \frac{1}{6} \xi^2 \right), \tag{11.8}$$

and the shear stress P_{12} and its dimensionless form Π are given as

$$P_{12} = \frac{p_0}{\sqrt{\pi}} \frac{u_w}{v_0}, \quad \Pi = \frac{1}{\sqrt{\pi}}. \tag{11.9}$$

where (11.3) and (11.4) have been used. Note that the temperature (11.8) contains only the quadratic term with respect to ξ, that is, it is equal to its equilibrium value in the linear approximation, while the shear stress (11.9) is always proportional to ξ at any value of the wall speed u_w.

11.1.3
Velocity Slip Regime

When the rarefaction parameter is large ($\delta \gg 1$), the problem is based on the momentum conservation law (3.27) and Newton's law (6.2). For the stationary one-dimensional flow, they read

$$\frac{\mathrm{d}P_{12}}{\mathrm{d}r_1} = 0, \quad P_{12} = -\mu \frac{\mathrm{d}u_2}{\mathrm{d}r_1}. \tag{11.10}$$

The slip boundary condition (10.2) is written taking into account the surface motion

$$u_2 = \mp \left(\frac{u_w}{2} - \sigma_\mathrm{P}\, \ell\, \frac{\mathrm{d}u_2}{\mathrm{d}r_1} \right) \quad \text{at} \quad r_1 = \pm \frac{a}{2}. \tag{11.11}$$

Then the velocity profile is obtained analytically

$$u_2 = -u_w \frac{r_1}{a} \left(1 + 2\frac{\sigma_\mathrm{P}}{\delta} \right)^{-1}, \quad \tilde{u} = -x \left(1 + 2\frac{\sigma_\mathrm{P}}{\delta} \right)^{-1}, \tag{11.12}$$

where the relation (5.13) of u_2 to $\tilde{u}$ has been used. The shear stress is obtained from Eq. (11.10) as

$$P_{12} = \mu \frac{u_w}{a} \left(1 + 2\frac{\sigma_\mathrm{P}}{\delta} \right)^{-1}, \quad \Pi = \frac{1}{\delta + 2\sigma_\mathrm{P}}, \tag{11.13}$$

where P_{12} is related to Π by (11.4). As it is expected, the velocity slip on the surface reduces the shear stress compared to that for the nonslip condition.

11.1.4
Kinetic Equation

The Couette flow can be solved by a linearization near both global and local Maxwellian. Choosing the first, the bulk source term $\tilde{g}$ is zero so that the linearized BE (5.20) is reduced to

$$c_1 \frac{\partial h}{\partial x} = \frac{a}{v_0} \hat{L} h. \tag{11.14}$$

Since we are interested only in the momentum transfer, the BGK model is appropriate for the problem in question. Thus, Eq. (7.47) with (7.48) and Table (7.1) takes the form

$$c_1 \frac{\partial h}{\partial x} = \delta(2c_2\tilde{u} - h). \tag{11.15}$$

The dimensionless velocity $\tilde{u}$ and shear stress Π are calculated via the perturbation according to (5.13) and (5.15)

$$\begin{bmatrix} \tilde{u}(x) \\ \Pi(x) \end{bmatrix} = \frac{1}{\pi^{3/2}} \int \begin{bmatrix} c_2 \\ 2c_1c_2 \end{bmatrix} e^{-c^2} h(x, \boldsymbol{c})\, \mathrm{d}^3\boldsymbol{c}. \tag{11.16}$$

According to the momentum conservation law, the shear stress Π is constant over the space, but such a property will be not used in the numerical scheme. This quantity is calculated in all nodes x_k, and then its variance is used as an additional criterion of the accuracy.

The boundary condition is obtained from (5.50) with the surface source term equal to

$$h_w = 2\boldsymbol{c} \cdot \tilde{\boldsymbol{u}}_w \quad \tilde{\boldsymbol{u}}_w = [0, \mp\frac{1}{2}, 0] \quad \text{at} \quad x \pm \frac{1}{2}. \tag{11.17}$$

Under assumption of the diffuse scattering, the boundary condition takes the form (5.70), where the quantity ϱ_r defined by (5.71) is zero because the perturbation function h is an odd function of the velocity component c_2 according to (11.15). Substituting (11.17) into (5.70) with $\varrho_r = 0$, the boundary condition becomes

$$h(x, \boldsymbol{c}) = \mp c_2 \quad \text{at} \quad x = \pm\frac{1}{2} \quad \text{for} \quad c_1 \lessgtr 0. \tag{11.18}$$

To solve numerically Eq. (11.15), the new perturbation function is introduced in order to reduce the number of variables

$$\Phi(x, c_1) = \frac{1}{\pi} \int_{-\infty}^{\infty} \int_{-\infty}^{\infty} e^{-c_2^2 - c_3^2} h(x, \boldsymbol{c}) c_2\, \mathrm{d}c_2\, \mathrm{d}c_3. \tag{11.19}$$

Multiplying Eq. (11.15) by $e^{-c_2^2-c_3^2} c_2/\pi$ and integrating it with respect to c_2 and c_3, the kinetic equation for the perturbation $\Phi(x, c_1)$ having two variables is obtained

$$c_1 \frac{\partial \Phi(x, c_1)}{\partial x} = \delta[\tilde{u}(x) - \Phi(x, c_1)]. \tag{11.20}$$

The boundary condition for the new perturbation function is obtained from (11.18) by the integration according to (11.19)

$$\Phi(x, c_1) = \mp\frac{1}{2} \quad \text{at} \quad x = \pm\frac{1}{2} \quad \text{for} \quad c_1 \lessgtr 0. \tag{11.21}$$

Note that the solution of the integro-differential equation (11.20) with the boundary condition (11.21) has a certain symmetry, namely,

$$\Phi(x, c_1) = -\Phi(-x, -c_1). \tag{11.22}$$

Such a symmetry allows us to calculate the perturbation Φ only for positive values of the velocity c_1, which is used to determine the half-moments defined as

$$\begin{bmatrix} \tilde{u}^{(1/2)}(x) \\ \Pi^{(1/2)}(x) \end{bmatrix} = \frac{1}{\sqrt{\pi}} \int_0^\infty \begin{bmatrix} 1 \\ 2c_1 \end{bmatrix} e^{-c_1^2} \Phi(x, c_1)\, \mathrm{d}c_1, \tag{11.23}$$

that is, the moments are calculated only for the positive velocity $c_1 > 0$. The full moments are calculated using the symmetry (11.22)

$$\tilde{u}(x) = \tilde{u}^{(1/2)}(x) - \tilde{u}^{(1/2)}(-x), \quad \Pi(x) = \Pi^{(1/2)}(x) + \Pi^{(1/2)}(-x). \tag{11.24}$$

11.1.5 Numerical Scheme

According to the DVM, some values c_i $(1 \le i \le N_c)$ of the velocity c_1 are chosen. Each value c_i has its own weight denoted as W_i so that the integral with respect to the velocity is expressed by the quadrature (9.15). The nodes and weights obtained by the Gauss rule are most appropriate for our task. Their values for $N_c = 4, 6$, and 8 are given in Appendix, Section B.1.1.

The simple way to distribute the nodes for the space coordinate x is as follows

$$x_k = \frac{k}{2N_x}, \quad -N_x \le k \le N_x, \tag{11.25}$$

where N_x is an integer. The perturbation function and the bulk velocity are discretized as

$$\Phi_{ki} = \Phi(x_k, c_i), \quad \tilde{u}_k = \tilde{u}(x_k). \tag{11.26}$$

Then the finite difference scheme (9.40) for Eq. (11.15) reads

$$\Phi_{ki} = \frac{(4N_x c_i - \delta)\Phi_{k-1,i} + \delta(\tilde{u}_k + \tilde{u}_{k-1})}{4N_x c_i + \delta}, \tag{11.27}$$

where the coefficient $C_k = 4N_x c_i$ is obtained from the definition (9.39) and the node distribution (11.25). Since the solution is symmetric (11.22), the sweeping of the node x_k can be done in one direction, for example, from $k = -N_x + 1$ to $k = N_x$ only for positive values of c_1.

When the scheme (11.27) is realized for all nodes of the velocity space, the new values of the bulk velocity are calculated by the quadrature (9.15) in accordance with the expression in Eq. (11.23) as

$$\begin{bmatrix} \tilde{u}_k^{(1/2)} \\ \Pi_k^{(1/2)} \end{bmatrix} = \sum_{i=1}^{N_c} \begin{bmatrix} 1 \\ 2c_i \end{bmatrix} \Phi_{ki} W_i. \tag{11.28}$$

In practice, the contribution of Φ_{ki} can be added into the moments just after its calculation avoiding the storage of the data. Since the sweeping with respect to the nodes x_k is done from $k = -N_x + 1$, the contribution of the boundary node

$k = -N_x$ is calculated separately substituting the boundary value of Φ given by (11.21) for $c_1 > 0$ into the definition (11.23)

$$\tilde{u}_k^{(1/2)} = \frac{1}{4}, \quad \Pi_k^{(1/2)} = \frac{1}{2\sqrt{\pi}}, \quad \text{for} \quad k = -N_x. \tag{11.29}$$

When the half-moments are known, the full moments are calculated as

$$\tilde{u}_k = \tilde{u}_k^{(1/2)} - \tilde{u}_{-k}^{(1/2)}, \quad \Pi_k = \Pi_k^{(1/2)} + \Pi_{-k}^{(1/2)}, \tag{11.30}$$

that follows from (11.24). Thus, the iteration is completed. Then all procedures are repeated up to the convergence, when the shear stress Π does not change in two successive iterations within a given error.

A listing of the numerical program couette_planar.for based on the scheme described earlier and written in FORTRAN language is given in Section B.2.

11.1.6 Numerical Results

Using the code couette_planar.for, it is verified that the shear stress Π is calculated with the numerical error less than 0.1% using a few number of nodes, namely, $N_x = 10$ and $N_c = 6$. The results obtained by this code are given in Table 11.1 and compared with those obtained by Siewert [91] from the linearized BE based on the HS potential and by the DSMC method based on the AI potential for argon in Ref. [75]. These data show that the discrepancy between results obtained by quite different methods does not exceed 1%. However, the computational effort to solve the BE or to apply the DSMC method is several orders larger than that to solve the model equation. The analytical limit results (11.9) and (11.13) given in Table 11.1 show that the flow is practically free-molecular at $\delta \leq 0.01$, and it is well described by the slip solution at $\delta \geq 20$.

Table 11.1 Dimensionless shear stress Π in planar Couette flow versus rarefaction δ obtained by various methods: BGK by the scheme (11.27), BE based on HS in Ref. [91], and DSMC based on AI for argon in Ref. [75].

δ	Π		
	BGK	BE	DSMC
0[a]	0.5642		
0.01	0.5593		0.5575
0.1	0.5220	0.5209	0.5167
1.0	0.3389	0.3405	0.3365
10	0.08311	0.08351	0.08320
20	0.04539		0.04531
20[b]	0.04538		

a) Free-molecular solution (11.9).
b) Slip solution (11.13).

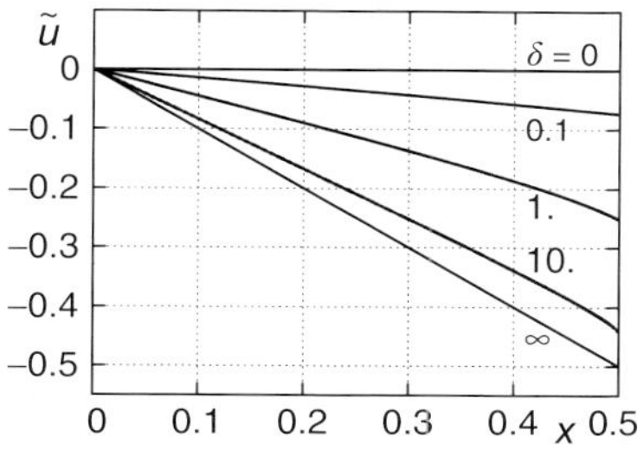

Figure 11.2 Dimensionless velocity $\tilde{u}$ versus coordinate x in planar Couette flow.

The velocity profiles for several values of δ are depicted in Figure 11.2. All profiles are very close to the straight lines. However, their slopes depend on the rarefaction parameters being zero in the free-molecular regime ($\delta = 0$) and smoothly varying to that given by the analytical solution (11.12) when the rarefaction δ increases.

Numerical data on the planar Couette for a large ratio u_w/v_0 obtained by the DSMC technique based on the AI potential reported in Ref. [75] show that such kind of flow is not sensitive to the intermolecular potential. Some results on the planar Couette flow for gaseous mixtures are reported in the papers [75, 113–115]. According to the results obtained in Ref. [113], the deviation of the shear stress Π from its value for a single gas is stronger for a mixture with a larger ratio of the atomic weight ratio. For a helium–xenon mixture, the deviation reaches 25%, while the deviation does not exceed 1.5% for a neon–argon mixture.

11.2 Planar Heat Transfer

A heat transfer between two parallel plates confining a gas is one more classical problem of fluid mechanics. The technique to solve it is very similar to that described for the planar Couette flow. Thus, many definitions of the previous section are also used in this one.

11.2.1 Definitions

Let us again consider two parallel plates fixed at $r_1 = \pm a/2$. In contrast to the Couette flow, now the plates are at rest, but the left plate is maintained at a temperature $T_0 + \Delta T/2$, while the right plate has another temperature equal to $T_0 - \Delta T/2$ as is depicted in Figure 11.3. The distance a is assumed to be the characteristic size, so that the rarefaction parameter δ is given by Eq. (7.46). The unique dimensionless coordinate x is introduced in the same way as in the Couette flow by (11.2). It is assumed that the temperature difference is small compared to the equilibrium temperature $\Delta T \ll T_0$. Then the ratio

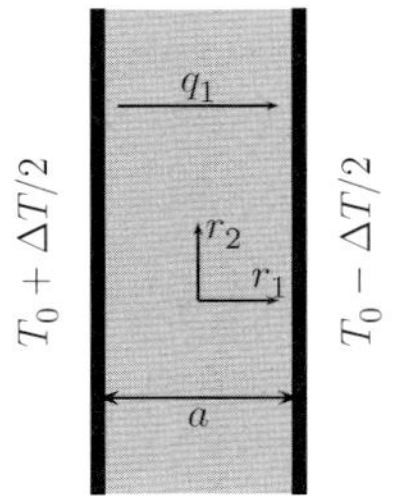

Figure 11.3 Scheme for heat transfer between two plates.

$$\xi = \frac{\Delta T}{T_0}, \quad |\xi| \ll 1 \tag{11.31}$$

can be used as the small parameter for the linearization. In this problem, the temperature distribution $T(x)$ and heat flux q_1 in the gap between the plates are calculated. These quantities are related to the corresponding dimensionless moments as

$$T = T_0(1 + \xi\tau), \quad q_1 = \xi p_0 v_0 \tilde{q} \tag{11.32}$$

in accordance with definitions (5.14) and (5.16), respectively. Since for the problem in question the vector $\tilde{\boldsymbol{q}}$ has the unique component $\tilde{q}_1$, its subscript "1" has been omitted.

11.2.2
Free-Molecular Regime

In the free molecular regime ($\delta \to 0$), the problem can be solved analytically for an arbitrary temperature drop ΔT. Let us denote the temperatures of the left and right surfaces as

$$T_\pm = T_0 \pm \frac{\Delta T}{2} \quad \text{at} \quad x = \mp\frac{1}{2}, \tag{11.33}$$

respectively. Under the assumption of the diffuse scattering on both surfaces, the VDF takes the form

$$f_\pm(\boldsymbol{v}) = n_\pm \left(\frac{m}{2\pi k_B T_\pm}\right)^{3/2} \exp\left(-\frac{mv^2}{2k_B T_\pm}\right), \quad \text{for} \quad v_1 \gtrless 0. \tag{11.34}$$

The quantities n_+ and n_- are obtained from two conditions. First, the total number density should be the same as in the equilibrium, that is,

$$\int_{v_1>0} f_+(\boldsymbol{v})\, d^3\boldsymbol{v} + \int_{v_1<0} f_-(\boldsymbol{v})\, d^3\boldsymbol{v} = n_0. \tag{11.35}$$

Physically it means that none of the temperature variation changes the gas quantity in the gap. The second condition expresses the fact that the gas does not move, that is,

$$\int_{v_1>0} v_1 f_+(\boldsymbol{v})\, d^3\boldsymbol{v} + \int_{v_1<0} v_1 f_-(\boldsymbol{v})\, d^3\boldsymbol{v} = 0. \tag{11.36}$$

In other words, the bulk velocity is zero. Substituting (11.34) into (11.35) and (11.36) and solving the system of algebraic equations, the quantities n_+ and n_- are found as

$$n_+ = 2n_0 \frac{T_-^{1/2}}{T_-^{1/2} + T_+^{1/2}}, \quad n_- = 2n_0 \frac{T_+^{1/2}}{T_+^{1/2} + T_-^{1/2}}. \tag{11.37}$$

Then the heat flow q_1 is calculated by the substitution of (11.34) into the definition (2.32) taking into account the gas immobility ($\boldsymbol{u} = 0$)

$$q_1 = \frac{p_0 v_0}{\sqrt{\pi}} \frac{\Delta T}{T_0} \frac{2[1 - (\Delta T/2T_0)^2]^{1/2}}{(1 + \Delta T/2T_0)^{1/2} + (1 - \Delta T/2T_0)^{1/2}}. \tag{11.38}$$

To find the temperature, the VDF (11.34) is substituted into the definition (2.26). Then the simple expression is obtained

$$T = (T_+ T_-)^{1/2} = T_0 \left[1 - \left(\frac{\Delta T}{2T_0} \right)^2 \right]^{1/2}. \tag{11.39}$$

In case of small temperature difference, $|\Delta T|/T_0 \ll 1$, the last fraction in the expression (11.38) becomes unity. Then the dimension q_1 and dimensionless heat flows (11.32) are reduced to

$$q_1 = \frac{p_0 v_0}{\sqrt{\pi}} \xi, \quad \tilde{q} = \frac{1}{\sqrt{\pi}}. \tag{11.40}$$

The temperature does not have terms of the first order with respect to ξ, but it is constant and equal to its equilibrium value T_0 in the linear approximation.

11.2.3
Temperature Jump Regime

In the hydrodynamic regime ($\delta \to \infty$), the energy balance equation (3.28) is solved together with Fourier's law (6.3)

$$\frac{\mathrm{d}q_1}{\mathrm{d}r_1} = 0, \quad q_1 = -\kappa \frac{\mathrm{d}T}{\mathrm{d}r_1}. \tag{11.41}$$

The combination of these equations leads to the differential equation of the second order

$$\frac{\mathrm{d}}{\mathrm{d}r_1} \kappa \frac{\mathrm{d}T}{\mathrm{d}r_1} = 0. \tag{11.42}$$

The temperature jump condition (10.39), which for the plane heat transfer reads

$$T = T_0 \pm \left(\frac{\Delta T}{2} + \zeta_T \ell \frac{\mathrm{d}T}{\mathrm{d}r_1} \right) \quad \text{at} \quad r_1 = \mp \frac{a}{2} \tag{11.43}$$

is used as the boundary condition to Eq. (11.42). Under the assumption of the small temperature difference (11.31), the thermal conductivity can be considered constant. Then the temperature distribution is obtained as

$$T(r_1) = T_0 - \Delta T \, \frac{r_1}{a}\left(1 + 2\frac{\zeta_{\mathrm{T}}}{\delta}\right)^{-1}, \quad \tau = -x\left(1 + 2\frac{\zeta_{\mathrm{T}}}{\delta}\right)^{-1}. \tag{11.44}$$

The heat flux is calculated from Fourier's equation (11.41) as

$$q_1 = \kappa \frac{\Delta T}{a}\left(1 + 2\frac{\zeta_{\mathrm{T}}}{\delta}\right)^{-1}, \quad \tilde{q} = \frac{5}{4\,\mathrm{Pr}}\frac{1}{\delta + 2\zeta_{\mathrm{T}}}. \tag{11.45}$$

It is evident that the heat flux magnitude decreases when the temperature jump condition is applied instead of the temperature continuity condition.

11.2.4
Kinetic Equation

Since the heat flux is significant for the problem under question, the BGK model is not recommended, but the S-model, see Eq. (7.47) with Table 7.1, is appropriate. For the specific case considered here, the S-model takes the form

$$c_1 \frac{\partial h}{\partial x} = \delta(H - h), \tag{11.46}$$

where the function H reads

$$H = \varrho + \left(c^2 - \frac{3}{2}\right)\tau + \frac{4}{15}\left(c^2 - \frac{5}{2}\right) c_1 \tilde{q}. \tag{11.47}$$

The moments are obtained from (5.17)

$$\begin{bmatrix} \varrho(x) \\ \tau(x) \\ \tilde{q}(x) \end{bmatrix} = \frac{1}{\pi^{3/2}} \int \begin{bmatrix} 1 \\ \frac{2}{3}c^2 - 1 \\ c^2 c_1 \end{bmatrix} e^{-c^2} h(x, \boldsymbol{c}) \,\mathrm{d}^3\boldsymbol{c}. \tag{11.48}$$

Note that the bulk velocity has been assumed to be zero in both (11.47) and (11.48). The energy conservation law expressed by the first equality of (11.41) is valid for any gas rarefaction, that is, $\tilde{q}$ does not depend on the coordinate x. However, the conservation law is not used in our scheme so that the heat flow will be calculated in all nodes x_k. The variance of $\tilde{q}$ from node to node is used as an additional criterion of the numerical accuracy.

The linearization near the global Maxwellian can be used for this problem. It means that the bulk source term is zero, $\tilde{g}(\boldsymbol{c}) = 0$. The surface source term is obtained from (5.52) as

$$h_w = \tau_w \left(c^2 - \frac{3}{2}\right), \quad \tau_w = \pm\frac{1}{2} \quad \text{at} \quad x = \mp\frac{1}{2}. \tag{11.49}$$

For the diffuse scattering law, the expression (5.70) is reduced to

$$h = \pm\left[\varrho_r + \frac{1}{2}\left(c^2 - 2\right)\right] \quad \text{at} \quad x = \mp\frac{1}{2} \quad \text{for} \quad c_1 \gtrless 0, \tag{11.50}$$

where the quantity ϱ_r defined by (5.71) can be calculated only for one surface. For the right surface, we have

$$\varrho_r = \frac{2}{\pi}\int_{c_1>0} c_1 e^{-c^2} h\left(\frac{1}{2}, \boldsymbol{c}\right) \,\mathrm{d}^3\boldsymbol{c}. \tag{11.51}$$

Since the left-hand side of this equation does not contain c_1, the integration variable c_1' in the right-hand side has been replaced by c_1. For the left surface, this quantity just changes its own sign

$$\varrho_r|_{x=-1/2} = -\varrho_r|_{x=1/2}. \tag{11.52}$$

To solve Eq. (11.46), two new perturbation functions are introduced

$$\Phi^{(1)}(x, c_1) = \frac{1}{\pi} \int_{-\infty}^{\infty} \int_{-\infty}^{\infty} e^{-c_2^2 - c_3^2} h(x, \mathbf{c}) \, dc_2 \, dc_3, \tag{11.53}$$

and

$$\Phi^{(2)}(x, c_1) = \frac{1}{\pi} \int_{-\infty}^{\infty} \int_{-\infty}^{\infty} e^{-c_2^2 - c_3^2} h(x, \mathbf{c})(c_2^2 + c_3^2 - 1) \, dc_2 \, dc_3. \tag{11.54}$$

Multiplying Eq. (11.46) by $e^{-c_2^2-c_3^2}/\pi$ and $e^{-c_2^2-c_3^2}(c_2^2 + c_3^2 - 1)/\pi$, then integrating it with respect to c_2 and c_3, the following equations are obtained for the new perturbations

$$c_1 \frac{\partial \Phi^{(\alpha)}(x, c_1)}{\partial x} = \delta(H^{(\alpha)} - \Phi^{(\alpha)}), \quad \alpha = 1, 2, \tag{11.55}$$

where the functions $H^{(\alpha)}$ are obtained from (11.47)

$$H^{(1)} = \varrho + \left(c_1^2 - \frac{1}{2}\right) \tau + \frac{4}{15} \left(c_1^2 - \frac{3}{2}\right) c_1 \tilde{q}, \quad H^{(2)} = \tau + \frac{4}{15} c_1 \tilde{q}. \tag{11.56}$$

The boundary condition (11.50) is given in terms of the new perturbations as

$$\Phi^{(1)} = \pm \left[\varrho_r + \frac{1}{2}\left(c_1^2 - 1\right)\right] \quad \text{at} \quad x = \mp\frac{1}{2} \quad \text{for} \quad c_1 \gtrless 0, \tag{11.57}$$

$$\Phi^{(2)} = \pm\frac{1}{2} \quad \text{at} \quad x = \mp\frac{1}{2} \quad \text{for} \quad c_1 \gtrless 0, \tag{11.58}$$

with the quantity ϱ_r calculated as

$$\varrho_r = 2 \int_0^{\infty} c_1 e^{-c_1^2} \Phi^{(1)}\left(\frac{1}{2}, c_1\right) \, dc_1 \tag{11.59}$$

that follows from (11.51).

The perturbations obeying Eq. (11.55) with the boundary conditions (11.57) and (11.58) have the following symmetry

$$\Phi^{(\alpha)}(x, c_1) = -\Phi^{(\alpha)}(-x, -c_1). \tag{11.60}$$

To employ this symmetry in the numerical scheme, the half-moments are defined as

$$\begin{bmatrix} \varrho^{(1/2)}(x) \\ \tau^{(1/2)}(x) \\ \tilde{q}^{(1/2)}(x) \end{bmatrix} = \frac{1}{\pi^{1/2}} \int_0^{\infty} \left\{ \begin{bmatrix} 1 \\ \frac{1}{3}(2c_1^2 - 1) \\ c_1^3 \end{bmatrix} \Phi^{(1)}(x, c_1) + \begin{bmatrix} 0 \\ \frac{2}{3} \\ c_1 \end{bmatrix} \Phi^{(2)}(x, c_1) \right\} e^{-c_1^2} \, dc_1, \tag{11.61}$$

where the integration is done only for the positive component c_1. Then the full moments are calculated as

$$\varrho(x) = \varrho^{(1/2)}(x) - \varrho^{(1/2)}(-x), \tag{11.62}$$

$$\tau(x) = \tau^{(1/2)}(x) - \tau^{(1/2)}(-x), \tag{11.63}$$

$$\tilde{q}(x) = \tilde{q}^{(1/2)}(x) + \tilde{q}^{(1/2)}(-x). \tag{11.64}$$

where the properties (11.60) have been used.

11.2.5 Numerical Scheme

Similar to the Couette flow, the nodes c_i and weight W_i given in Section B.1.1 and the nodes defined by (11.25) are used for the problem under consideration. To use the scheme (9.40), the notations

$$\Phi_{ki}^{(\alpha)} = \Phi^{(\alpha)}(x_k, c_i), \quad H_{ki}^{(\alpha)} = H^{(\alpha)}(x_k, c_i), \quad \alpha = 1, 2, \tag{11.65}$$

$$\varrho_k = \varrho(x_k), \quad \tau_k = \tau(x_k), \quad \tilde{q}_k = \tilde{q}(x_k) \tag{11.66}$$

are introduced. Since both $\Phi^{(1)}$ and $\Phi^{(2)}$ obey the equations of the same structure (11.55), the numerical scheme is the same, therefore, the superscript α is omitted below. For the problem in question, the kinetic equation (11.55) is quite similar to (11.20) so that its numerical scheme is exactly as (11.27) replaying $\tilde{u}_k$ by H_{ki},

$$\Phi_{ki} = \frac{(4N_x c_i - \delta)\Phi_{k-1,i} + \delta(H_{ki} + H_{k-1,i})}{4N_x c_i + \delta}. \tag{11.67}$$

Because of the symmetry (11.60), it is enough to know the perturbations Φ only for positive values of the velocity nodes c_i, sweeping all nodes in the physical space x_k. In other words, the half-moments defined by (11.61) are calculated as

$$\begin{bmatrix} \varrho_k^{(1/2)} \\ \tau_k^{(1/2)} \\ \tilde{q}_k^{(1/2)} \end{bmatrix} = \sum_{i=1}^{N_c} \left\{ \begin{bmatrix} 1 \\ \frac{1}{3}(2c_i^2 - 1) \\ c_i^3 \end{bmatrix} \Phi_{ki}^{(1)} + \begin{bmatrix} 0 \\ \frac{2}{3} \\ c_i \end{bmatrix} \Phi_{ki}^{(2)} \right\} W_i \tag{11.68}$$

sweeping the nodes x_k from $-N_1 + 1$ to N_x. The half-moments in the first node x_k $(k = -N_x)$ are calculated using the boundary perturbation functions (11.57) and (11.58) for $c_1 > 0$ and $\varrho_r^{(n)}$ obtained in the previous iteration

$$\varrho_k^{(1/2)} = \frac{\varrho_r}{2} - \frac{1}{8}, \quad \tau_k^{(1/2)} = \frac{1}{4}, \quad \tilde{q}_k^{(1/2)} = \frac{\varrho_r + 1}{2\pi^{1/2}} \quad \text{for} \quad k = -N_x. \tag{11.69}$$

The quantity ϱ_r is calculated from (11.59) using only positive values of the velocity

$$\varrho_r = 2\pi^{1/2} \sum_{i=1}^{N_c} c_i \Phi_{ki} W_i \quad \text{for} \quad k = N_x. \tag{11.70}$$

Once the half-moments are known, the full moments are calculated employing the symmetry

$$\varrho_k = \varrho_k^{(1/2)} - \varrho_{-k}^{(1/2)}, \quad \tau_k = \tau_k^{(1/2)} - \tau_{-k}^{(1/2)}, \quad \tilde{q}_k = \tilde{q}_k^{(1/2)} + \tilde{q}_{-k}^{(1/2)} \tag{11.71}$$

that follow from (11.62) and (11.64).

The numerical codes heat_planar.for in FORTRAN and an example with output data are given in Section B.3.

11.2.6 Numerical Results

Numerical values of the heat flow calculated by the code heat_planar.for using the values $N_x = 10$ and $N_c = 6$ are given in Table 11.2. It is verified that these scheme parameters of the heat flow is calculated with the numerical error less than 0.1%. The variance of the heat flow from node to node has the order 10^{-6} and smaller, that is, the conservation law is fulfilled with a high precision. The values of heat flow $\tilde{q}$ calculated by the DSMC technique with the AI potential [76] are also given in Table 11.2. The difference of these values from those obtained from the S-model by the DVM does not exceed 1%. However, the computational effort to apply the DSMC method is significantly larger than that to solve the model equation. The analytical solutions in the free-molecular (11.40) and hydrodynamic (11.45) regimes are also given in Table 11.2. These solutions show that the flow is free-molecular when $\delta \leq 0.01$ and it is well described by the jump solution at $\delta \geq 20$.

The temperature profiles τ depicted in Figure 11.4 are quite similar to the velocity profile in the planar Couette flow, see Figure 11.2, namely, the profiles are practically the straight lines and their slopes being zero in the free-molecular regime

Table 11.2 Dimensionless heat flow $\tilde{q}$ between two plates versus rarefaction δ obtained from the S-model by the scheme (11.67) and that obtained by the DSMC method based on AI for argon in Ref. [76].

δ	$\tilde{q}$	
	S-model	DSMC
0[a)]	0.5642	
0.01	0.5607	0.5577
0.1	0.5348	0.5294
1	0.4001	0.3971
10	0.1348	0.1347
20	0.07842	0.07843
20[b)]	0.07843	

a) Free-molecular solution (11.40).
b) Jump solution (11.45).

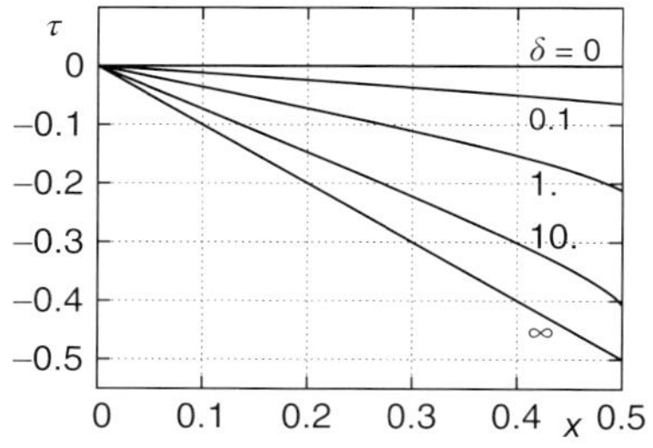

Figure 11.4 Temperature deviation τ versus coordinate x in planar heat transfer.

($\delta = 0$) smoothly vary to that given by the analytical solution (11.44) by increasing the rarefaction δ.

Numerical data obtained from the nonlinearized BE based on the HS potential were reported by Ohwada [116]. Graur and Polikarpov [117] calculated the heat flow $\tilde{q}$ from the BGK, S-model, and ellipsoidal model without the linearization. Their results show that the heat flow calculated on the basis of the S and ellipsoidal models deviate for 3% from that obtained on the basis of the BE [116]. The heat flow based on the BGK model is smaller than that based on the BE for about 30%. Thus, the choice of the kinetic model with the correct Prandtl number is very important for flows involving heat transfer phenomena.

The heat flow between two plates confining a polyatomic gas was studied by Pazooki and Loyalka [118]. In addition to the translational energy, a polyatomic molecule has the rotational and vibrational energies. They create additional mechanism of the energy transfer and can significantly change the total heat flow.

Numerical results for heat transfer through a mixture can be found in Refs [76, 119, 120]. According to the paper [119], the deviation of the heat flow $\tilde{q}$ from its value for a single gas is stronger for a mixture with a larger ratio of the atomic weights. For instance, the heat flow through a helium–xenon mixture can be twice larger than that for a single gas, while the deviation does not exceed 6% for neon–argon mixture.

11.3 Planar Poiseuille and Thermal Creep Flows

11.3.1 Definitions

Now, let us consider a gas flow confined between two infinite plates fixed at $r_1 = \pm a/2$ because of longitudinal small gradients of the pressure p and temperature T along the coordinate r_3

$$\xi_{\mathrm{P}} = \frac{a}{p}\frac{\mathrm{d}p}{\mathrm{d}r_3}, \quad |\xi_{\mathrm{P}}| \ll 1, \quad \xi_{\mathrm{T}} = \frac{a}{T}\frac{\mathrm{d}T}{\mathrm{d}r_3}, \quad |\xi_{\mathrm{T}}| \ll 1, \tag{11.72}$$

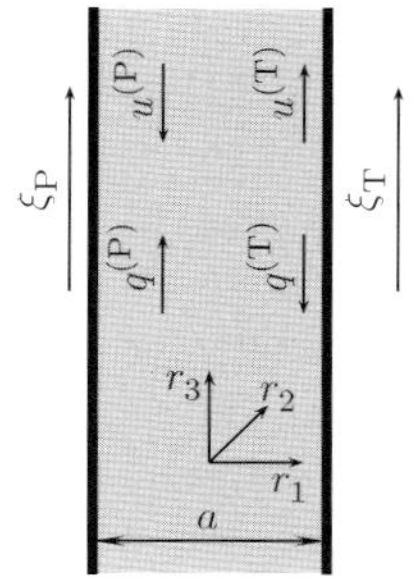

Figure 11.5 Scheme and coordinates of planar Poiseuille and creep flows.

as is shown in Figure 11.5. In other words, we assume that the pressure and temperature depend on the longitudinal coordinate as

$$p(r_3) = p_0\left(1 + \xi_P \frac{r_3}{a}\right), \quad T(r_3) = T_0\left(1 + \xi_T \frac{r_3}{a}\right). \tag{11.73}$$

Since the plates are infinite, the bulk velocity and the heat flux have the longitudinal component only depending on the transversal coordinate r_1

$$\boldsymbol{u} = [0, 0, u_3], \quad u_3 = u_3(r_1), \quad \boldsymbol{q} = [0, 0, q_3], \quad q_3 = q_3(r_1). \tag{11.74}$$

We are going to calculate the mass and heat flow rates through a cross section of the channel defined as

$$\dot{M} := mn_0 \int_{-a/2}^{a/2} u_3(r_1)\,\mathrm{d}r_1, \quad \dot{E} := \int_{-a/2}^{a/2} q_3(r_1)\,\mathrm{d}r_1, \tag{11.75}$$

respectively. Physically, these definitions mean that the flow rates are calculated per length unity in the direction r_2.

The distance between the plates a is used as the characteristic size so that the rarefaction parameter is defined by (7.46) and the dimensionless coordinates are introduced by (7.43).

For the problem in question, the linearization is done near the local Maxwellian. The pressure and temperature gradients can be used as small parameters of the linearization. Since we are considering two driving forces, the two reference perturbations $h_R^{(P)}$ and $h_R^{(T)}$ are introduced. The first reference perturbation $h_R^{(P)}$ is calculated assuming a constant pressure gradient along the channel with the bulk velocity being zero and the temperature equal to its equilibrium value, that is,

$$\varrho_R^{(P)} = x_3, \quad \boldsymbol{u}_R^{(P)} = 0, \quad \tau_R^{(P)} = 0. \tag{11.76}$$

The second reference perturbation $h_R^{(T)}$ corresponds to a constant temperature gradient when the pressure is equal to its equilibrium values with the bulk velocity being zero. To maintain the constant pressure, the density should have a gradient opposite to that of the temperature. Then the reference moments take the form

$$\varrho_R^{(T)} = -x_3, \quad \boldsymbol{u}_R^{(T)} = 0, \quad \tau_R^{(T)} = x_3. \tag{11.77}$$

Equations (11.76) and (11.77) mean that the distributions (11.73) are used as the reference pressure and temperature. Substituting (11.76) and (11.77) into (5.30), the reference perturbations are obtained as

$$h_{\mathrm{R}}^{(\mathrm{P})} = x_3, \quad h_{\mathrm{R}}^{(\mathrm{T})} = x_3 \left(c^2 - \frac{5}{2} \right). \tag{11.78}$$

Then, the bulk velocity and heat flow will have two dimensionless terms, that is, Eqs. (5.13) and (5.16) are written down as

$$u_3 = v_0 (\tilde{u}^{(\mathrm{P})} \xi_{\mathrm{P}} + \tilde{u}^{(\mathrm{T})} \xi_{\mathrm{T}}), \quad q_3 = p_0 v_0 (\tilde{q}^{(\mathrm{P})} \xi_{\mathrm{P}} + \tilde{q}^{(\mathrm{T})} \xi_{\mathrm{T}}). \tag{11.79}$$

Since each of the dimensionless quantities $\tilde{u}$ and $\tilde{q}$ has the unique component, the subscript "3" is omitted. A substitution of (11.79) into the definitions (11.75) leads to a decoupling of the flow rates

$$\dot{M} = \frac{a p_0}{v_0} (-G_{\mathrm{P}}^* \xi_{\mathrm{P}} + G_{\mathrm{T}}^* \xi_{\mathrm{T}}), \quad \dot{E} = \frac{1}{2} a p_0 v_0 (Q_{\mathrm{P}}^* \xi_{\mathrm{P}} - Q_{\mathrm{T}}^* \xi_{\mathrm{T}}), \tag{11.80}$$

where the following coefficients have been introduced

$$G_{\mathrm{P}}^* := -2 \int_{-1/2}^{1/2} \tilde{u}^{(\mathrm{P})}(x_1) \, \mathrm{d}x_1, \quad G_{\mathrm{T}}^* := 2 \int_{-1/2}^{1/2} \tilde{u}^{(\mathrm{T})}(x_1) \, \mathrm{d}x_1, \tag{11.81}$$

$$Q_{\mathrm{P}}^* := 2 \int_{-1/2}^{1/2} \tilde{q}^{(\mathrm{P})}(x_1) \, \mathrm{d}x_1, \quad Q_{\mathrm{T}}^* := -2 \int_{-1/2}^{1/2} \tilde{q}^{(\mathrm{T})}(x_1) \, \mathrm{d}x_1. \tag{11.82}$$

The coefficients are defined as that all of them are positive. Their physical meaning is as follows: G_{P}^* describes the Poiseuille flow, that is, mass flow rate caused by a pressure gradient ξ_{P}; G_{T}^* describes the thermal creep, that is, mass flow rate caused by a temperature gradient ξ_{T}; Q_{P}^* corresponds to the mechanocaloric effect, that is, the heat flow caused by a pressure gradient ξ_{P}; and Q_{T}^* is the ordinary heat flux caused by a temperature gradient ξ_{T}.

11.3.2
Slip Solution

Combining the momentum conservation law (3.27) and Newton's law (6.2) for the problem under consideration, we obtain

$$\mu \frac{\mathrm{d}^2 u_3}{\mathrm{d}r_1^2} = \frac{\mathrm{d}p}{\mathrm{d}r_3}. \tag{11.83}$$

The boundary conditions are written down taking into account the viscous (10.2) and thermal (10.17) slips

$$u_3 = \pm \sigma_{\mathrm{P}} \, \ell \, \frac{\mathrm{d}u_3}{\mathrm{d}r_1} + \sigma_{\mathrm{T}} \frac{\mu}{\rho T} \frac{\mathrm{d}T}{\mathrm{d}r_3} \quad \text{at} \quad r_1 = \pm \frac{a}{2}. \tag{11.84}$$

The solution of Eq. (11.83) subject to this boundary condition reads

$$u_3 = v_0 \left[\frac{\delta}{2} \left(x_1^2 - \frac{1}{4} \right) - \frac{\sigma_{\mathrm{P}}}{\delta} \right] \xi_{\mathrm{P}} + v_0 \frac{\sigma_{\mathrm{T}}}{2\delta} \xi_{\mathrm{T}}. \tag{11.85}$$

Comparing it with (11.79), the two dimensionless terms are derived as

$$\tilde{u}^{(\mathrm{P})} = \frac{\delta}{2} \left(x_1^2 - \frac{1}{4} \right) - \frac{\sigma_{\mathrm{P}}}{2}, \quad \tilde{u}^{(\mathrm{T})} = \frac{\sigma_{\mathrm{T}}}{2\delta}. \tag{11.86}$$

Then the coefficients G_P^* and G_T^* are obtained from (11.81)

$$G_P^* = \frac{\delta}{6} + \sigma_P, \quad G_T^* = \frac{\sigma_T}{\delta}. \tag{11.87}$$

To obtain the coefficient Q_T^*, Fourier's law (6.3) is written as

$$q_3 = -\kappa \frac{dT}{dr_3}. \tag{11.88}$$

Combining Eqs. (6.4), (7.46), and (11.72), this equation is transformed into

$$q_3 = -\frac{5}{4\delta\,\mathrm{Pr}} p_0 v_0 \xi_T. \tag{11.89}$$

Then the dimensionless heat flow takes the form

$$\tilde{q}^{(T)} = -\frac{5}{4\delta\,\mathrm{Pr}} \approx -\frac{15}{8\delta}. \tag{11.90}$$

Its substitution into (11.82) leads to the coefficient Q_T^*

$$Q_T^* = \frac{5}{2\delta\,\mathrm{Pr}} \approx \frac{15}{4\delta}. \tag{11.91}$$

A derivation of the coefficient Q_P^* involves a constitutive equation obtained from the so-called Burnett equations [4] and the heat flux in the Knudsen layer (10.15). The details of such a derivation are given in Ref. [80]. Later in this book, it will be shown that this coefficient is equal to G_T^* so that the second expression of (11.87) provides the quantity Q_P^*.

11.3.3
Kinetic Equation

Since two driving forces are considered here, ξ_P and ξ_T, two perturbation functions $h^{(P)}$ and $h^{(T)}$ are obtained. However, the kinetic equation and its boundary condition have the same structure for both perturbations so that the superscripts "P" and "T" are omitted, understanding that an expression is valid for both perturbations $h^{(P)}$ and $h^{(T)}$.

Since we are interested in mass and heat transfer, the S-model would be appropriate. For the problem in question, the general model equation (7.47) with $A_1 = 1$, $A_2 = 0$, and $A_3 = 1$ (see Table 7.1) is reduced to

$$c_1 \frac{\partial h}{\partial x_1} = \delta(H - h) + \tilde{g}, \tag{11.92}$$

where

$$H = 2c_3\tilde{u} + \frac{4}{15} c_3 \left(c^2 - \frac{5}{2}\right) \tilde{q}. \tag{11.93}$$

Here, the density ϱ and temperature τ deviations are assumed to be zero. The source terms are calculated by substituting the reference perturbations (11.78) into (5.34) and then into (7.49)

$$\tilde{g}^{(P)} = -c_3, \quad \tilde{g}^{(T)} = -c_3 \left(c^2 - \frac{5}{2}\right). \tag{11.94}$$

The dimensionless velocity $\tilde{u}$ and heat flow $\tilde{q}$ are expressed via the perturbation as (5.17)

$$\begin{bmatrix} \tilde{u}(x_1) \\ \tilde{q}(x_1) \end{bmatrix} = \frac{1}{\pi^{3/2}} \int \begin{bmatrix} c_3 \\ \left(c^2 - \frac{5}{2}\right) c_3 \end{bmatrix} e^{-c^2} h(x_1, \mathbf{c})\, \mathrm{d}^3\mathbf{c}. \tag{11.95}$$

The surface source term used in the boundary condition (5.50) is zero, $h_w = 0$, because the surface temperature is equal to the reference temperature. Assuming the diffuse scattering, see the operator (5.67), we conclude that the perturbation function of reflected particles is zero

$$h = 0 \quad \text{at} \quad x_1 = \pm\frac{1}{2}, \quad \text{for} \quad c_1 \lessgtr 0, \tag{11.96}$$

because the incident perturbation function substituted into (5.67) linearly depends on the component c_3. Such a dependence follows from Eq. (11.92) where all terms contain c_3 except those containing h.

To eliminate the variables c_2 and c_3, new perturbation functions are introduced as

$$\Phi^{(1)}(x_1, c_1) = \frac{1}{\pi} \int_{-\infty}^{\infty} \int_{-\infty}^{\infty} e^{-c_2^2 - c_3^2} h\left(x_1, \mathbf{c}\right)\ c_3\, \mathrm{d}c_2\, \mathrm{d}c_3, \tag{11.97}$$

$$\Phi^{(2)}(x_1, c_1) = \frac{1}{\pi} \int_{-\infty}^{\infty} \int_{-\infty}^{\infty} e^{-c_2^2 - c_3^2} h\left(x_1, \mathbf{c}\right) \left(c_2^2 + c_3^2 - 2\right) c_3\, \mathrm{d}c_2\, \mathrm{d}c_3. \tag{11.98}$$

Multiplying Eq. (11.92) by $e^{-c_2^2-c_3^2} c_3/\pi$ and by $e^{-c_2^2-c_3^2}(c_2^2 + c_3^2 - 2)c_3/\pi$, then integrating it with respect to c_2 and c_3, we obtain the kinetic equations for the new perturbations

$$c_1 \frac{\partial \Phi^{(\alpha)}}{\partial x_1} = \delta(H^{(\alpha)} - \Phi^{(\alpha)}) + \tilde{g}^{(\alpha)}, \quad \alpha = 1, 2, \tag{11.99}$$

where the functions $H^{(\alpha)}$ are obtained from (11.93)

$$H^{(1)} = \tilde{u} + \frac{2}{15}\tilde{q}\left(c_1^2 - \frac{1}{2}\right), \quad H^{(2)} = \frac{4}{15}\tilde{q}. \tag{11.100}$$

Using the expressions (11.94), these sources take the form

$$\tilde{g}^{(1,\mathrm{P})} = -\frac{1}{2}, \quad \tilde{g}^{(1,\mathrm{T})} = -\frac{1}{2}\left(c_1^2 - \frac{1}{2}\right), \quad \tilde{g}^{(2,\mathrm{P})} = 0, \quad \tilde{g}^{(2,\mathrm{T})} = -1. \tag{11.101}$$

The velocity and heat flow are calculated via the new perturbations as

$$\begin{bmatrix} \tilde{u}(x_1) \\ \tilde{q}(x_1) \end{bmatrix} = \frac{1}{\sqrt{\pi}} \int_{-\infty}^{\infty} \left\{ \begin{bmatrix} 1 \\ c_1^2 - \frac{1}{2} \end{bmatrix} \Phi^{(1)}(x_1, c_1) + \begin{bmatrix} 0 \\ 1 \end{bmatrix} \Phi^{(2)}(x_1, c_1) \right\} e^{-c_1^2}\, \mathrm{d}c_1. \tag{11.102}$$

The boundary conditions for the new perturbations are the same as that for the original perturbation

$$\Phi^{(\alpha)} = 0 \quad \text{at} \quad x_1 = \pm\frac{1}{2}, \quad \text{for} \quad c_1 \lessgtr 0. \tag{11.103}$$

The perturbation functions obey the following symmetry:

$$\Phi^{(\alpha)}(x_1, c_1) = \Phi^{(\alpha)}(-x_1, -c_1), \tag{11.104}$$

that allows us to define the half-moments

$$\begin{bmatrix} \tilde{u}^{(1/2)}(x_1) \\ \tilde{q}^{(1/2)}(x_1) \end{bmatrix} = \frac{1}{\sqrt{\pi}} \int_0^{\infty} \left\{ \begin{bmatrix} 1 \\ c_1^2 - \frac{1}{2} \end{bmatrix} \Phi^{(1)}(x_1, c_1) + \begin{bmatrix} 0 \\ 1 \end{bmatrix} \Phi^{(2)}(x_1, c_1) \right\} e^{-c_1^2}\, \mathrm{d}c_1 \tag{11.105}$$

and to calculate the full moments as

$$\tilde{u}(x_1) = \tilde{u}^{(1/2)}(x_1) + \tilde{u}^{(1/2)}(-x_1), \quad \tilde{q}(x_1) = \tilde{q}^{(1/2)}(x_1) + \tilde{q}^{(1/2)}(-x_1). \tag{11.106}$$

11.3.4
Reciprocal Relation

A consideration of two driving forces allows us to obtain the reciprocal relation. The time-reversed kinetic coefficients (5.85) are obtained by using the source terms (11.94)

$$\Lambda^t_{\mathrm{P}m} = ((\hat{T}\tilde{g}^{(\mathrm{P})}, h^{(m)})) = \int_{-1/2}^{1/2} (c_3, h^{(m)})\, \mathrm{d}x_1, = \int_{-1/2}^{1/2} \tilde{u}^{(m)}\, \mathrm{d}x_1 \tag{11.107}$$

$$\Lambda^t_{\mathrm{T}m} = ((\hat{T}\tilde{g}^{(\mathrm{T})}, h^{(m)})) = \int_{-1/2}^{1/2} \left(c_3 \left(c^2 - \frac{5}{2} \right), h^{(m)} \right) \mathrm{d}x_1 = \int_{-1/2}^{1/2} \tilde{q}^{(m)}\, \mathrm{d}x_1, \tag{11.108}$$

where $m = \mathrm{P}$ or $m = \mathrm{T}$. These coefficients are expressed via the dimensionless flow rates defined by (11.81) and (11.82)

$$\Lambda^t_{\mathrm{PP}} = -\frac{1}{2} G^*_{\mathrm{P}}, \quad \Lambda^t_{\mathrm{PT}} = \frac{1}{2} G^*_{\mathrm{T}}, \quad \Lambda^t_{\mathrm{TP}} = \frac{1}{2} Q^*_{\mathrm{P}}, \quad \Lambda^t_{\mathrm{TT}} = -\frac{1}{2} Q^*_{\mathrm{T}}. \tag{11.109}$$

The equality (5.83) between the coefficient Λ^t_{PT} and Λ^t_{TP} leads to the coupling between the thermal creep and mechanocaloric heat flow

$$G^*_{\mathrm{T}} = Q^*_{\mathrm{P}}. \tag{11.110}$$

Initially, this relation was obtained by Loyalka [54] for the particular case of gas flow through a long capillary.

11.3.5
Numerical Scheme

Since the physical space of the problem in question is the same as that in the Couette flow, the same nodes defined by (11.25) can be used for the coordinate x_1. To approximate Eq. (11.99), the discrete functions are defined as

$$\Phi^{(\alpha)}_{ki} = \Phi^{(\alpha)}(x_k, c_i), \quad H^{(\alpha)}_{ki} = H(x_k, c_i), \quad \tilde{g}_i = \tilde{g}^{(\alpha)}(c_i). \tag{11.111}$$

Then the schemes (9.40) for the perturbations Φ reads

$$\Phi_{ki} = \frac{(4N_x c_i - \delta)\Phi_{k-1,i} + \delta(H_{ki} + H_{k-1,i}) + 2\tilde{g}_i}{4N_x c_i + \delta}. \tag{11.112}$$

As previously, the half-moments (11.105) are calculated as

$$\begin{bmatrix} \tilde{u}_k^{(1/2)} \\ \tilde{q}_k^{(1/2)} \end{bmatrix} = \sum_{i=1}^{N_c} \left\{ \begin{bmatrix} 1 \\ c_i^2 - \frac{1}{2} \end{bmatrix} \Phi_{ki}^{(1)} + \begin{bmatrix} 0 \\ 1 \end{bmatrix} \Phi_{ki}^{(2)} \right\} W_i \tag{11.113}$$

and then the full moments are obtained according to the symmetry (11.106) as

$$\tilde{u}_k = \tilde{u}_k^{(1/2)} + \tilde{u}_{-k}^{(1/2)}, \quad \tilde{q}_k = \tilde{q}_k^{(1/2)} + \tilde{q}_{-k}^{(1/2)}. \tag{11.114}$$

When the scheme converges, the flow rates are calculated using Simpson's rule

$$G_{\rm P}^* = -\sum_{k=0}^{N_x} \tilde{u}_k^{(\rm P)} W_k, \quad G_{\rm T}^* = \sum_{k=0}^{N_x} \tilde{u}_k^{(\rm T)} W_k, \tag{11.115}$$

$$Q_{\rm P}^* = \sum_{k=0}^{N_x} \tilde{q}_k^{(\rm P)} W_k, \quad Q_{\rm T}^* = -\sum_{k=0}^{N_x} \tilde{q}_k^{(\rm T)} W_k, \quad W_k = \frac{2}{N_x} W_k^S \tag{11.116}$$

with W_k^S defined by (9.14).

11.3.6 Splitting Scheme

If we try to use the Gauss nodes in the velocity space to the problem in question, we find that the number of these nodes drastically increases by decreasing the rarefaction parameter. To understand such a behavior, let us analyze Eq. (11.100) in the limit $\delta \to 0$. The estimation of the derivative in this limit shows that

$$\frac{\partial \Phi}{\partial x_1} \to \frac{\tilde{g}}{c_1} \quad \text{at} \quad \delta \to 0, \tag{11.117}$$

that is, the derivative tends to infinity for small values of the velocity c_1. The function $\Phi^{(\rm P)}$ in the middle point of the channel is plotted versus the velocity in Figure 11.6. This figure shows that for the small values of the rarefaction ($\delta = 0.1$ and 0.01), the perturbation drastically increases when the velocity c_1 tends to zero, but it does not happen for the intermediate ($\delta = 1$) and large rarefaction ($\delta = 10$). It means that the Gauss works well only in the range $\delta \geq 1$, but not near the free-molecular regime $\delta < 1$. In the last case, the density of nodes c_i should be larger near zero in order to reach a reasonable accuracy. As a consequence, the computational time drastically increases when δ tends to zero. To reduce the computational effort, the splitting scheme is suggested.

Let us decompose the perturbation function h into two parts:

$$h(x_1, \boldsymbol{c}) = h_0(x_1, \boldsymbol{c}) + h'(x_1, \boldsymbol{c}), \tag{11.118}$$

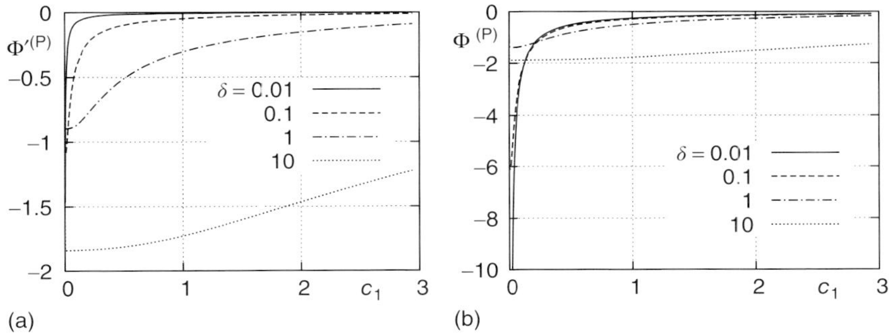

Figure 11.6 Perturbations $\Phi^{(P)}$ and $\Phi'^{(P)}$ in the middle point ($x_1 = 0$) versus velocity c_1.

where the first term obeys the equation without the moments but containing the source term

$$c_1 \frac{\partial h_0}{\partial x_1} + \delta h_0 = \tilde{g}, \tag{11.119}$$

while the second term satisfies the kinetic equation containing the moments but without the source term

$$c_1 \frac{\partial h'}{\partial x_1} = \delta \left[2c_3 \tilde{u} + \frac{4}{15} c_3 \left(c^2 - \frac{5}{2} \right) \tilde{q} - h' \right]. \tag{11.120}$$

Both terms h_0 and h' are subject to the same boundary condition (11.96). Equation (11.119) can be solved analytically by integrating it from $-1/2$ to x_1 so that its solution reads

$$h_0(x_1, c_1) = \frac{\tilde{g}}{\delta} \left[1 - \exp\left(-\frac{\delta(x_1 + \frac{1}{2})}{c_1} \right) \right], \quad c_1 > 0. \tag{11.121}$$

Its expression for the negative velocity $c_1 < 0$ is obtained from the symmetrical property

$$h_0(x_1, c_1, c_2, c_3) = h(-x_1, -c_1, c_2, c_3). \tag{11.122}$$

The half-moments are split too

$$\tilde{u}^{(1/2)} = u_0^{(1/2)} + \tilde{u}'^{(1/2)}, \quad \tilde{q}^{(1/2)} = q_0^{(1/2)} + \tilde{q}'^{(1/2)}, \tag{11.123}$$

where each term is calculated via the corresponding perturbation using the half-moment concept

$$\begin{bmatrix} \tilde{u}'^{(1/2)}(x_1) \\ \tilde{q}'^{(1/2)}(x_1) \end{bmatrix} = \frac{1}{\pi^{3/2}} \int_{c_1>0} \begin{bmatrix} c_3 \\ c_3 \left(c^2 - \frac{5}{2} \right) \end{bmatrix} e^{-c^2} h'(x_1, \mathbf{c}) \, \mathrm{d}^3\mathbf{c}, \tag{11.124}$$

$$\begin{bmatrix} u_0^{(1/2)}(x_1) \\ q_0^{(1/2)}(x_1) \end{bmatrix} = \frac{1}{\pi^{3/2}} \int_{c_1>0} \begin{bmatrix} c_3 \\ c_3 \left(c^2 - \frac{5}{2} \right) \end{bmatrix} e^{-c^2} h_0(x_1, \mathbf{c}) \, \mathrm{d}^3\mathbf{c}. \tag{11.125}$$

Then the full moments are obtained using the symmetry (11.106).

Substituting the solution (11.121) into (11.125), the first terms in (11.123) are obtained as

$$u_0^{(1/2,\mathrm{P})}(x_1) = -\frac{1}{2\delta}\left(\frac{1}{2} - \frac{1}{\sqrt{\pi}}I_0\right), \tag{11.126}$$

$$q_0^{(1/2,\mathrm{P})}(x_1) = u_0^{(1/2,\mathrm{T})}(x_1) = \frac{1}{2\sqrt{\pi}\delta}\left(I_2 - \frac{1}{2}I_0\right), \tag{11.127}$$

$$q_0^{(1/2,\mathrm{T})}(x_1) = -\frac{1}{2\delta}\left[\frac{5}{4} - \frac{1}{\sqrt{\pi}}\left(I_4 - I_2 + \frac{9}{4}I_0\right)\right], \tag{11.128}$$

where the source terms (11.101) has been used. The notation I_n $(n = 0, 2, 4)$ means the function defined by (A.29). All these functions in (11.126)–(11.128) have the same argument, namely,

$$I_n = I_n(\delta(x_1 + 1/2)). \tag{11.129}$$

The perturbation h' obeys the same kinetic equation as h, but without the source term. Eliminating this term, we avoided the large derivative for the small velocity (11.117). Such a "bad" behavior has been separated analytically into h_0. The numerical scheme for h' is exactly the same as that for h omitting the source term. Figure 11.6 shows the perturbation $\Phi'^{(\mathrm{P})}$ defined by (11.97) with h' instead h. We see that the splitting scheme does not eliminate completely the problem of the large derivative for the small velocity, but it reduces the sharpness of the perturbation at c_1. Moreover, the contribution of $\Phi'^{(\mathrm{P})}$ into the moments decreases by decreasing the rarefaction so that its accuracy becomes less important when the rarefaction tends to zero.

Since the perturbation continues to be sharp near the point $c_1 = 0$, it makes sense to introduce the velocity nodes so that their density would be larger near zero and smaller far from zero. It is reasonable to introduce two velocity intervals

$$0 \le c_1 \le c_{01} \quad \text{and} \quad c_{01} \le c_1 \le c_0 \tag{11.130}$$

with the same number of nodes regularly distributed in each interval so that the nodes are defined as

$$c_i = \begin{cases} 2c_{01}\frac{i}{N_c} & \text{for} \quad 0 \le i \le \frac{N_c}{2}, \\ c_{01} + 2(c_0 - c_{01})\frac{i}{N_c} & \text{for} \quad \frac{N_c}{2} < i \le N_c. \end{cases} \tag{11.131}$$

The weights are calculated using Simpson's rule (9.14) and including the factor $e^{-c_i^2}/\sqrt{\pi}$

$$W_i = \frac{2}{N_c}\frac{e^{-c_i^2}}{\sqrt{\pi}} W_i^S \begin{cases} c_{01} & \text{for} \quad 0 \le i < \frac{N_c}{2}, \\ c_0/2 & \text{for} \quad i = N_c/2, \\ c_0 - c_{01} & \text{for} \quad \frac{N_c}{2} < i \le N_c. \end{cases} \tag{11.132}$$

According to Figure 11.6, the sharp variation happens in the range $0 \leq c_1 \leq 0.1$, then the value $c_{01} = 0.1$ is appropriate for the upper limit of the first interval in (11.130). The upper limit of the second interval c_0 is chosen considering that the contribution of each node is proportional to $e^{-c_1^2}$. The value $c_0 = 3$ would be enough for high-accuracy calculations.

For the intermediate ($\delta = 1$) and large ($\delta = 10$) rarefaction parameters, the behaviors of Φ' is smooth and similar to that of Φ. In this case, the splitting becomes useless. Since the perturbation function is not sharp, the Gauss nodes and weights corresponding to the weighting function (9.16) are used.

11.3.7
Free-Molecular Limit

The splitting scheme helps us to obtain the asymptotic behavior of the moments and flow rates in the free-molecular regime $\delta \to 0$. Under the condition $\delta \to 0$, the first terms in (11.123) become dominant, while the second terms vanish so that

$$\tilde{u}(x_1) \to u_0^{(1/2)}(x_1) + u_0^{(1/2)}(-x_1), \quad \tilde{q}(x_1) \to q_0^{(1/2)}(x_1) + q_0^{(1/2)}(-x_1). \tag{11.133}$$

The asymptotic behaviors of $u_0^{(1/2)}$ and $q_0^{(1/2)}$ are obtained using the representations (A.31) and (A.30). Retaining only the main terms containing $\ln \delta$, the velocity and heat flow take the form

$$\tilde{u}^{(P)} = \frac{\ln \delta}{2\pi^{1/2}}, \quad \tilde{u}^{(T)} = \tilde{q}^{(P)} = -\frac{\ln \delta}{4\pi^{1/2}}, \quad \tilde{q}^{(T)} = \frac{9 \ln \delta}{8\pi^{1/2}}. \tag{11.134}$$

A substitution of these expressions into the definitions (11.81) and (11.82) provides the flow rates

$$G_P^* = -\frac{\ln \delta}{\pi^{1/2}}, \quad G_T^* = Q_P^* = -\frac{\ln \delta}{2\pi^{1/2}}, \quad Q_T^* = -\frac{9 \ln \delta}{4\pi^{1/2}}. \tag{11.135}$$

This is a very curious result, namely, the bulk velocity $\tilde{u}$, heat flow $\tilde{q}$ and consequently the flow rates through the cross section of finite area tend to infinity in the free-molecular regime. This unphysical behavior is explained by the degenerate geometry: the channel is infinite in two directions. Furthermore, it will be shown that if we restrict the channel at least in one direction (length or width), the bulk velocity and heat flow become finite quantities.

Since the splitting scheme is just slightly different from the simple scheme, a unique numerical code "poiseuille_creep_planar.for" was elaborated and presented in Section B.4. The variable "lg" determines if the simple or splitting scheme is used.

11.3.8
Numerical Results

The scheme parameters and results obtained by the code "poiseuille_creep_planar.for" are given in Table 11.3. The same values of the flow rates are reported in papers published by many researchers, see example, Refs

Table 11.3 Numerical scheme parameters and coefficients G^*_P, G^*_T, Q^*_P, Q^*_T for gas flow between plates versus rarefaction δ based on the code given in Section B.4.

δ	N_x	N_c	Iters.	G^*_P	$G^*_T = Q^*_P$	Q^*_T
0.001	10	40	6	4.278	1.858	9.593
0.01	10	40	8	3.052	1.247	6.734
0.1	10	60	15	2.040	0.7328	4.054
0.2	10	60	19	1.817	0.6075	3.312
1	20	6	47	1.554	0.3654	1.754
4	40	4	177	1.862	0.1891	0.7348
10	40	4	614	2.781	0.0983	0.3408
100	100	4	32974	17.70	0.0116	0.0372
100[a]				17.68	0.0117	0.0375

Slip solution (11.87), (11.91).

[121]–[124]. First of all, the reciprocal relation (11.110) is fulfilled within the numerical accuracy that is 0.1%. Second, the scheme parameters vary significantly by variation of the rarefaction parameter. The main problem arising in the hydrodynamic regime ($\delta \to \infty$) is the drastic increase in the iteration number, in other words, the convergence becomes very slow that is why the convergence criterion, variable "e" in the code, must be small enough, for example, 10^{-10}. The values of the coefficients G^*_P, G^*_T, and Q^*_T calculated in the slip flow regime by (11.87) and (11.91) at $\delta = 100$ are also given in Table 11.3. These values are very close to those obtained from the kinetic equation.

It is useful to construct an interpolating formula for the Poiseuille coefficient G^*_P, which is frequently used in practice. It is difficult to find a unique formula for the whole range of the rarefaction because the behavior of G^*_P near the free-molecular is quite different from that in the hydrodynamic limit. The following combination of two formulas provides the numerical of Poiseuille coefficient from Table 11.3 with an accuracy of 0.3%

$$G^*_P = \begin{cases} -\frac{\ln\ \delta}{\sqrt{\pi}} + 0.376 - (1.77 \ln \delta + 0.584)\delta + 2.12\delta^2 & \text{for} \quad \delta \leq 0.45, \\ \frac{\delta}{6} + 1.081 + \frac{0.61}{\delta^{0.7}} - \frac{0.24}{\delta} & \text{for} \quad \delta > 0.45. \end{cases} \tag{11.136}$$

It is interesting to compare the results based on the S-model with those obtained from the BE, which was solved for both HS and LJ potentials in the work [94]. The coefficients G^*_P, G^*_T, and Q^*_T based on this work are plotted in Figure 11.7. We see that the difference of the flow rates based on the S-model from those obtained on the basis of the BE with HS potential is about 5% for G^*_P and Q^*_T, while it is about 9% for G^*_T and Q^*_P. The same difference from the data based on the BE with LJ potential reaches 18%.

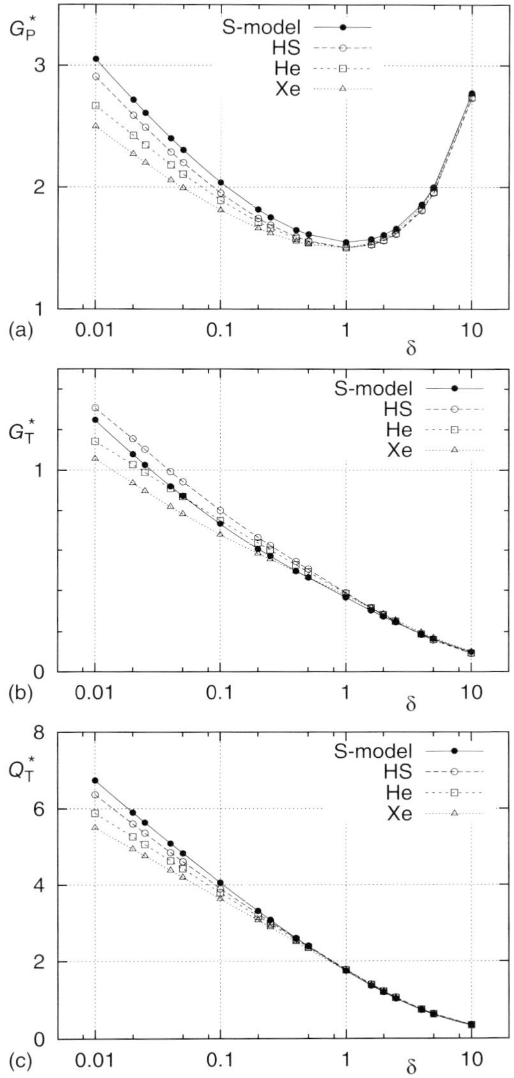

Figure 11.7 Coefficients G_P^*, G_T^*, and Q_T^* for gas flow between two plates versus rarefaction parameter δ. Comparison of data based on S-model with those obtained in Ref. [94] applying the BE with HS and LJ potentials for helium (He) and xenon (Xe).

The influence of the ACs α_t and α_n on the flow rates was studied in the papers [124, 125]. The coefficients G_P^*, G_T^*, and Q_T^* based on the S-model and for various values of the ACs are plotted in Figure 11.8. These data show that the Poiseuille

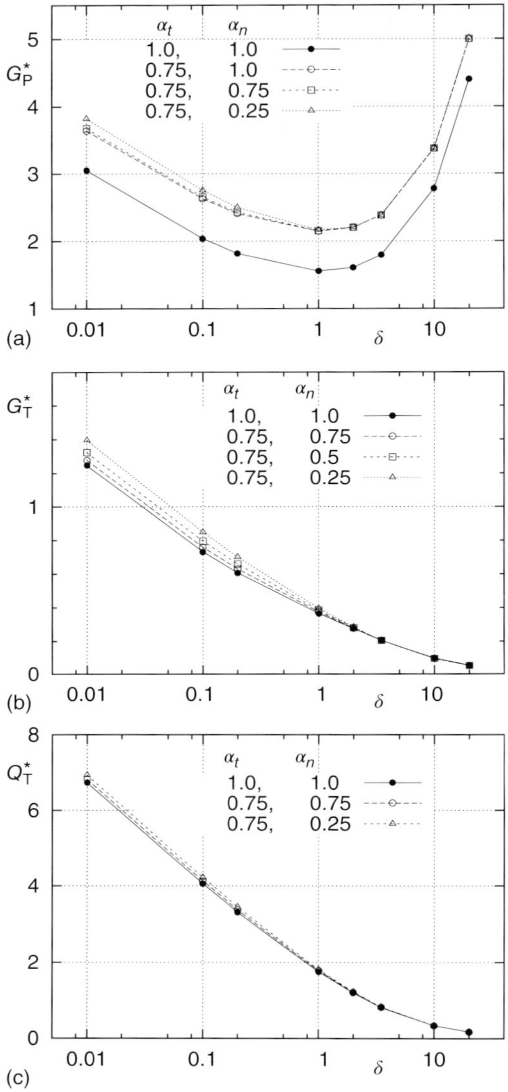

Figure 11.8 Coefficients G_P^*, G_T^*, and Q_T^* for gas flow between two plates versus rarefaction parameter δ. Comparison of data based on diffuse scattering with those obtained in Ref. [124] applying the CL scattering kernel.

coefficient G_P^* significantly depends on the AC α_t in the whole range of the rarefaction parameter δ. It always decreases by increasing α_t. The dependence of the Poiseuille coefficient G_P^* on the AC α_n is very weak. The thermal creep coefficient G_T^* depends on both ACs α_t and α_n. It is unexpected, but the heat flow

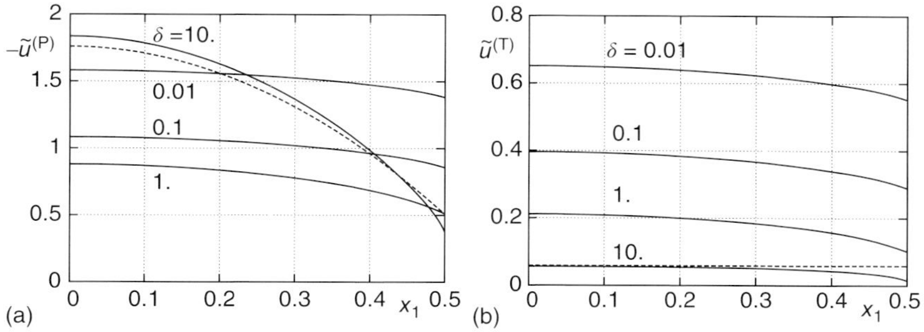

Figure 11.9 Dimensionless velocities $\tilde{u}^{(P)}$ and $\tilde{u}^{(T)}$ for gas flow between plates versus coordinate x_1: solid lines–kinetic equation, dashed line–slip solution (11.86).

rate Q_T^* weakly depends on the AC α_n. The dependence practically vanishes near the hydrodynamic regime.

The velocity profiles $\tilde{u}^{(P)}$ and $\tilde{u}^{(T)}$ are depicted in Figure 11.9. Note that the profile due to the pressure gradient $\tilde{u}^{(P)}$ is negative. For the large value of the rarefaction parameter ($\delta = 10$), the numerical velocity profile of the Poiseuille flow $\tilde{u}^{(P)}$ is very close to that obtained analytically using the slip condition (11.86), which is plotted by the dashed line. The profile becomes flatter by decreasing the rarefaction parameter δ, and the magnitude of the bulk velocity increases when δ tends to zero. The velocity profiles of thermal creep $\tilde{u}^{(T)}$ are more or less flat, and their magnitudes decrease by increasing the rarefaction parameter δ. For $\delta = 10$, the velocity magnitude obtained numerically is close to that obtained from the slip solutions (11.86) only in the middle point, while near the channel wall, the numerical profiles differ from the corresponding slip solution.

The heat flow profiles $\tilde{q}^{(P)}$ and $\tilde{q}^{(T)}$ are presented in Figure 11.10. It is curious that for the large value of the rarefaction parameter ($\delta = 10$), the heat flow of the

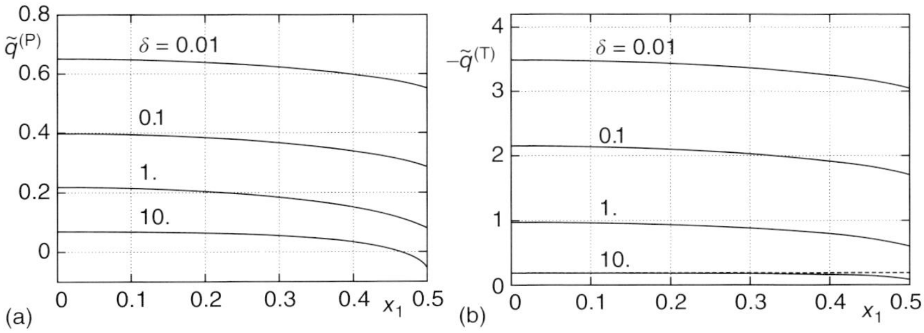

Figure 11.10 Dimensionless heat flows $\tilde{q}^{(P)}$ and $\tilde{q}^{(T)}$ for gas flow between plates versus coordinate x_1: solid lines–kinetic equation, dashed line–hydrodynamic solution (11.90).

mechanocaloric effect $\tilde{q}^{(P)}$ is positive over practically whole range of the coordinate x_1 and it is negative near the plates. It is an evidence that the $\tilde{u}^{(P)}$ and $\tilde{q}^{(P)}$ are different, but their integrals are the same according to the reciprocal relation (11.110). The heat flow $\tilde{q}^{(P)}$ increases by decreasing the rarefaction parameter and becomes equal to that of the thermal creep $\tilde{u}^{(P)}$. The heat flow due to the temperature gradient $\tilde{q}^{(T)}$ is always negative. For the large value of δ, it is equal to the analytical expression (11.90) plotted by the dashed line practically over the whole range of x_1 except the small region near the plates.

The planar Poiseuille and creep flows of polyatomic gases were studied by Loyalka and Storvick [126] and by Titarev and Shakhov [127], who concluded that the Poiseuille coefficient is not affected by the internal structure of molecules, while the thermal creep coefficient is affected within 10%. Some results on mixture flows between two parallel plates were reported in the papers [128, 129].

Exercises

11.1 Obtain (11.9).

11.2 What will change in the Couette flow problem if the linearization is done near the local Maxwellian?

11.3 What will change in the heat flow problem if the linearization is done near the local Maxwellian?

11.4 Obtain (11.37).

11.5 Obtain (11.134).
Hint: Use the limit expressions at $x \to 0$

$$I_0(x) \to \frac{\sqrt{\pi}}{2} - x \ln\, x, \quad I_2(x) \to \frac{\sqrt{\pi}}{4}, \quad I_4(x) \to \frac{3\sqrt{\pi}}{8}.$$

11.6 Consider two square parallel plates having the area of $A = 1\ \text{cm}^2$ separated by a distance of $a = 5.0\ \mu\text{m}$. The gap between the plates is filled by helium. One of the plate is kept at the temperature $T_1 = 295$ K and the other one is kept at $T_2 = 305$ K. Calculate the heat rate transfer Q between the plates equal to $Q = Aq_1$, at: (a) $p = 1$ atm. (see Table A.1); (b) $p = 4.5$ kPa.
Solution: (a) Since $\delta = 23$, Eq. (11.45) is used for q_1. Then $Q = 27$ W; (b) Since $\delta = 1$, Eq. (11.32) with $\tilde{q} = 0.4001$ is used for q_1. Then $Q = 6.7$ W.

11.7 Consider the same plates of the previous exercise confining the helium gas. The plates are kept at the same temperature $T = 300$ K, but there is a longitudinal pressure gradient equal to 10 Pa/m. Calculate the mass flow rate per width unity at: (a) $p = 1$ atm. (see Table A.1); (b) $p = 4.5$ kPa.
Solution: (a) Since $\delta = 23$, Eqs. (11.80), (11.87) are used. Then $\dot{M} = 1.1 \times 10^{-9}$g/s m; (b) Since $\delta = 1$, Eq. (11.80) with $G_\text{p}^* = 1.554$ is used. Then $\dot{M} = 3.5 \times 10^{-10}$ g/s m.

11.8 Consider the previous exercise, but with a temperature gradient of 10 K/m instead of the pressure gradient.
Solution: (a) Since $\delta = 23$, Eqs. (11.80), (11.87) are used. Then $\dot{M} = 3.9 \times 10^{-9}$ g/s m; (b) Since $\delta = 1$, Eq. (11.80) with $G_{\mathrm{T}}^{*} = 0.3654$ is used. Then $\dot{M} = 1.2 \times 10^{-9}$ g/s m.

12
One-Dimensional Axisymmetrical Flows

This chapter describes problems similar to those solved in the previous chapter. However, flows considered here are not planar, but they are axisymmetrical. This geometrical configuration requires some modifications in numerical schemes and, in some cases, it changes behaviors of some quantities.

12.1
Cylindrical Couette Flow

12.1.1
Definitions

Consider a gas flow between two coaxial cylinders. The radius of the internal cylinder is denoted as a, while the radius of the external cylinder is β times larger and equal to $a\beta$. The cylinder axis is directed along the coordinate r_3 as depicted in Figure 12.1. The cylinders are assumed to be so long that the end effects can be neglected and the solution does depend on the coordinate r_3. The gas is perturbed by a rotation of the internal cylinder with an angular speed ω. Our purpose is to calculate the velocity profile and the shear stress between the cylinders. Since the problem is axisymmetrical, we are considering the azimuthal bulk velocity u_ϕ and the radial–azimuthal component of the shear stress $P_{r\phi}$, which are calculated via the distribution function as

$$u_\phi(r) = \frac{1}{n}\int v_\phi f(\boldsymbol{r}, \boldsymbol{v})\, \mathrm{d}v_r\, \mathrm{d}v_\phi\, \mathrm{d}v_3, \tag{12.1}$$

$$P_{r\phi}(r) = m\int v_r(v_\phi - u_\phi) f(\boldsymbol{r}, \boldsymbol{v})\, \mathrm{d}v_r\, \mathrm{d}v_\phi\, \mathrm{d}v_3, \tag{12.2}$$

where $r = \sqrt{r_1^2 + r_2^2}$ is the radial coordinate, v_r and v_ϕ are radial and azimuthal components of molecular velocity, respectively. Equations (12.1) and (12.2) are the definitions (2.16) and (2.28) written down in the cylindrical coordinates. In contrast to the planar Couette flow, the shear stress $P_{r\phi}$ is not constant but the

Rarefied Gas Dynamics: Fundamentals for Research and Practice, First Edition. Felix Sharipov.

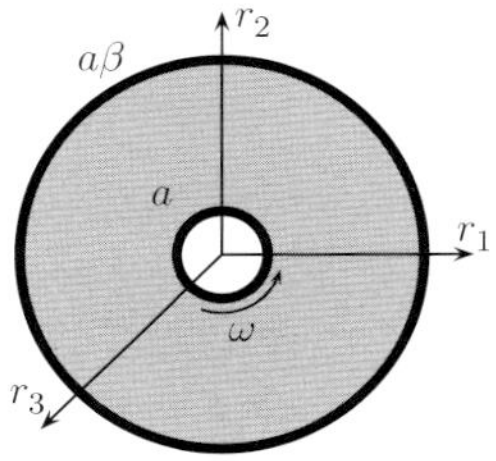

Figure 12.1 Scheme of cylindrical Couette flow.

momentum conservation law in this case is written as

$$P_{r\phi}r^2 = \text{const.} \tag{12.3}$$

We assume that the surface speed ωa is small in comparison with the MPS so that the ratio

$$\xi = \frac{\omega a}{v_0}, \quad |\xi| \ll 1 \tag{12.4}$$

will be used as the small parameter for the linearization. Then the dimensionless velocity $\tilde{u}$ and shear stress Π are defined as

$$u_\phi = \xi v_0 \tilde{u}, \quad P_{r\phi} = \xi p_0 \Pi \tag{12.5}$$

that follow from Eqs. (5.13) and (5.15), respectively.

Denoting the internal cylinder radius as a, we automatically adopted this value as the characteristic size. Then the rarefaction parameter is given by (7.46) and the dimensionless coordinates $[x_1, x_2, x_3]$ are introduced by (7.43).

12.1.2
Slip Flow Regime

In the viscous regime, the Navier–Stokes equation written down in the cylindrical coordinates is solved. The conservation law is given by (12.3), while Newton's law (6.2) is written as

$$P_{r\phi}(r) = -\mu\left(\frac{\mathrm{d}u_\phi}{\mathrm{d}r} - \frac{u_\phi}{r}\right). \tag{12.6}$$

The slip boundary condition (10.2) must also be written in the cylindrical variables (r, ϕ) as

$$u_\phi = \begin{cases} \omega a + \sigma_\mathrm{P}\,\ell\left(\frac{\mathrm{d}u_\phi}{\mathrm{d}r} - \frac{u_\phi}{r}\right), & \text{at} \quad r = a, \\ -\sigma_\mathrm{P}\,\ell\left(\frac{\mathrm{d}u_\phi}{\mathrm{d}r} - \frac{u_\phi}{r}\right), & \text{at} \quad r = a\beta. \end{cases} \tag{12.7}$$

Solving Eqs. (12.3) and (12.6) with (12.7), the velocity profile is obtained as

$$u_\phi(r) = \omega a\left[\frac{a}{r} - \frac{r}{a\beta^2}\left(1 - 2\frac{\sigma_\mathrm{P}}{\delta\beta}\right)\right] B, \tag{12.8}$$

where

$$\mathcal{B} = \left[1 - \frac{1}{\beta^2} + 2\frac{\sigma_{\rm P}}{\delta}\left(\frac{1}{\beta^3} + 1\right)\right]^{-1}. \tag{12.9}$$

The dimensionless velocity defined by (12.5) is given by

$$\tilde{u}(x) = \left[\frac{1}{x} - \frac{x}{\beta^2}\left(1 - 2\frac{\sigma_{\rm P}}{\delta\beta}\right)\right]\mathcal{B}. \tag{12.10}$$

The shear stress is derived from (12.8) and its dimensionless form is obtained from (12.5),

$$P_{r\phi}(r) = 2\,\mu\,\omega\,\frac{a^2}{r^2}\mathcal{B}, \quad \Pi = \frac{2\mathcal{B}}{\delta x^2}. \tag{12.11}$$

In the limit of high radii ratio, $\beta \to \infty$, this solution yields

$$P_{r\phi}(r) = 2\,\mu\,\omega\,\frac{a^2}{r^2}\left(1 + 2\frac{\sigma_{\rm P}}{\delta}\right)^{-1}, \quad \Pi(x) = \frac{2}{\delta x^2}\left(1 + 2\frac{\sigma_{\rm P}}{\delta}\right)^{-1}. \tag{12.12}$$

Physically, it means that the internal cylinder is not affected by the external cylinder.

12.1.3
Kinetic Equation

As is shown in Section 9.4.4, the axial symmetry allows us to reduce the number of variables and consider x, c_p, and θ instead of x_1, x_2, c_1, and c_2. The relation between these coordinates is given by Eqs. (9.59), (9.62), and Figure 9.2. Similar to the planar Couette flow, the axisymmetrical one is solved by linearizing the BGK model near the global Maxwellian (5.6), that is, Eq. (7.47) is written for $A_1 = 1$, $A_2 = 0$, and $A_3 = 0$ without the source term. The density ϱ and temperature τ deviations for this problem are zero so that the function H defined by (7.48) is reduced to

$$H = 2c_p\tilde{u}\sin\theta. \tag{12.13}$$

To eliminate one more variable, namely, c_3, the new perturbation is introduced as

$$\Phi(x,\theta,c_p) = \frac{1}{\pi^{1/2}}\int_{-\infty}^{\infty} e^{-c_3^2}h(x,\theta,c_p,c_3)\,\mathrm{d}c_3, \tag{12.14}$$

which obeys the following kinetic equation:

$$c_p\cos\theta\frac{\partial\Phi}{\partial x} - \frac{c_p\sin\theta}{x}\frac{\partial\Phi}{\partial\theta} = \delta(H - \Phi), \tag{12.15}$$

where the left-hand side of the kinetic equation has been written in the cylindrical coordinates (9.63). Using the same reasonings as those to obtain (11.18) and assuming the diffuse scattering, the boundary conditions for the new perturbation are obtained as

$$\Phi = \begin{cases} 2c_p\sin\theta & \text{at} \quad x = 1 \quad \text{and} \quad 0 \le \theta \le \frac{\pi}{2}, \\ 0 & \text{at} \quad x = \beta \quad \text{and} \quad \frac{\pi}{2} \le \theta \le \pi. \end{cases} \tag{12.16}$$

Analyzing the equation (12.15) and boundary conditions (12.16), we conclude that the perturbation is antisymmetric

$$\Phi(x, \theta, c_p) = -\Phi(x, -\theta, c_p). \tag{12.17}$$

The moments $\tilde{u}$ and Π are obtained from Eqs. (12.1), (12.2), and (12.5) replacing the velocity coordinates c_x and c_ϕ by c_p and θ according to (9.62)

$$\begin{bmatrix} \tilde{u}(x) \\ \Pi(x) \end{bmatrix} = \frac{2}{\pi} \int_0^\pi \int_0^\infty \begin{bmatrix} c_p \sin\theta, \\ c_p^2 \sin 2\theta \end{bmatrix} \Phi(x, \theta, c_p) e^{-c_p^2} c_p \, dc_p \, d\theta, \tag{12.18}$$

where the property (12.17) has been used.

12.1.4
Free-Molecular Regime

To obtain the solution in the free-molecular regime, the left-hand side of the kinetic equation is written in the form (9.53), while the right-hand side is zero. Integrating such an equation with respect to the variable s from a point with the radial coordinate x up to a boundary, the perturbation Φ is obtained as

$$\Phi = \begin{cases} 2xc_p \sin\theta & \text{at} \quad 0 \le \theta \le \theta_0, \\ 0 & \text{at} \quad \theta_0 \le \theta \le \pi, \end{cases} \qquad \sin\theta_0 = \frac{1}{x}. \tag{12.19}$$

The geometrical meaning of the angle θ_0 is shown in Figure 12.2. The range $0 \le \theta \le \theta_0$ corresponds to the situation when gaseous particles come to the point x from the internal cylinder. Their perturbation is determined by the cylinder surface motion. The other range $\theta_0 < \theta \le \pi$ corresponds to the particles coming from the external cylinder which are not perturbed $\Phi = 0$.

After the substitution of (12.19) into (12.18), we obtain the moments as

$$\tilde{u}(x) = \frac{1}{\pi}(x\theta_0 - \cos\theta_0), \quad \Pi(x) = \frac{1}{\sqrt{\pi}x^2}. \tag{12.20}$$

It is interesting that this solution does not depend on the ratio of radii β, but it is determined only by the internal cylinder. Note that at the internal cylinder surface $x = 1$, the expression for Π coincides with that for the plane Couette flow given by Eq. (11.9).

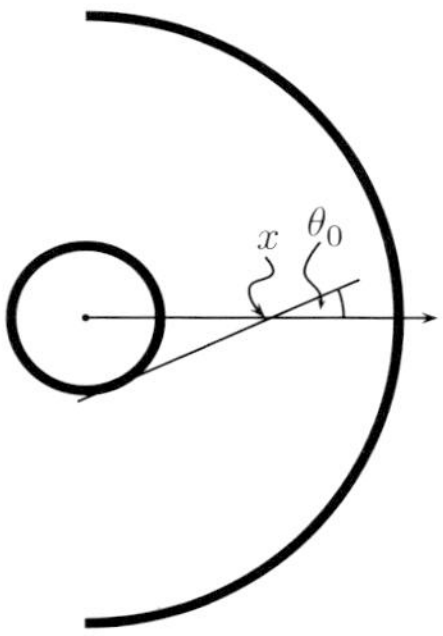

Figure 12.2 Scheme to obtain (12.19).

12.1.5
Numerical Scheme

Let us introduce the regularly distributed grid for the radial x as

$$x_k = \frac{k}{N_x}, \quad N_x \leq k \leq \beta N_x, \tag{12.21}$$

where N_x is an integer. Note that the number of nodes is equal to $(\beta - 1)N_x$, which must be an integer too. The Gauss nodes described in Section B.1.2 will be used for the speed c_p. Thus, the perturbation Φ, the function H defined by (12.13), and the moments are discretized as

$$\Phi_{kji} = \Phi(x_k, \theta_j, c_i), \quad H_{k,j+1/2,i} = 2c_i\tilde{u}_k \sin\theta_{j+1/2}, \tag{12.22}$$

$$\tilde{u}_k = \tilde{u}(x_k), \ \Pi_k = \Pi(x_k). \tag{12.23}$$

Since all expressions given in this section are the same for all values of the speed c_i, the subscript i will be omitted here.

For the problem considered here, the perturbation Φ for the first $\theta_0 = 0$ and last θ_{N_θ} nodes is equal to zero that follows from (12.13), (12.15), and boundary condition (12.16) so that we skip the scheme (9.70) and use directly (9.75), which takes the form

$$\begin{aligned}\Phi_{kj} = &\left[\left(C_j^x - \frac{C_j^\theta}{k - l/2} - \delta\right)\Phi_{k-l,j} + \left(-C_j^x + \frac{C_j^\theta}{k - l/2} - \delta\right)\Phi_{k,j+1}\right.\\ &+ \left(C_j^x + \frac{C_j^\theta}{k - l/2} - \delta\right)\Phi_{k-l,j+1}\\ &\left.+ 2\delta\left(H_{k,j+1/2} + H_{k-l,j+1/2}\right)\right]\left(C_j^x + \frac{C_j^\theta}{k - l/2} + \delta\right)^{-1},\end{aligned} \tag{12.24}$$

where the coefficients C_j^x and C_j^θ are those defined by (9.76) with a small modification

$$C_j^x = 2c_iN_x\left|\cos\left(\pi\frac{j + 1/2}{N_\theta}\right)\right|, \quad C_j^\theta = 2c_iN_x\frac{N_\theta}{\pi}\sin\left(\pi\frac{j + 1/2}{N_\theta}\right). \tag{12.25}$$

The angle nodes θ_j are swept from $j = N_\theta - 1$ to 1, while the nodes x_k are swept in the following directions:

$$\begin{aligned}&\text{from} \quad N_x + 1 \quad &\text{to} \quad &\beta N_x \quad &\text{for} \quad &l = 1,\\ &\text{from} \quad \beta N_x - 1 \quad &\text{to} \quad &N_x \quad &\text{for} \quad &l = -1,\end{aligned} \tag{12.26}$$

where l is defined by (9.73).

To calculate the moment $\tilde{u}$ and Π, we should retrieve the subscript i attached to Φ as has been defined by (12.22). Then the moment defined by (12.18) and discretized by (12.23) is written in the form of quadrature

$$\begin{bmatrix} \tilde{u}_k \\ \Pi_k \end{bmatrix} = \sum_{j=1}^{N_\theta - 1} \sum_{i=1}^{N_c} \begin{bmatrix} c_i \sin\theta_j \\ c_i^2 \sin 2\theta_j \end{bmatrix} \Phi_{kji} W_i^c W_j^\theta, \tag{12.27}$$

where W_i^c are the weights of the speed nodes c_i and W_j^θ are the weights of the angle nodes. Using Simpson's rule, the weights W_j^θ are defined as

$$W_j^\theta = \frac{2\pi}{N_\theta} W_j^S, \tag{12.28}$$

being W_j^S given by (9.14). In practical calculations, the contribution of each Φ_{kji} is added to the moments immediately after its calculation that allows to avoid the storage of all its values.

12.1.6
Splitting Scheme

One of the difficulties to apply a finite difference scheme to the kinetic equation is the discontinuity of the perturbation function. In the free-molecular regime, see Eq. (12.19), the discontinuity is evident in the point $\theta = \theta_0$. The discontinuity also exists near the same point for small values of the rarefaction δ. Mathematically it means that the derivative $\partial\Phi/\partial\theta$ near this point is large that causes a significant numerical error when it is approximated by a finite difference. Such an error can distort the perturbation function not only near the discontinuity point but also far from it. The typical behavior of the perturbation Φ calculated by the above-described scheme ($N_x = 200$, $N_\theta = 100$, and $N_c = 6$) is plotted on Figure 12.3 (a) as a function of the angle θ in the middle point between the cylinders, $x = (\beta + 1)/2$, for the radii ratio $\beta = 2$, and for the third speed node $c_i \approx 1$ ($i = 3$). The graph shows that for the small values of rarefaction (δ=0.1 and 1), the perturbation Φ drastically changes near the point $\theta_0 = 0.23\pi$ and it has the sawtooth behavior in the range $\theta < \theta_0$. For larger values of the rarefaction, the teeth become smaller and the function Φ is smooth for $\delta = 5$.

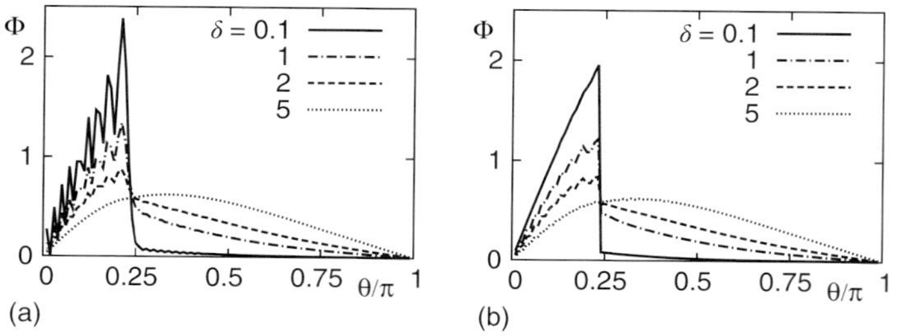

Figure 12.3 Perturbations Φ versus angle θ at the middle point between cylinders $x = (\beta + 1)/2$ for $\beta = 2$ and $c_i \approx 1$: (a) without split, (b) with split.

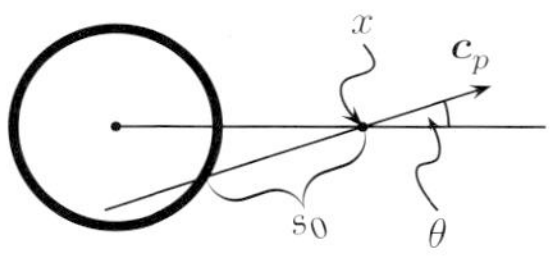

Figure 12.4 Scheme to calculate s_0 (see Eq. (12.32)).

To reduce the numerical error related to the discontinuity, the perturbation function is usually split, see for example, Refs [35, 130],

$$\Phi = \Phi_0 + \Phi', \tag{12.29}$$

where the term Φ_0 obeys the following equation:

$$-c_p \frac{\partial \Phi_0}{\partial s} + \delta \Phi_0 = 0 \tag{12.30}$$

subject to the boundary condition (12.16). The form (9.53) of the derivatives has been used in (12.30). Its integration with respect to the variable s leads to the following result:

$$\Phi_0 = \begin{cases} 2xc_p \sin\theta \exp\left(-\frac{\delta s_0}{c_p}\right) & \text{at} \quad 0 \le \theta \le \theta_0, \\ 0 & \text{at} \quad \theta_0 \le \theta \le \pi, \end{cases} \tag{12.31}$$

where the angle θ_0 is defined by (12.19), while s_0 is given as

$$s_0 = x\cos\theta - \sqrt{1 - (x\sin\theta)^2} \tag{12.32}$$

and represents the distance from the point x to the internal cylinder surface in the direction $-\mathbf{c}_p$ as is shown in Figure 12.4. The moments u_0 and Π_0 are calculated via Φ_0 by the definition (12.18) as

$$u_0 = \frac{4x}{\pi} \int_0^{\theta_0} I_3(\delta s_0)\sin^2\theta \, d\theta, \quad \Pi_0 = \frac{8x}{\pi} \int_0^{\theta_0} I_4(\delta s_0)\sin^2\theta \cos\theta \, d\theta, \tag{12.33}$$

where I_3 and I_4 are defined by (A.29).

The second term Φ' obeys the original equation (12.15),

$$c_p \cos\theta \frac{\partial \Phi'}{\partial x} - \frac{c_p \sin\theta}{x} \frac{\partial \Phi'}{\partial \theta} = \delta(H - \Phi'), \tag{12.34}$$

assuming its value to be zero at the boundary so that this equation is solved by the same numerical scheme as Eq. (12.15). The function H given by (12.13) contains the velocity $\tilde{u}$, which is split together with the shear stress

$$\tilde{u} = \tilde{u}_0 + \tilde{u}', \quad \Pi = \Pi_0 + \Pi', \tag{12.35}$$

where $\tilde{u}_0$ and Π_0 are given by (12.33) and $\tilde{u}'$ and Π' are calculated via Φ' according to their definition (12.27).

The perturbation function Φ calculated by the splitting scheme (12.29) is plotted on Figure 12.3 (b), which shows that it becomes completely smooth at $\delta = 0.1$, slightly smoother at $\delta = 1$ and 2. However, the split does not change the perturbation Φ for the large rarefaction $\delta = 5$ so that it makes sense to apply the splitting scheme only in the range $0 < \delta \le 2$.

The numerical codes couette_axisym.for based on the scheme described earlier written in FORTRAN and an example with output data are given in Section B.5.

Table 12.1 Numerical scheme parameters and shear stress Π at $x = 1$ for cylindrical Couette flow versus rarefaction δ and radii ratio β based on the code given in Section B.5.

				Π			Iters.	Var. (%)
δ	N_x	N_θ	N_c	$\beta = 2$	3	5	5	5
0[a]				0.5642	0.5642	0.5642		
0.01	10	20	4	0.5635	0.5634	0.5634	7	0.1
0.1	10	20	4	0.5573	0.5565	0.5562	15	0.8
1	20	200	4	0.4968	0.4884	0.4837	110	1.
2	200	200	4	0.4363	0.4206	0.4120	377	1.
5	400	400	4	0.3059	0.2822	0.2706	1365	1.
10	500	500	4	0.1983	0.1768	0.1670	4673	1.
10[b]				0.2044	0.1819	0.1717		

a) Free-molecular solution (12.20).
b) Slip solution (12.11).

12.1.7 Results

The shear stress Π and numerical scheme parameters providing its accuracy of 0.1% are presented in Table 12.1, that is, the variation of the shear stress will be within 0.1% if we increase the number of all nodes. These data show that both N_x and N_θ drastically increases by increasing the rarefaction parameter, while the number of speed nodes N_c is extremely small because the function $\Phi(c_p)$ is rather smooth. The number of iterations corresponding to $\beta = 5$ is given in the eighth column of Table 12.1. For the other values of β, this number is smaller. Usually, the number of iterations becomes very large in the hydrodynamic regime ($\delta = 10$).

The shear stress Π presented in Table 12.1 is referred to the position $x = 1$ corresponding to the internal cylinder surface. In this point, the influence of the perturbation function discontinuity is smallest and then the accuracy of Π at $x = 1$ is best. The value of Π in the other points is usually calculated with a worse accuracy. It is interesting to check the momentum conservation law (12.3) in its dimensionless form $\Pi x^2 =$ constant. The relative variance of Πx^2 for $\beta = 5$ given in the ninth column of Table 12.1 is about 1% for $\delta \geq 0.1$. It can be reduced by increasing the numbers N_x and N_θ.

The data presented in Table 12.1 show that the difference between the shear stress Π at $\beta = 3$ and that for $\beta = 5$ is rather small. In practice, it means that the results corresponding to the radii ratio $\beta = 5$ can be successfully applied for larger values of this ratio. The last row of Table 12.1 presents the shear stress Π calculated in the slip regime by Eq. (12.11). Comparing it with the numerical results of Π for $\delta = 10$, we verify that the expression (12.11) yields the shear stress with an error less than 3%. Naturally, this error decreases for larger values of δ.

The velocity profiles $\tilde{u}$ corresponding to the radii ratio $\beta = 2$ are plotted in Figure 12.5. In the hydrodynamic regime ($\delta \to \infty$), the velocity $\tilde{u}$ is close to a

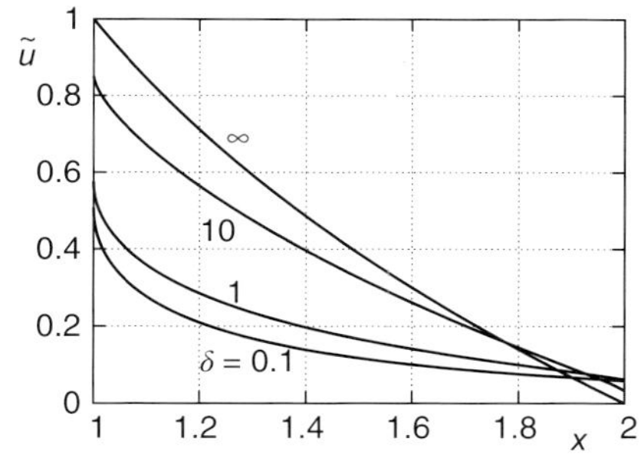

Figure 12.5 Dimensionless velocity $\tilde{u}$ in cylindrical Couette flow versus radial coordinate x at $\beta = 2$.

linear function and varies from 1 to 0. At $\delta = 10$, the behavior is the same, but the velocity slip is observed near the cylinders. In the transitional ($\delta = 1$) and near free-molecular ($\delta = 0.1$) regimes, the velocity drastically decreases near the internal cylinder and it is practically flat near the external cylinder.

Cercignani and Sernagiotto [131] calculated the cylindrical Couette flow applying the integro-moment method to the linearized BGL model. A nonlinearized flow was calculated by the DSMC method by Nanbu [132]. The situation when both cylinders rotate was analyzed in the papers [133, 134].

12.2
Heat Transfer between Two Cylinders

12.2.1
Definitions

Consider again two coaxial cylinders with radii a and βa as is drawn in Figure 12.6. The external cylinder is maintained at temperature T_0, while the internal cylinder has a different temperature $T_0 + \Delta T$. The radius a of the internal cylinder is assumed to be the characteristic size, so that the rarefaction parameter is given by Eq. (7.46). The quantities of our interest are the density $n(r)$ and temperature $T(r)$ distributions, and the radial heat flux q_r expressed via the distribution function as

$$n(r) = \int f(r, \boldsymbol{v})\, \mathrm{d}v_r\, \mathrm{d}v_\phi\, \mathrm{d}v_3, \tag{12.36}$$

$$T(r) = \frac{m}{3nk_\mathrm{B}} \int v^2 f(r, \boldsymbol{v})\, \mathrm{d}v_r\, \mathrm{d}v_\phi\, \mathrm{d}v_3, \tag{12.37}$$

$$q_r(r) = \frac{m}{2} \int v^2 v_r f(r, \boldsymbol{v})\, \mathrm{d}v_r\, \mathrm{d}v_\phi\, \mathrm{d}v_3 \tag{12.38}$$

that follow from (2.12), (2.26), and (2.32), respectively. Here, $r = \sqrt{r_1^2 + r_2^2}$ is the radial coordinate, and v_r and v_ϕ are radial and azimuthal components of molecular

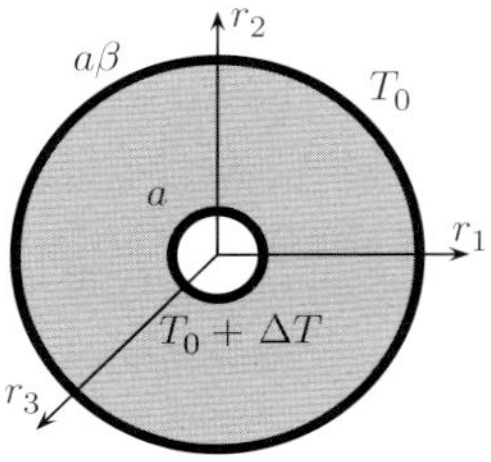

Figure 12.6 Scheme of heat transfer between two cylinders.

velocity. The energy conservation law for the problem in question reads

$$q_r\, r = \text{const.} \tag{12.39}$$

If we assume that the temperature difference is small, $\Delta T \ll T_0$, then the relative temperature difference can be used as the small parameter of linearization

$$\xi = \frac{\Delta T}{T_0}, \quad |\xi| \ll 1. \tag{12.40}$$

Then the moments are linearized as

$$n = n_0(1 + \xi\varrho), \quad T = T_0(1 + \xi\tau), \quad q_r = \xi p_0 v_0 \tilde{q}. \tag{12.41}$$

that follow from Eqs. (5.12), (5.14), and (5.16), respectively.

12.2.2
Temperature Jump Solution

In the hydrodynamic regime ($\delta \to \infty$), Fourier's law in the cylindrical coordinates is applied

$$q_r = -\kappa \frac{\mathrm{d}T}{\mathrm{d}r}. \tag{12.42}$$

Together with the energy conservation law (12.39) and the temperature jump boundary condition (10.39) written for the problem in question as

$$T - T_0 = \begin{cases} \Delta T + \zeta_{\mathrm{T}} \ell \, \frac{\mathrm{d}T}{\mathrm{d}r}, & \text{at} \quad r = a, \\ -\zeta_{\mathrm{T}} \ell \, \frac{\mathrm{d}T}{\mathrm{d}r}, & \text{at} \quad r = \beta a, \end{cases} \tag{12.43}$$

Fourier's law leads to the following temperature profile:

$$T(r) = T_0 + \tau(x)\Delta T, \quad \tau(x) = 1 - \left(\ln x + \frac{\zeta_{\mathrm{T}}}{\delta}\right) \mathcal{B}, \tag{12.44}$$

where (12.41) has been used. Here, the quantity $\mathcal{B}$ reads

$$\mathcal{B} = \left[\ln \beta + \frac{\zeta_{\mathrm{T}}}{\delta}\left(1 + \frac{1}{\beta}\right)\right]^{-1}. \tag{12.45}$$

The density distribution is obtained from the fact that the pressure should be constant in the gap between cylinders, $\varrho(x) = -\tau(x)$.

The heat flux is calculated substituting (12.44) into Eq. (12.42)

$$q_r = \kappa \frac{\mathcal{B}}{r} \Delta T, \quad \tilde{q} = \frac{5}{4\mathrm{Pr}} \frac{\mathcal{B}}{\delta x} \approx \frac{15}{8} \frac{\mathcal{B}}{\delta x}, \tag{12.46}$$

where (12.41) has been used and the Prandtl number Pr defined by (6.4) has been assumed to be equal to 2/3.

When the external cylinder radius is significantly larger than that of the internal cylinder, then the heat flow (12.46) takes the form

$$\tilde{q} = \frac{15}{8} \frac{1}{x(\delta \ln \beta + \zeta_\mathrm{T})} \quad \text{at} \quad \beta \gg 1. \tag{12.47}$$

It means that in contrast to the cylindrical Couette flow, the influence of the external cylinder does not vanish for the heat transfer.

12.2.3
Kinetic Equation

Since the main objective of this problem is the heat flow, the kinetic equation should provide the correct Prandtl number, hence the ellipsoidal or S-model would be appropriate. Let us apply the last one, that is, Eq. (7.47) with $A_1 = 1$, $A_2 = 0$, $A_3 = 1$ according to Table 7.1. For the problem in question, the function H takes the form

$$H = \varrho + \left(c^2 - \frac{3}{2}\right)\tau + \frac{4}{15} c_p \cos\theta \left(c^2 - \frac{5}{2}\right)\tilde{q}, \tag{12.48}$$

where the moments are calculated via the perturbation linearizing the expressions (12.36)–(12.38)

$$\begin{bmatrix} \varrho(x) \\ \tau(x) \\ \tilde{q}(x) \end{bmatrix} = \frac{2}{\pi^{3/2}} \int_{-\infty}^{\infty} \int_0^{\pi} \int_0^{\infty} \begin{bmatrix} 1 \\ \frac{2}{3}c^2 - 1 \\ c_p \cos\theta \left(c^2 - \frac{5}{2}\right) \end{bmatrix} \times h(x, \theta, c_p, c_3) e^{-c^2} c_p \, \mathrm{d}c_p \, \mathrm{d}\theta \, \mathrm{d}c_3, \tag{12.49}$$

where the symmetry $h(\theta) = h(-\theta)$ has been considered.

The source term $\tilde{g}$ is zero because the linearization is done near the global Maxwellian, while the surface source term has the form (5.52),

$$h_w = \tau_w \left(c^2 - \frac{3}{2}\right), \quad \tau_w = \begin{cases} 1 & \text{at} \quad x = 1, \\ 0 & \text{at} \quad x = \beta. \end{cases} \tag{12.50}$$

Under the diffuse scattering assumption, the perturbation of reflected particles is given by (5.70) and particularly expressed as

$$h = \begin{cases} \varrho_r + (c^2 - 2) & \text{at} \quad x = 1, \\ 0 & \text{at} \quad x = \beta, \end{cases} \tag{12.51}$$

where the value ϱ_r is obtained from (5.71)

$$\varrho_r = -\frac{4}{\pi} \int_{-\infty}^{\infty} \int_{\pi/2}^{\pi} \int_0^{\infty} e^{-c^2} h(1, \theta, c_p, c_3) c_p^2 \cos\theta \, \mathrm{d}c_p \, \mathrm{d}\theta \, \mathrm{d}c_3. \tag{12.52}$$

To eliminate the variable c_3, two new perturbations are defined as

$$\Phi^{(1)}\left(x,\theta,c_p\right) = \frac{1}{\pi^{1/2}}\int_{-\infty}^{\infty} e^{-c_3^2} h\left(x,\theta,c_p,c_3\right)\, \mathrm{d}c_3, \tag{12.53}$$

$$\Phi^{(2)}\left(x,\theta,c_p\right) = \frac{1}{\pi^{1/2}}\int_{-\infty}^{\infty} e^{-c_3^2} h\left(x,\theta,c_p,c_3\right)\left(c_3^2-\frac{1}{2}\right)\, \mathrm{d}c_3. \tag{12.54}$$

Each of them obeys the same equation

$$c_p\cos\theta\frac{\partial\Phi^{(\alpha)}}{\partial x} - \frac{c_p\sin\theta}{x}\frac{\partial\Phi^{(\alpha)}}{\partial\theta} = \delta(H^{(\alpha)}-\Phi^{(\alpha)}), \quad \alpha=1,2, \tag{12.55}$$

with the following H functions

$$H^{(1)} = \varrho + \left(c_p^2-1\right)\tau + \frac{4}{15}c_p\cos\theta\left(c_p^2-2\right)\tilde{q}, \tag{12.56}$$

$$H^{(2)} = \frac{1}{2}\tau + \frac{2}{15}c_p\cos\theta\;\tilde{q}. \tag{12.57}$$

The moments (12.49) are expressed in terms of the new perturbations as

$$\begin{bmatrix}\varrho(x)\\ \tau(x)\\ \tilde{q}(x)\end{bmatrix} = \frac{2}{\pi}\int_0^{\pi}\int_0^{\infty}\left\{\begin{bmatrix}1\\ \frac{2}{3}(c_p^2-1)\\ c_p\cos\theta\left(c_p^2-2\right)\end{bmatrix}\Phi^{(1)}(x,\theta,c_p)\right.$$
$$\left.+\begin{bmatrix}0\\ 2/3\\ c_p\cos\theta\end{bmatrix}\Phi^{(2)}\left(x,\theta,c_p\right)\right\}e^{-c_p^2}c_p\,\mathrm{d}c_p\,\mathrm{d}\theta. \tag{12.58}$$

The boundary conditions (12.51) take the form

$$\Phi^{(1)} = \begin{cases}\varrho_r + (c_p^2-\frac{3}{2}),\\ 0,\end{cases} \quad \Phi^{(2)} = \begin{cases}\frac{1}{2} & \text{at} \quad x=1, \quad 0\le\theta<\frac{\pi}{2},\\ 0 & \text{at} \quad x=\beta, \quad \frac{\pi}{2}\le\theta\le\pi,\end{cases} \tag{12.59}$$

where the expression (12.52) is transformed into

$$\varrho_r = -\frac{4}{\pi^{1/2}}\int_{\pi/2}^{\pi}\int_0^{\infty} e^{-c_p^2}\Phi^{(1)}\left(1,\theta,c_p\right)c_p^2\cos\theta\,\mathrm{d}c_p\,\mathrm{d}\theta. \tag{12.60}$$

12.2.4
Free-Molecular Regime

In the free-molecular regime ($\delta \ll 1$), the perturbation in (12.60) is zero because it corresponds to the external cylinder, $\varrho_r = 0$. The perturbations in an arbitrary point is equal to its boundary values (12.59) in the angle $0 \le \theta \le \theta_0$,

$$\Phi^{(1)} = \begin{cases}(c_p^2-3/2),\\ 0,\end{cases} \quad \Phi^{(2)} = \begin{cases}\frac{1}{2} & \text{at} \quad 0\le\theta<\theta_0,\\ 0 & \text{at} \quad \theta_0\le\theta\le\pi,\end{cases} \tag{12.61}$$

where the angle θ_0 is depicted on Figure 12.2 and calculated as $\theta_0 = \arcsin(1/x)$. Substituting Eqs. (12.61) into the moment definitions (12.58), the density, temperature, and radial heat flux $\tilde{q}$ are calculated analytically as

$$\varrho(x) = -\frac{\theta_0(x)}{2\pi}, \quad \tau(x) = \frac{\theta_0(x)}{\pi}, \quad \tilde{q}(x) = \frac{1}{\sqrt{\pi}x}. \tag{12.62}$$

Note that the free-molecular solution is not affected by the radii ratio β. At the internal cylinder $x = 1$, the expression of the radial heat flux $\tilde{q}$ at $x = 1$, coincides with the plane heat flux given by Eq. (11.40).

12.2.5
Numerical Scheme

The cylindrical heat transfer problem has the same geometrical configuration as that of the cylindrical Couette flow; therefore, the nodes (12.21) for the radial variable x and those (9.65) for the angle θ are used also for the flow in consideration. The functions H are discretized as

$$H_{kji}^{(1)} = \varrho_k + (c_i^2 - 1)\tau_k + \frac{4}{15}c_i \cos\theta_j(c_i^2 - 2)\tilde{q}_k, \tag{12.63}$$

$$H_{kji}^{(2)} = \frac{1}{2}\tau_k + \frac{2}{15}c_i \cos\theta_j\, \tilde{q}_k, \tag{12.64}$$

where the discretized moments are introduced

$$\varrho_k = \varrho(x_k), \quad \tau_k = \varrho(x_k), \quad \tilde{q}_k = \tilde{q}(x_k). \tag{12.65}$$

Since both $\Phi^{(1)}$ and $\Phi^{(2)}$ obey the same equation (12.55), the numerical scheme is the same so that the superscript is omitted below. Moreover, the subscript i is also omitted because the finite difference scheme does not change from one node c_i to the other.

In contrast to the Couette flow, the perturbation functions for the first $j = 0$ and last $j = N_\theta$ nodes of the angle θ_j are not zero anymore so that they must be calculated first using the scheme (9.70), which takes the form

$$\Phi_{kj} = \frac{(C^x - \delta)\Phi_{k-l,j} + \delta(H_{kj} + H_{k-l,j})}{C^x + \delta}, \quad j = 0, N_\theta, \tag{12.66}$$

where the index l is defined by (9.68) and the coefficient C^x is the simplified expression of (9.71)

$$C^x = 2c_i N_x. \tag{12.67}$$

For the other nodes of θ_j $(1 \le j \le N_\theta - 1)$, the scheme is exactly the same as that for the Couette flow (12.24).

Now let us retrieve the superscript and subscript and write down the quadratures for the moments based on their expressions (12.58)

$$\begin{bmatrix} \varrho_k \\ \tau_k \\ \tilde{q}_k \end{bmatrix} = \sum_{j=0}^{N_\theta} \sum_{i=1}^{N_c} \left\{ \begin{bmatrix} 1 \\ \frac{2}{3}\left(c_i^2 - 1\right) \\ c_i \cos\theta_j (c_i^2 - 2) \end{bmatrix} \Phi_{kji}^{(1)} + \begin{bmatrix} 0 \\ \frac{2}{3} \\ c_i \cos\theta_j \end{bmatrix} \Phi_{kji}^{(2)} \right\} W_i^c W_j^\theta. \tag{12.68}$$

The quantity ϱ_r defined by (12.60) is approximated by the following quadrature:

$$\varrho_r = -2\pi^{1/2} \sum_{j=N_\theta/2}^{N_\theta} \sum_{i=1}^{N_c} \Phi_{N_x,ji}^{(1)} c_i \cos\theta_j W_i^c W_j^\theta \tag{12.69}$$

using only the half of the nodes θ_j. The weights W_i^c and nodes c_i are given in Section B.1.2. The angle weights W_j^θ correspond to Simpson's rule

$$W_j^\theta = \frac{2\pi}{N_\theta} W_j^S, \tag{12.70}$$

where W_j^S means (9.14).

Finally, we should calculate the contribution of the half-moments at the internal surface cylinder ($x = 1$), which is not counted by the numerical scheme. The perturbations in this point are given by the boundary conditions (12.59). Substituting them into the definitions (12.58) and integrating only in the angle range $[0, \pi/2]$, we obtain

$$\begin{bmatrix} \varrho_{N_x}^{(1/2)} \\ \tau_{N_x}^{(1/2)} \\ \tilde{q}_{N_x}^{(1/2)} \end{bmatrix} = \frac{2}{\pi} \int_0^{\pi/2} \int_0^\infty \left\{ \begin{bmatrix} 1 \\ \frac{2}{3}\left(c_p^2 - 1\right) \\ (c_p^2 - 2) c_p \cos\theta \end{bmatrix} \left[\varrho_r + \left(c_p^2 - \frac{3}{2}\right)\right] + \begin{bmatrix} 0 \\ \frac{2}{3} \\ c_p \cos\theta \end{bmatrix} \frac{1}{2} \right\} e^{-c_p^2} c_p \, dc_p \, d\theta = \begin{bmatrix} \frac{\varrho_r}{2} - \frac{1}{4} \\ \frac{1}{2} \\ \frac{1}{\sqrt{\pi}}\left(1 - \frac{\varrho_r}{4}\right) \end{bmatrix}. \tag{12.71}$$

These values must be added to the corresponding moments in the point $k = N_x$ at each iteration.

12.2.6 Splitting Scheme

As in case of the Couette flow, the perturbation functions $\Phi^{(1)}$ and $\Phi^{(2)}$ are discontinuous at the point $\theta = \theta_0$ when the rarefaction parameter δ is small. To reduce the influence of this discontinuity on the numerical accuracy, each perturbation is also split as (12.29), where both $\Phi_0^{(1)}$ and $\Phi_0^{(2)}$ obey (12.30) with the boundary conditions (12.61), while the perturbations $\Phi_0^{\prime(1)}$ and $\Phi_0^{\prime(2)}$ obey Eq. (12.34) with the respective H functions defined by (12.56) and (12.57) assuming their value at the boundary to be zero. The moments are also split as

$$\varrho = \varrho_0 + \varrho', \quad \tau = \tau_0 + \tau', \quad \tilde{q} = \tilde{q}_0 + \tilde{q}', \tag{12.72}$$

where each part is calculated by (12.58) via the corresponding perturbation.

The perturbations $\Phi_0^{(1)}$ and $\Phi_0^{(2)}$ are obtained analytically

$$\Phi_0^{(2)} = \begin{cases} \frac{1}{2}\exp\left(-\frac{\delta s_0}{c_p}\right) & \text{at} \quad 0 \le \theta \le \theta_0, \\ 0 & \text{at} \quad \theta_0 \le \theta \le \pi. \end{cases} \tag{12.73}$$

Substituting these expressions into (12.58), the corresponding terms of the moments are obtained

$$\varrho_0 = \frac{2}{\pi}\int_0^{\theta_0}\left(\varrho_r I_1 + I_3 - \frac{3}{2}I_1\right)\,\mathrm{d}\theta, \tag{12.74}$$

$$\tau_0 = \frac{4}{3\pi}\int_0^{\theta_0}\left[\varrho_r(I_3 - I_1) + I_5 - \frac{5}{2}I_3 + 2I_1\right]\,\mathrm{d}\theta, \tag{12.75}$$

$$\tilde{q}_0 = \frac{2}{\pi}\int_0^{\theta_0}\left[\varrho_r(I_4 - 2I_2) + I_6 - \frac{7}{2}I_4 + \frac{7}{2}I_2\right]\cos\theta\,\mathrm{d}\theta, \tag{12.76}$$

where the functions I_n defined by (A.29) have the same argument δs_0 with s_0 given by (12.32). The small difference of these expressions from (12.33) is that they contain the unknown quantity ϱ_r so that only the integrals are calculated before the iterations, while the final expressions of the moments ϱ_0, τ_0, and $\tilde{q}_0$ are calculated in each iteration. It makes sense to apply the splitting scheme in the range $0 < \delta \le 2$, while it is useless for a larger value of the rarefaction δ.

The numerical codes heat_axisym.for in FORTRAN and an example with output data are given in Section B.6.

12.2.7
Numerical Results

The radial heat flux $\tilde{q}$ for the radii ratio $\beta = 2$ calculated by the code heat_axisym.for is presented in Table 12.2. The numerical scheme parameters providing its accuracy of 0.1% are also given in Table 12.2, that is, the heat flow $\tilde{q}$ will not vary more than 0.1% by increasing the number of nodes N_x, N_θ, and N_c. Similar to the Couette flow, the node numbers N_x and N_θ drastically increase in the hydrodynamic regime, while the number of speed nodes N_c is rather small. The number of iterations given in the sixth column of Table 12.2 again increases by increasing the rarefaction parameter. The solution for $\delta = 10$ based on the temperature jump condition given in the last row of Table 12.2 differs from the numerical value of $\tilde{q}$ for 1%, that is, it provides reasonable values of the heat flow in the range $\delta \ge 10$.

The numerical values of heat flow $\tilde{q}$ for larger ratios of radii are reported in Refs [35, 136]. Some of these data are plotted on Figure 12.7, which confirms that the

Table 12.2 Numerical scheme parameters and dimensionless heat flow $\tilde{q}$ at $x = 1$ and $\beta = 2$ versus rarefaction δ for cylindrical heat transfer based on the code given in Section B.6.

δ	N_x	N_θ	N_c	q	Iters.	Var. (%)
0[a)]				0.5642		
0.01	10	20	4	0.5631	7	0.1
0.1	10	20	4	0.5537	13	0.8
1	20	80	4	0.4734	61	0.4
2	100	250	6	0.4076	132	0.5
5	200	300	6	0.2851	491	0.7
10	200	400	6	0.1879	1468	0.4
10[b)]				0.1901		

a) Free-molecular solution (12.62).
b) Jump solution (12.46).

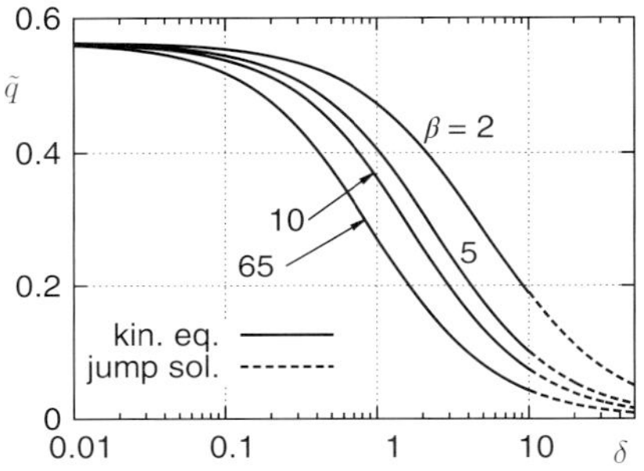

Figure 12.7 Dimensionless heat flow $\tilde{q}$ between two cylinders versus rarefaction δ for several values of radii ratio β. Numerical data are reported in Ref. [35]. Reproduced from Ref. [136] with permission.

temperature jump solution (12.46) yields a reasonable approximation for $\delta \geq 10$. Unlike the cylindrical Couette flow, the dependence of the heat flow $\tilde{q}$ on the radii ratio β is strong. According to the asymptotic formula (12.47), the heat flow always decreases by increasing the ratio β.

The works [35, 136] also report numerical data for the heat flow based on the CL scattering kernel (4.17). A comparison of these numerical results with experimental data reported in Ref. [37] is performed in Figure 12.8. The comparison shows that heavy gases such as argon, krypton, and xenon interact diffusely, $\alpha_t = 1$ and $\alpha_n = 1$, with a surface, while light gases such as helium and neon represent a significant deviation from the complete accommodation. The corresponding values of the accommodation coefficients α_t and α_n are given in Table 4.2.

The temperature profiles at $\beta = 2$ are shown in Figure 12.9 for some values of the rarefaction parameter δ. They are very similar to those for the velocity profiles, see Figure 12.5 in the Couette flow. In the hydrodynamic regime ($\delta \to \infty$), the profile $\tau(x)$ is practically linear from 1 to 0. For $\delta = 10$, the temperature profile is quite similar to that in the hydrodynamic regime, but it is subject to the temperature

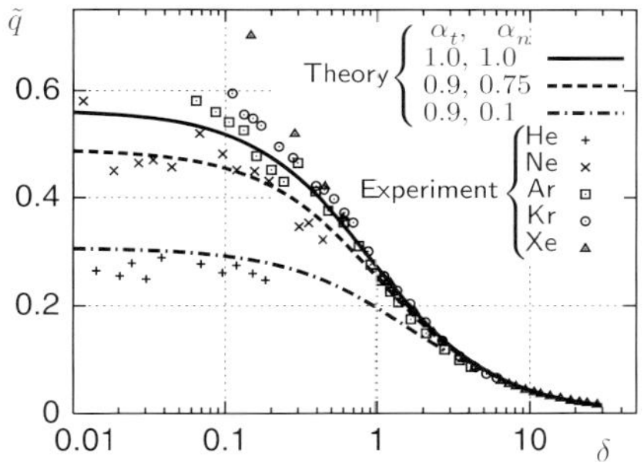

Figure 12.8 Dimensionless heat flow $\tilde{q}$ between two cylinders versus rarefaction δ for $\beta = 65$: curves–theoretical results [35] based on CL kernel (4.17); symbols–experimental data [37]. Reproduced from Ref. [35] with permission.

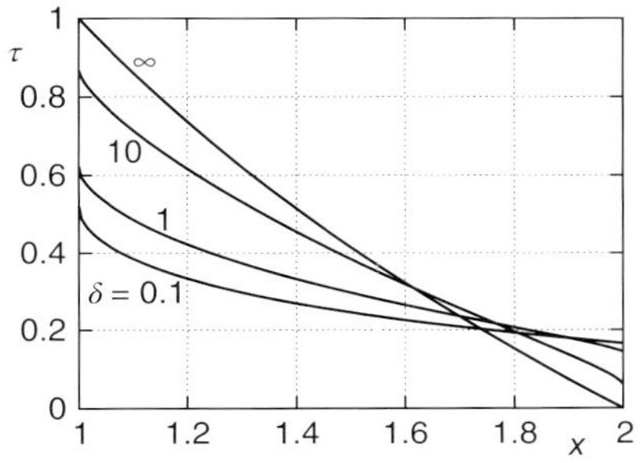

Figure 12.9 Temperature deviation $\tau(x)$ between two cylinders at $\beta = 2$.

jump at the cylinder surfaces. In the transitional ($\delta = 1$) and near free-molecular ($\delta = 0.1$) regimes, the temperature sharply decreases near the internal cylinder and weakly varies near the external cylinder.

Tantos *et al.* [135] studied the cylindrical heat transfer through various kinds of polyatomic gases. The heat transfer between two cylinders based on the nonlinearized S-model was investigated by Pantazis and Valougeorgis [137]. Some data on the heat transfer through a gas confined between two rotating cylinders are reported in Refs [138–141].

12.3 Cylindrical Poiseuille and Thermal Creep Flows

12.3.1 Definitions

A gas flow through a long cylindrical tube is very similar to that between two plates considered in Section 11.3 so that many formulas are the same. To avoid repetitions, only different expressions will be given here.

Consider a monatomic rarefied gas flowing through a long tube due to small longitudinal gradients of pressure ξ_P and temperature ξ_T defined by (11.72). As is shown in Figure 12.10, the coordinate r_3 is the longitudinal coordinate coinciding

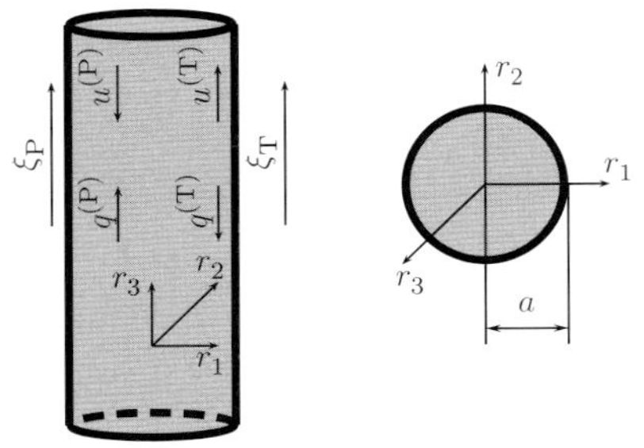

(a) longitudinal section (b) cross section

Figure 12.10 Scheme of cylindrical Poiseuille and creep flows.

with the tube axis. The tube radius a is chosen as the characteristic size so that the dimensionless coordinates are given by (7.43) and the rarefaction parameter is defined by (7.46).

We assume that the tube is so long that the end effects can be neglected and the flow is considered to be one-dimensional, that is, the pressure and temperature depend linearly only on the coordinate r_3 and is given by (11.73). The bulk velocity and the heat flux have the longitudinal component only depending on the radial coordinate r only

$$\boldsymbol{u} = [0, 0, u_3], \quad u_3 = u_3(r), \quad \boldsymbol{q} = [0, 0, q_3], \quad q_3 = q_3(r), \tag{12.77}$$

where $r = \sqrt{r_1^2 + r_2^2}$ is the radial coordinate.

Our aim is to calculate the velocity $u_3(r)$, heat flow vector $q_3(r)$, mass flow rate and heat flow rates defined as

$$\dot{M} := 2\pi m n_0 \int_0^a u_3(r) r \, \mathrm{d}r, \quad \dot{E} := 2\pi \int_0^a q_3(r) r \, \mathrm{d}r. \tag{12.78}$$

Similar to the planar Poiseuille flow (see Section 11.3), the linearization is done near the local Maxwellian using the pressure and temperature gradients as the small parameters. Then the reference perturbations $h_R^{(P)}$ and $h_R^{(T)}$ are the same and given by (11.78). Moreover, the bulk velocity u_3 and heat flow q_3 are decomposed into two terms (11.79). The mass and heat flow rates are also decomposed and written as

$$\dot{M} = \frac{\pi a^2 p_0}{v_0}(-G_P^* \xi_P + G_T^* \xi_T), \quad \dot{E} = \frac{1}{2}\pi a^2 p_0 v_0 (Q_P^* \xi_P - Q_T^* \xi_T), \tag{12.79}$$

where the following coefficients have been introduced

$$G_P^* := -4 \int_0^1 \tilde{u}^{(P)} x \, \mathrm{d}x, \quad G_T^* := 4 \int_0^1 \tilde{u}^{(T)} x \, \mathrm{d}x, \tag{12.80}$$

$$Q_P^* := 4 \int_0^1 \tilde{q}^{(P)} x \, \mathrm{d}x, \quad Q_T^* := -4 \int_0^1 \tilde{q}^{(T)} x \, \mathrm{d}x, \tag{12.81}$$

with $x = r/a$ being the radial dimensionless coordinate. The coefficients G_P^*, G_T^*, Q_P^*, and Q_T^* have the same physical meaning as those introduced for the planar Poiseuille flow by (11.81) and (11.82).

12.3.2
Slip Solution

A combination of the momentum conservation law (3.27) and Newton's law in the cylindrical coordinates reads

$$\frac{\mu}{r}\frac{\partial}{\partial r}\left(r\frac{\partial u_3}{\partial r}\right) = \frac{\mathrm{d}p}{\mathrm{d}r_3}. \tag{12.82}$$

The viscous (10.2) and thermal slip conditions (10.17) for the problem in question take the form

$$u_3 = -\sigma_\mathrm{P}\ell\frac{\mathrm{d}u_3}{\mathrm{d}r} + \sigma_\mathrm{T}\frac{\mu}{\rho T}\frac{\mathrm{d}T}{\mathrm{d}r_3}, \quad \text{at} \quad r = a. \tag{12.83}$$

Then solving Eq. (12.82) subject to this condition, we obtain

$$u_3 = v_0\left\{\left[\frac{\delta}{4}\left(r^2 - 1\right) - \frac{\sigma_\mathrm{P}}{2}\right]\xi_\mathrm{P} + \frac{\sigma_\mathrm{T}}{2\delta}\xi_\mathrm{T}\right\}. \tag{12.84}$$

Comparing it with Eq. (12.79), the dimensionless bulk velocity is derived as

$$\tilde{u}^{(\mathrm{P})}(x) = \frac{\delta}{4}(x^2 - 1) - \frac{\sigma_\mathrm{P}}{2}, \quad \tilde{u}^{(\mathrm{T})}(x) = \frac{\sigma_\mathrm{T}}{2\delta}. \tag{12.85}$$

Then the coefficients defined by (12.80) are given as

$$G_\mathrm{P}^* = \frac{\delta}{4} + \sigma_\mathrm{P}, \quad G_\mathrm{T}^* = \frac{\sigma_\mathrm{T}}{\delta}. \tag{12.86}$$

To obtain the heat flow $\tilde{q}^{(\mathrm{T})}$, the derivations (11.88) and (11.89) done for the planar flow should be repeated, which leads to the same expressions of the heat flow $\tilde{q}$ and coefficient Q_T^*, that is, Eqs. (11.90) and (11.91), respectively.

The derivation of the coefficient Q_P^* is omitted here because of its complexity, but as will be shown later, its expression is the same as that for G_T^* due to the reciprocal relation (see Eq. (12.100)).

12.3.3
Kinetic Equation

Under the adopted assumptions and reference perturbations (11.78), the unknown perturbation h function does not depend on the longitudinal coordinate x_3 and is axisymmetric so that the left-hand side of the linearized kinetic equation can be written in the form of Eq. (9.63). If we apply the S-model, the right-hand side of Eq. (7.47) is written with $A_1 = 1$, $A_2 = 0$, and $A_3 = 1$. The density ϱ and temperature τ deviations disappear because their local values are equal to the reference values. Finally, the S-model for the cylindrical Poiseuille flow reads

$$c_p\cos\theta\frac{\partial h}{\partial x} - \frac{c_p\sin\theta}{x}\frac{\partial h}{\partial\theta} = \delta(H - h) + \tilde{g}, \tag{12.87}$$

where the function H and source terms are exactly the same as those for the planar flow, that is, (11.93) and (11.94), respectively. The dimensionless velocity $\tilde{u}$ and heat flux $\tilde{q}$ are related to the perturbation function as

$$\begin{bmatrix} \tilde{u}(x) \\ \tilde{q}(x) \end{bmatrix} = \frac{1}{\pi^{3/2}} \int_{-\infty}^{\infty} \int_0^{2\pi} \int_0^{\infty} \begin{bmatrix} c_3 \\ \left(c^2 - \frac{5}{2}\right) c_3 \end{bmatrix} e^{-c^2} \times h\left(x, \theta, c_p, c_3\right) c_p \, \mathrm{d}c_p \, \mathrm{d}\theta \, \mathrm{d}c_3 \tag{12.88}$$

that follow from (5.17) replacing the Cartesian coordinate $[c_1, c_2, c_3]$ by the cylindrical ones $[c_p, \theta, c_3]$ in accordance with (9.18).

Let us assume the diffuse scattering on the tube surface given by (5.70). The second and third terms of this expression are zero because the wall is at rest $\tilde{\boldsymbol{u}}_w = 0$. The last term is also zero because the wall temperature is equal to the reference temperature $\tau_w = 0$. The first term of (5.70) is expressed via the incident perturbation function by (5.71), which is an odd function of c_3. This fact follows from Eq. (12.87) where all terms, namely, H and $\tilde{g}$, contain c_3 except those containing h, hence, the integral in (5.71) is zero $\varrho_r = 0$. Thus, we conclude that the perturbation of the reflected particles on the tube wall is zero,

$$h = 0 \quad \text{at} \quad x = 1, \quad \text{for} \quad \frac{\pi}{2} \leq \theta \leq \pi. \tag{12.89}$$

To eliminate the variable c_3, Eq. (12.87) is split introducing new perturbations as

$$\Phi^{(1)}(x, \theta, c_p) = \frac{1}{\pi^{1/2}} \int_{-\infty}^{\infty} e^{-c_3^2} h(x, \theta, c_p, c_3) c_3 \, \mathrm{d}c_3, \tag{12.90}$$

$$\Phi^{(2)}(x, \theta, c_p) = \frac{1}{\pi^{1/2}} \int_{-\infty}^{\infty} e^{-c_3^2} h(x, \theta, c_p, c_3) \left(c_3^2 - \frac{3}{2}\right) c_3 \, \mathrm{d}c_3. \tag{12.91}$$

Multiplying Eq. (12.87) by $c_3/\pi^{1/2}$ and integrating it with respect to c_3, the equation for $\Phi^{(1)}$ is obtained. Doing the same with the factor $c_3(c_3^2 - 3/2)/\pi^{1/2}$, the equation for $\Phi^{(2)}$ is derived. Both equations have the same structure and read

$$c_p \cos\theta \frac{\partial \Phi^{(\alpha)}}{\partial x} - \frac{c_p \sin\theta}{x} \frac{\partial \Phi^{(\alpha)}}{\partial \theta} = \delta(H^{(\alpha)} - \Phi^{(\alpha)}) + \tilde{g}^{(\alpha)}, \quad \alpha = 1, 2, \tag{12.92}$$

where the functions $H^{(\alpha)}$ are given as

$$H^{(1)} = \tilde{u} + \frac{2}{15}(c_p^2 - 1)\tilde{q}, \quad H^{(2)} = \frac{1}{5}\tilde{q}, \tag{12.93}$$

which are obtained from H defined by (11.93). The source terms $\tilde{g}$ are obtained by substituting (11.78) into (7.49) and now they have two superscripts

$$\tilde{g}^{(1,\mathrm{P})} = -\frac{1}{2}, \quad \tilde{g}^{(1,\mathrm{T})} = -\frac{1}{2}(c_p^2 - 1), \quad \tilde{g}^{(2,\mathrm{P})} = 0, \quad \tilde{g}^{(2,\mathrm{T})} = -\frac{3}{4}. \tag{12.94}$$

The dimensionless moments (12.88) are expressed via the new perturbations as

$$\begin{bmatrix} \tilde{u}(x) \\ \tilde{q}(x) \end{bmatrix} = \frac{2}{\pi} \int_0^{\pi} \int_0^{\infty} \left\{ \begin{bmatrix} 1 \\ c_p^2 - 1 \end{bmatrix} \Phi^{(1)}(x, \theta, c_p) + \begin{bmatrix} 0 \\ 1 \end{bmatrix} \Phi^{(2)}(x, \theta, c_p) \right\} e^{-c_p^2} c_p \, \mathrm{d}c_p \, \mathrm{d}\theta. \tag{12.95}$$

Note that the integral has been multiplied by the factor 2, and the integration with respect of the angle θ is done over the interval $[0, \pi]$ because of the symmetry.

Both perturbations satisfy the boundary condition (12.89)

$$\Phi^{(1)} = 0 \quad \text{and} \quad \Phi^{(2)} = 0 \quad \text{at} \quad x = 1, \quad \text{for} \quad \frac{\pi}{2} \le \theta \le \pi. \tag{12.96}$$

12.3.4
Reciprocal Relation

A consideration of two driving forces allows us to obtain the reciprocal relation. The time-reversed kinetic coefficients (5.85) are obtained by using the source terms (11.94)

$$\begin{aligned}\Lambda^t_{\mathrm{P}m} &= \left(\left(\hat{T}\tilde{g}^{(\mathrm{P})}, h^{(m)}\right)\right) = 2\pi \int_0^1 (c_3, h^{(m)}) x \, \mathrm{d}x \\ &= 2\pi \int_0^1 \tilde{u}^{(m)} x \, \mathrm{d}x,\end{aligned} \tag{12.97}$$

$$\begin{aligned}\Lambda^t_{\mathrm{T}m} &= \left(\left(\hat{T}\tilde{g}^{(\mathrm{T})}, h^{(m)}\right)\right) = 2\pi \int_0^1 \left(c_3 \left(c^2 - \frac{5}{2}\right), h^{(m)}\right) x \, \mathrm{d}x \\ &= 2\pi \int_0^1 \tilde{q}^{(m)} x \, \mathrm{d}x,\end{aligned} \tag{12.98}$$

where $m = \mathrm{P}$ or $m = \mathrm{T}$. These coefficients are expressed via the dimensionless flow rates defined by (11.81) and (11.82)

$$\Lambda^t_{\mathrm{PP}} = -\frac{\pi}{2} G^*_{\mathrm{P}}, \quad \Lambda^t_{\mathrm{PT}} = \frac{\pi}{2} G^*_{\mathrm{T}}, \quad \Lambda^t_{\mathrm{TP}} = \frac{\pi}{2} Q^*_{\mathrm{P}}, \quad \Lambda^t_{\mathrm{TT}} = -\frac{\pi}{2} Q^*_{\mathrm{T}}. \tag{12.99}$$

The equality (5.83) between the coefficient Λ^t_{PT} and Λ^t_{TP} leads to the coupling between the thermal creep and mechanocaloric heat flow

$$G^*_{\mathrm{T}} = Q^*_{\mathrm{P}}, \tag{12.100}$$

which is the same as in case of the planar flow (11.110).

12.3.5
Free-Molecular Regime

To obtain the perturbation function h in the free-molecular regime, the left-hand side of the kinetic equation (12.87) is written in the form (9.53), while the right-hand side contains only the source term

$$-c_p \frac{\partial h}{\partial s} = \tilde{g}. \tag{12.101}$$

Integrating it with respect to the variable s from 0 to boundary, the perturbation is obtained as

$$h = \frac{s_0}{c_p} \tilde{g}, \quad s_0 = x \cos\theta + \sqrt{1 - (x \sin\theta)^2}, \tag{12.102}$$

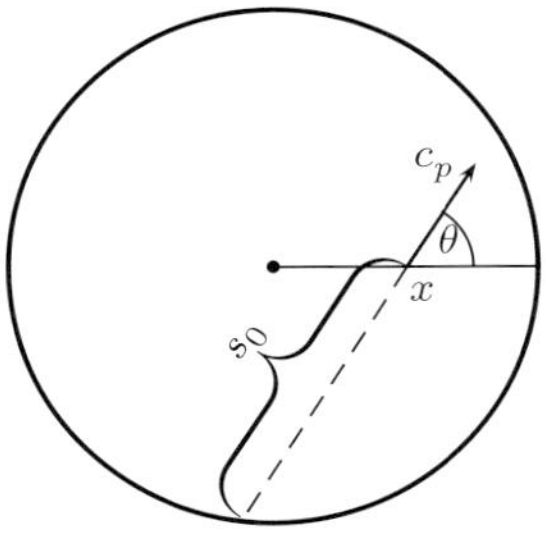

Figure 12.11 Scheme to obtain (12.102).

where s_0 is the distance from the position x to the tube wall in the direction $-\boldsymbol{c}_p$ as is shown in Figure 12.11. Substituting this into (5.13) and (5.16) and taking into account the source term expressions (12.94), the dimensionless bulk velocity and heat flow are obtained as

$$\tilde{u}^{(\mathrm{P})}(x) = -\frac{1}{\sqrt{\pi}} \int_0^{\pi/2} [1 - (x \sin\theta)^2]^{1/2}\, \mathrm{d}\theta, \tag{12.103}$$

$$\tilde{u}^{(\mathrm{T})}(x) = -\frac{1}{2}\tilde{u}^{(\mathrm{P})}(x), \quad \tilde{q}^{(\mathrm{P})}(x) = \tilde{u}^{(\mathrm{T})}(x), \quad \tilde{q}^{(\mathrm{T})}(x) = \frac{9}{4}\tilde{u}^{(\mathrm{P})}(x). \tag{12.104}$$

A substitution of these expressions into the definitions (12.80) and (12.81) leads to the results

$$G_{\mathrm{P}}^{*} = \frac{8}{3\sqrt{\pi}}, \quad G_{\mathrm{T}}^{*} = Q_{\mathrm{P}}^{*} = \frac{4}{3\sqrt{\pi}}, \quad Q_{\mathrm{T}}^{*} = \frac{6}{\sqrt{\pi}} \tag{12.105}$$

that obeys the reciprocal relation (12.100).

12.3.6
Numerical Scheme

Since Eq. (12.92) is the same for both $\Phi^{(\alpha)}$ ($\alpha = 1, 2$), the finite difference scheme will be given without the superscript α. Let us consider some discrete values c_i of the speed c_p and their weights W_i^c. Regularly distributed nodes for the radial coordinate x varying in the range [0, 1] can be used in this problem,

$$x_k = \frac{k}{N_x}, \quad 0 \le k \le N_x, \tag{12.106}$$

where N_x is an even integer. Similar to the two previous problems, the angle nodes defined by (9.65) will be used here. The perturbation functions Φ, functions H, and source terms $\tilde{g}$ are discretized with respect to the three variables,

$$\Phi_{kji} = \Phi(x_k, \theta_j, c_i), \quad H_{ki} = H(x_k, c_i), \quad \tilde{g}_i = \tilde{g}(c_i), \tag{12.107}$$

where H_{ki} is expressed by (12.93) via the discretized moments

$$\tilde{u}_k = \tilde{u}(x_k), \quad \tilde{q}_k = \tilde{q}(x_k). \tag{12.108}$$

The finite difference scheme is the same for all values of c_i so that the subscript i will be omitted in the following Eqs. (12.109)–(12.111).

The nodes θ_j are swept from N_θ to 0. The first $j = 0$ and last nodes $j = N_\theta$ are treated using the scheme (9.70),

$$\Phi_{kj} = \frac{(C^x - \delta)\Phi_{k-l,j} + \delta(H_k + H_{k-l}) + 2\tilde{g}}{C^x + \delta}, \quad j = 0, N_\theta. \tag{12.109}$$

This scheme differs from (12.66) only by the source term $\tilde{g}$ so that the coefficient C^x is the same and is given by (12.67). When the sweeping for $j = N_\theta$ is done, the value Φ_{kj} at $k = 0$ is stored and it is further used as the boundary condition for the angle node θ_j in the range $0 \le j < N_\theta/2$. Moreover, this value is used to calculate the moments in the central point

$$\begin{bmatrix} \tilde{u}_0 \\ \tilde{q}_0 \end{bmatrix} = 2\pi \sum_{i=1}^{N_c} \left\{ \begin{bmatrix} 1 \\ c_i^2 - 1 \end{bmatrix} \Phi^{(1)}_{0,N_\theta,i} + \begin{bmatrix} 0 \\ 1 \end{bmatrix} \Phi^{(2)}_{0,N_\theta,i} \right\} W_i^c. \tag{12.110}$$

The rest values Φ_{kj} $(1 \le j \le N_\theta - 1)$ are calculated using the scheme (9.75)

$$\begin{aligned} \Phi_{kj} = & \left[\left(C_j^x - \frac{C_j^\theta}{k - l/2} - \delta \right) \Phi_{k-l,j} + \left(-C_j^x + \frac{C_j^\theta}{k - l/2} - \delta \right) \Phi_{k,j+1} \right. \\ & \left. + \left(C_j^x + \frac{C_j^\theta}{k - l/2} - \delta \right) \Phi_{k-l,j+1} + 2\delta(H_k + H_{k-l}) + 4\tilde{g} \right] \\ & \times \left(C_j^x + \frac{C_j^\theta}{k - l/2} + \delta \right)^{-1}, \end{aligned} \tag{12.111}$$

where the coefficients C_j^x and C_j^θ are the same as in the scheme (12.24) and are defined by (12.25). The index l defined by (9.73) determines the sweeping direction as follows:

$$\begin{array}{llllll} \text{from} & N_x - 1 & \text{to} & 1 & \text{for} & l = 1, \\ \text{from} & 1 & \text{to} & N_x & \text{for} & l = -1. \end{array} \tag{12.112}$$

Note that since the contribution of the node $k = 0$ has been calculated by Eqs. (12.109) and (12.110), it is not treated by the scheme (12.111).

To calculate the moment $\tilde{u}$ and $\tilde{q}$, we should retrieve the subscript i to Φ as has been defined by (12.107). The superscript α also should be restored according to Eqs. (12.90) and (12.91). Then the moment definitions (12.95) are written in the form of quadrature

$$\begin{bmatrix} \tilde{u}_k \\ \tilde{q}_k \end{bmatrix} = \sum_{j=0}^{N_\theta} \sum_{i=1}^{N_c} \left\{ \begin{bmatrix} 1 \\ c_i^2 - 1 \end{bmatrix} \Phi^{(1)}_{kji} + \begin{bmatrix} 0 \\ 1 \end{bmatrix} \Phi^{(2)}_{kji} \right\} W_i^c W_j^\theta, \tag{12.113}$$

where the weight W_j^θ is chosen according to Simpson's rule (12.70),

$$W_j^\theta = \frac{2\pi}{N_\theta} W_j^S \tag{12.114}$$

with W_j^S given by (9.14). The behavior of the perturbation as a function of the molecular speed c_p is similar to that for the planar Poiseuille flow, that is, it changes sharply near the point $c_p = 0$, so that the density of the nodes near zero should be larger than that far from zero. Again, it can be considered two velocity intervals (11.130) with the same number of nodes regularly distributed in each of them, that is, c_i are given by (11.131). The first interval could vary in the range $0 \leq c_p \leq 0.1$, while the second interval could be stretched from 0.1. to 4. Simpson's rule is applied in each interval with the weights including the factor $1/\pi$ (see the integral in (12.95)). Moreover, the value c_i appearing as c_p in (12.95) before dc_p should be taken into account. Then the weights W_i^c in (12.113) take the form

$$W_i = \frac{2}{N_c}\frac{c_i e^{-c_i^2}}{\pi} W_i^S \begin{cases} c_{01} & \text{for} \quad 1 \leq i < \frac{N_c}{2}, \\ c_0/2 & \text{for} \quad i = N_c/2, \\ c_0 - c_{01} & \text{for} \quad \frac{N_c}{2} < i \leq N_c. \end{cases} \tag{12.115}$$

For $\delta \geq 1$, the behavior of the perturbation is smooth, and the Gauss nodes and weights corresponding to the weighting function (9.22) given in Section B.1.2 can be used.

When the scheme converges, the flow rates are calculated using, for instance, Simpson's rule as

$$G_\mathrm{P}^* = -\sum_{k=1}^{N_x} \tilde{u}_k^{(\mathrm{P})} x_k W_k, \quad G_\mathrm{T}^* = \sum_{k=1}^{N_x} \tilde{u}_k^{(\mathrm{T})} x_k W_k, \tag{12.116}$$

$$Q_\mathrm{P}^* = \sum_{k=1}^{N_x} \tilde{q}_k^{(\mathrm{P})} x_k W_k, \quad Q_\mathrm{T}^* = -\sum_{k=1}^{N_x} \tilde{q}_k^{(\mathrm{T})} x_k W_k, \tag{12.117}$$

where

$$W_k = \frac{2}{N_x} W_k^S, \tag{12.118}$$

with W_k^S given by (9.14).

A listing of the numerical program "poiseuille_creep_axisym.for" based on the scheme described earlier and written in FORTRAN language is presented in Section B.7.

12.3.7 Results

The scheme parameters and results obtained by the numerical code "poiseuille_creep_axisym.for" with the numerical error within 0.1% are given in Table 12.3. The same values are reported in papers published by many researchers (see for example, Refs [34, 142, 143]). Note that the reciprocal relation (11.110) is fulfilled within the numerical accuracy. The scheme parameters vary significantly by the variation of the rarefaction parameter. As in the planar Poiseuille flow, the number of iterations drastically increases in the hydrodynamic regime ($\delta \to \infty$).

The numerical data on the Poiseuille coefficient G_P^* are most used in practice so that it would be convenient to use some interpolating formula based on the

Table 12.3 Numerical scheme parameters and coefficients G_P^*, G_T^*, Q_P^*, and Q_T^* for gas flow through a circular tube versus δ based on the code given in Section B.7.

δ	N_x	N_c	N_θ	Iters.	G_P^*	$G_T^* = Q_P^*$	Q_T^*
0[a)]					1.505	0.7523	3.385
0.001	20	40	20	5	1.499	0.74601	3.368
0.01	20	40	20	7	1.477	0.7207	3.285
0.1	20	60	20	13	1.409	0.6203	2.880
1	40	6	20	47	1.476	0.3967	1.675
2	40	4	20	92	1.677	0.3026	1.180
5	40	4	20	294	2.364	0.1764	0.6188
10	50	4	40	866	3.576	0.1021	0.3413
20	50	4	40	2858	6.048	0.0549	0.1791
20[b)]					6.018	0.0588	0.1875

a) Free-molecular solution, Eq. (12.105).
b) Slip solution, Eqs. (11.91), (12.86).

numerical results given in Table 12.3. Among several formulas proposed in the literature, the following one obtained in Ref. [136] combines the simplicity and accuracy

$$G_P^* = \frac{8}{3\sqrt{\pi}}\frac{1+0.04\,\delta^{0.7}\ln\delta}{1-0.78\,\delta^{0.8}} + \left(\frac{\delta}{4}+1.018\right)\frac{\delta}{1+\delta}. \tag{12.119}$$

It provides the limit expressions (12.86) and (12.105), and reproduces the numerical results based on the S-model, see Table 12.3, with a discrepancy of less than 0.2%.

The thermal creep coefficient G_T^* is also frequently used in practical calculations. Considering the limit solutions (12.86) and (12.105), the following formula can be proposed

$$G_T^* = \begin{cases} \frac{4}{3\sqrt{\pi}} + 0.825(1+\ln\delta)\delta - (1.18-0.61\ln\delta)\delta^2 & \text{for} \quad \delta \le 1, \\ \frac{1.175}{\delta} - \frac{1.75}{\delta^2} + \frac{1.47}{\delta^3} - \frac{0.5}{\delta^4} & \text{for} \quad \delta > 1, \end{cases} \tag{12.120}$$

which interpolates the numerical data of Table 12.3 and those reported in Refs [34, 142, 143] within the error of 0.6%.

It is interesting to compare the results based on the S-model with those obtained from the BE, which was solved by Loyalka and Hamoodi [144] for the HS potential. The coefficient G_P^* based on this work are plotted in Figure 12.12. We see that the difference of the flow rates obtained from the S-model from those obtained from the BE with HS potential does not exceed 2% for G_P^*.

The influence of the AC α_t and α_n on the flow rates was studied in the papers [34]. The coefficients G_P^* and G_T^* based on the S-model and for various values of the AC are plotted in Figure 12.13. Similar to the planar flow, these data show that the Poiseuille coefficient G_P^* significantly depends on the AC α_t in the whole

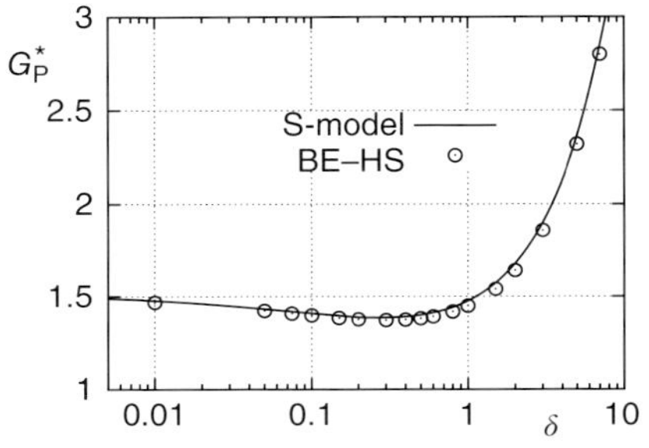

Figure 12.12 Poiseuille coefficient G_P^* for gas flow through tube versus rarefaction δ. Comparison of data based on S-model with those obtained in Ref. [144] applying the BE with HS potential.

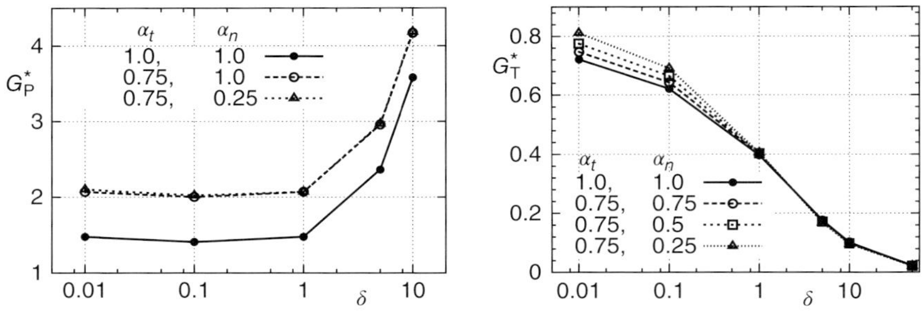

Figure 12.13 Coefficients G_P^* and G_T^* for gas flow through tube versus rarefaction δ. Comparison of data based on diffuse scattering with those obtained in Ref. [34] applying the CL scattering kernel.

range of the rarefaction parameter δ. It always decreases by increasing α_t. The dependence of the Poiseuille coefficient G_P^* on the AC α_n is very weak. The thermal creep coefficient G_T^* depends on both ACs α_t and α_n. The dependence practically vanishes near the hydrodynamic regime.

The velocity profiles $\tilde{u}^{(P)}$ and $\tilde{u}^{(T)}$ are depicted in Figure 12.14. Note that the profile due to the pressure gradient $\tilde{u}^{(P)}$ is negative. For the large value of the rarefaction parameter ($\delta = 10$), the numerical velocity profile of the Poiseuille flow $\tilde{u}^{(P)}$ is very close to that obtained analytically using the slip condition (12.85), which is plotted by the dashed line. The profile becomes flatter by decreasing the rarefaction parameter δ. The velocity profiles of thermal creep $\tilde{u}^{(T)}$ are more or less flat and their magnitudes decrease by increasing the rarefaction parameter δ. For $\delta = 10$, the velocity magnitude obtained numerically is close to that obtained from the slip solutions (12.85) only in the middle point, while near the tube wall, the numerical profiles differ from the corresponding slip solution.

The heat flow profiles $\tilde{q}^{(P)}$ and $\tilde{q}^{(T)}$ are presented in Figure 12.15. As in case of the planar flow, the heat flow of the mechanocaloric effect $\tilde{q}^{(P)}$ is positive practically over the whole range of the coordinate x and negative near the wall when the rarefaction parameter is large ($\delta = 10$). It is evident that the profiles $\tilde{u}^{(P)}$ and $\tilde{q}^{(P)}$ are different, but their integrals are the same according to the reciprocal relation (12.100). The heat flow $\tilde{q}^{(P)}$ increases by decreasing the rarefaction parameter and

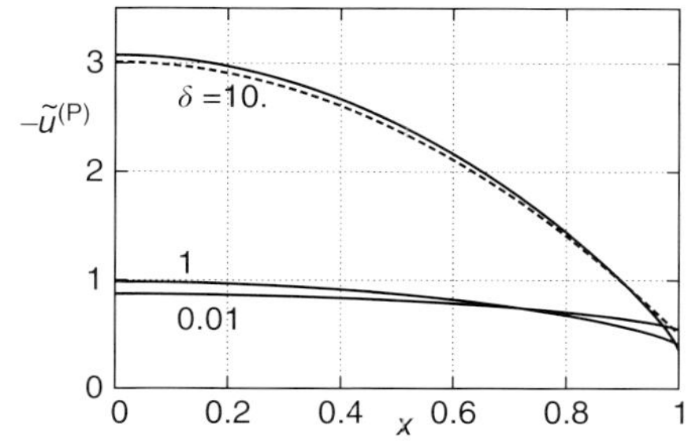

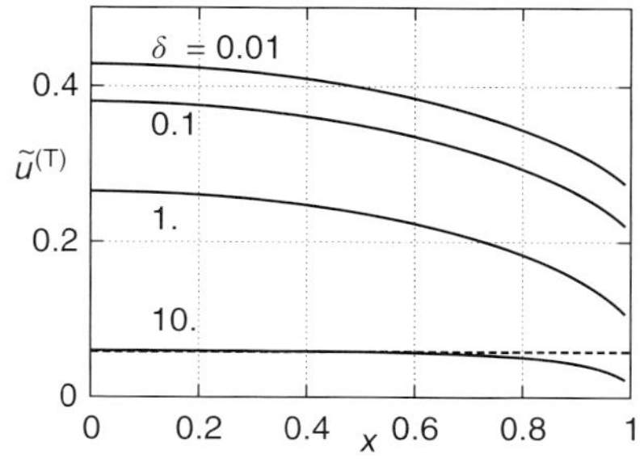

Figure 12.14 Dimensionless velocities $\tilde{u}^{(P)}$ and $\tilde{u}^{(T)}$ for gas flow through tube versus coordinate x: solid lines – kinetic equation, dashed line – slip solution (12.85).

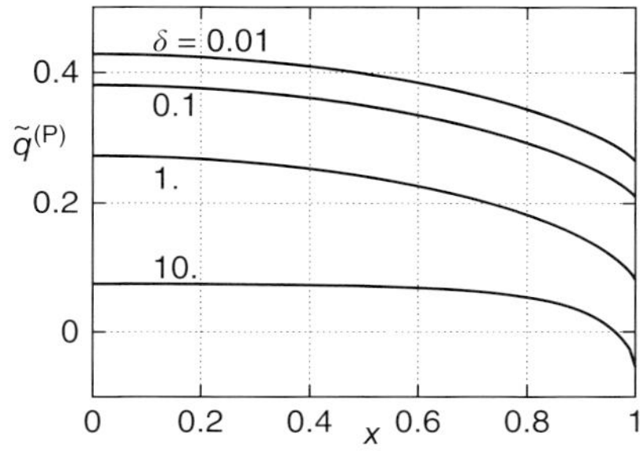

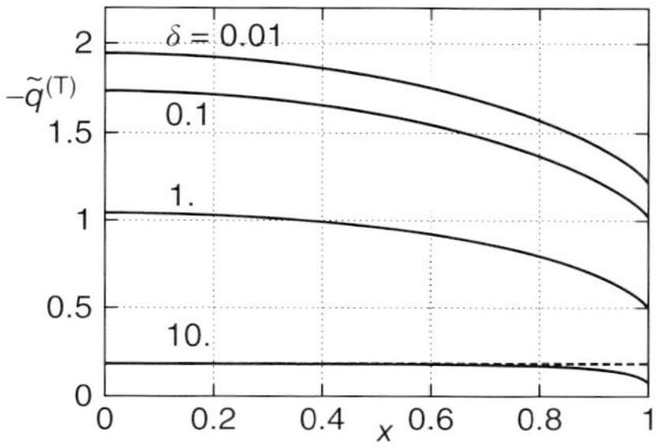

Figure 12.15 Dimensionless heat flows $\tilde{q}^{(P)}$ and $\tilde{q}^{(T)}$ for gas flow through tube versus coordinate x: solid lines – kinetic equation, dashed line – hydrodynamic solution (11.90).

becomes equal to that of the thermal creep $\tilde{u}^{(P)}$. The heat flow due to the temperature gradient $\tilde{q}^{(T)}$ is always negative. For the large value of $\delta = 10$, it is equal to the analytical expression (11.90) plotted by the dashed line practically over the whole range of x except the small region near the wall.

Flows of polyatomic gases through circular tubes were studied by Loyalka *et al.* [145] and by Titarev and Shakhov [127]. According to these data, the Poiseuille coefficient is weakly sensitive to the internal structure of molecules, while the thermal creep coefficient for polyatomic gas can be about 30% smaller than that for monatomic gas. Some results of flows of gaseous mixtures through tubes are reported in the papers [146, 147].

Exercises

12.1 Consider helium confined between two long coaxial cylinders. Calculate its velocity u_ϕ and shear stress $P_{r\phi}$ at the internal cylinders under the following conditions: radius $a = 0.5$ cm, radii ratio $\beta = 2$, angular speed $\omega = 10^3$ Hz, gas temperature $T = 300$ K. Assume the diffuse gas–surface interaction and consider two values of the gas pressure: (a) $p = 50$ Pa and (b) $p = 0.1$ Pa.

Solution: (a) Since $\delta = 11$, Eqs. (12.8) and (12.11) are used. Then $u_\phi = 4.1$ m/s, $P_{r\phi} = 4.2 \times 10^{-2}$ Pa; (b) Since $\delta = 2.2 \times 10^{-2}$, Eqs. (12.20) and (12.5) are used. Then $u_\phi = 2.5$ m/s, $P_{r\phi} = 2.5 \times 10^{-4}$ Pa.

12.2 Consider the same gas, cylinders, and pressures as in the previous exercise. However, the external cylinder is kept at the temperature 300 K, while the internal cylinder is hotter for $\Delta T = 10$ K. Calculate the temperature of the gas and heat flux at the internal cylinders.
Solution: (a) Equations (12.44) and (12.46) are used. Then $T = 308$ K, $q_r = 326$ W/m^2; (b) Equations (12.62) and (12.41) are used. Then $T = 305$ K, $q_r = 2.1$ W/m^2.

12.3 Consider helium flowing through a circular tube of radius $a = 0.5$ mm and length l =10 cm. Calculate the mass flow rate $\dot{M}$ under the following conditions: the temperature is constant and equal to 300 K, the average pressure is p =45 Pa, and the pressure drop $\Delta p = 1$ Pa.
Solution: Since $\delta = 1$, Eq. (12.79) is used with $G_\mathrm{P}^* = 1.476$ (Table 12.3) and $\xi_\mathrm{P} = a\Delta p/lp$. Then $\dot{M} = 5.2 \times 10^{-9}$ g/s.

12.4 Consider the same gas, cylinders, and average pressures as in the previous exercise. Calculate the mass flow rate driven by a temperature gradient 100 K/m instead of the pressure drop.
Solution: Since $\delta = 1$, Eq. (12.79) is used with $G_\mathrm{T}^* = 0.3967$ (Table 12.3) and $\xi_\mathrm{T} = 1.67 \times 10^{-4}$. Then $\dot{M} = 2.1 \times 10^{-9}$ g/s.

13
Two-Dimensional Planar Flows

13.1
Flows Through a Long Rectangular Channel

In this section, the flow between two parallel plates considered in Section 11.3 is generalized to that through a long channel with a rectangular cross section. The main difference of the geometrical configuration considered here is that the channel is not infinite in the r_2 direction, but it is restricted.

13.1.1
Definitions

Consider a channel with a rectangular cross section as is shown in Figure 13.1. The channel size in the r_1 direction continues to be a, but the size in the r_2 direction is β times larger than a. As in the planar flow considered in Section 11.3, the distance a is adopted as the characteristic size so that the dimensionless coordinates are introduced by (7.43) and the rarefaction parameter is defined by Eq. (7.46).

Similar to the one-dimensional planar flow, here we also consider the gas flow caused by small longitudinal gradients of pressure ξ_{P} and temperature ξ_{T} defined by (11.72). Since the channel is long in the r_3-direction, the pressure and temperature are constant over a cross section and depend linearly only on the longitudinal coordinate according to Eq. (11.73). Under such conditions, the bulk velocity and heat flow have only longitudinal component depending on the two coordinates as

$$\boldsymbol{u} = [0, 0, u_3], \quad u_3 = u_3(r_1, r_2), \quad \boldsymbol{q} = [0, 0, q_3], \quad q_3 = q_3(r_1, r_2). \tag{13.1}$$

In practice, we are interested in the mass and heat flow rates, defined as

$$\dot{M} := m\, n_0 \int_{-\beta a/2}^{\beta a/2} \int_{-a/2}^{a/2} u_3(r_1, r_2)\, \mathrm{d}r_1\, \mathrm{d}r_2, \tag{13.2}$$

$$\dot{E} := \int_{-\beta a/2}^{\beta a/2} \int_{-a/2}^{a/2} q_3(r_1, r_2)\, \mathrm{d}r_1\, \mathrm{d}r_2. \tag{13.3}$$

Following the same reasonings as for the planar flow, see Section 11.3, the linearization is done near the local Maxwellian using the pressure ξ_{P} and temperature

Rarefied Gas Dynamics: Fundamentals for Research and Practice, First Edition. Felix Sharipov.

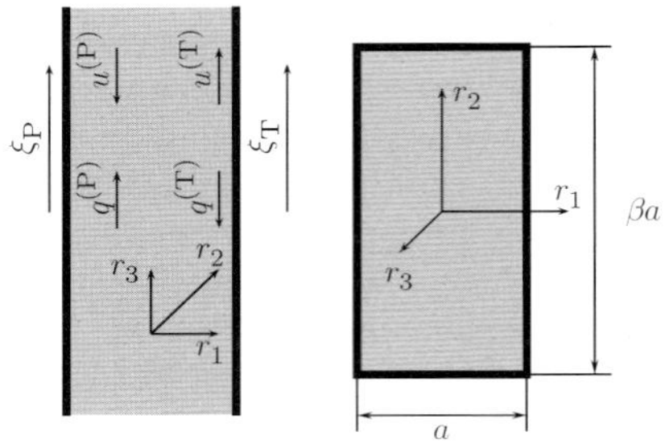

(a) Longitudinal section (b) Cross section

Figure 13.1 Scheme for gas flow through a channel of rectangular cross section.

ξ_T gradients as the small parameters. Then the corresponding reference perturbations $h_R^{(P)}$ and $h_R^{(T)}$ are considered. They are exactly the same as those for the planar flow (11.78). The bulk velocity u_3 and heat flow q_3 are decomposed according to (11.79). The mass and heat flow rates are decomposed similar to (12.85),

$$\dot{M} = \frac{\beta a^2\, p_0}{v_0}\left(-G_P^*\, \xi_P + G_T^*\, \xi_T\right), \quad \dot{E} = \frac{\beta a^2 p_0 v_0}{2}\left(-Q_P^*\, \xi_P + Q_T^*\, \xi_T\right), \tag{13.4}$$

where the dimensionless coefficients G_P^*, G_T^*, Q_P^*, and Q_T^* have the same physical meaning as those for the planar and cylindrical flows. For the flow under consideration, they are defined via the dimensionless velocity and heat flow as

$$G_P^* := -\frac{2}{\beta}\int_{-\beta/2}^{\beta/2}\int_{-1/2}^{1/2} \tilde{u}^{(P)}(x_1, x_2)\, dx_1\, dx_2, \tag{13.5}$$

$$G_T^* := \frac{2}{\beta}\int_{-\beta/2}^{\beta/2}\int_{-1/2}^{1/2} \tilde{u}^{(T)}(x_1, x_2)\, dx_1\, dx_2, \tag{13.6}$$

$$Q_P^* := \frac{2}{\beta}\int_{-\beta/2}^{\beta/2}\int_{-1/2}^{1/2} \tilde{q}^{(P)}(x_1, x_2)\, dx_1\, dx_2, \tag{13.7}$$

$$Q_T^* := -\frac{2}{\beta}\int_{-\beta/2}^{\beta/2}\int_{-1/2}^{1/2} \tilde{q}^{(T)}(x_1, x_2)\, dx_1\, dx_2. \tag{13.8}$$

13.1.2
Slip Solution

The velocity profile $u_3(r_1, r_2)$ and coefficients G_P^* and G_T^* in the viscous regime ($\delta \to \infty$) are calculated on the basis of the Navier–Stokes equation. A combination of the conservation law (3.27) and Newton's law (6.2) for two-dimensional flow leads to the following relation:

$$\mu\left(\frac{\partial^2 u_3}{\partial r_1^2} + \frac{\partial^2 u_3}{\partial r_2^2}\right) = \frac{dp}{dr_3}. \tag{13.9}$$

Table 13.1 Coefficients $G_\mathrm{P}^*(\delta = 0)$, $\mathcal{H}$, and $\mathcal{S}$ versus aspect ratio β.

	$\beta = 1$	2	5	10	50	100	∞
$G_\mathrm{P}^*(\delta = 0)$	0.839	1.152	1.618	1.991	2.884	3.273	∞
$\mathcal{H}$	0.422	0.686	0.874	0.937	0.989	0.994	1.0
$\mathcal{S}$	0.562	0.749	0.899	0.949	0.990	0.994	1.0

The slip boundary conditions (10.2) and (10.17) are posed on all four surfaces of the channel, that is, additionally to (11.84), the following condition is applied

$$u_3 = \pm\sigma_\mathrm{P}\,\ell\,\frac{\partial u_3}{\partial r_2} + \sigma_\mathrm{T}\frac{\mu}{\rho T}\frac{\mathrm{d}T}{\mathrm{d}r_3} \quad \text{at} \quad r_2 = \pm\frac{\beta a}{2}. \tag{13.10}$$

The analytical solution of (13.9) with the boundary condition (13.10) is very cumbersome so that only the flow rates are provided here, while the flow field can be found in Refs [148–150]. The Poiseuille coefficient obtained from (13.9) reads

$$G_\mathrm{P}^* = \frac{\delta}{6}\mathcal{H} + \sigma_\mathrm{P}\mathcal{S}, \tag{13.11}$$

where the first term corresponds to the solution of (13.9) with the no-slip boundary condition and reads

$$\mathcal{H} = 1 - \frac{6}{\beta}\sum_{n=0}^{\infty}\frac{\tanh(\beta\lambda_n)}{\lambda_n^5}, \qquad \lambda_n = \pi(n + 1/2). \tag{13.12}$$

The hyperbolic tangent $\tanh(x)$ is defined by (A.28). The second term $\mathcal{S}$ representing the slip correction calculated numerically in Ref. [148] and analytically in Refs [149, 150] has the following expression:

$$\mathcal{S} = \frac{4}{3}\mathcal{H} - 2\left(1 - \frac{1}{\beta}\right)\sum_{n=0}^{\infty}\frac{\tanh(\beta\lambda_n)}{\lambda_n^4}. \tag{13.13}$$

The quantities $\mathcal{H}$ and $\mathcal{S}$ for some values of β are given in Table 13.1. If the channel is wide, $\beta \gg 1$, then the expression (13.11) is reduced to that obtained for the planar flow (11.87).

The thermal creep coefficient G_T^* does not depend on the value of β, and it is the same (11.87) as that obtained in Section 11.3.

13.1.3
Kinetic Equation

The kinetic equation (7.47) with the S-model collision operator, that is, $A_1 = 1$, $A_2 = 0$, and $A_3 = 1$, reads

$$c_p \cos\theta\frac{\partial h}{\partial x_1} + c_p \sin\theta\frac{\partial h}{\partial x_2} = \delta(H - h) + \tilde{g}, \tag{13.14}$$

where the function H is defined by (11.93) and the source terms are given by (11.94). The moments are calculated by (12.88).

The boundary condition for the problem in question is similar to that for the planar (11.96) and axisymmetrical (12.89) flows, that is, the perturbation of particles reflected by the walls is zero.

To eliminate the variable c_3, new perturbation functions $\Phi^{(1)}$ and $\Phi^{(2)}$ are introduced

$$\Phi^{(1)}(x_1, x_2, \theta, c_p) = \frac{1}{\pi^{1/2}} \int_{-\infty}^{\infty} e^{-c_3^2} h(x_1, x_2, \theta, c_p, c_3)\, c_3\, \mathrm{d}c_3, \tag{13.15}$$

$$\Phi^{(2)}(x_1, x_2, \theta, c_p) = \frac{1}{\pi^{1/2}} \int_{-\infty}^{\infty} e^{-c_3^2} h(x_1, x_2, \theta, c_p, c_3)(c_3^2 - 1) c_3\, \mathrm{d}c_3, \tag{13.16}$$

which satisfy the following kinetic equation:

$$c_p \cos\theta \frac{\partial \Phi^{(\alpha)}}{\partial x_1} + c_p \sin\theta \frac{\partial \Phi^{(\alpha)}}{\partial x_2} = \delta(H^{(\alpha)} - \Phi^{(\alpha)}) + \tilde{g}^{(\alpha)}, \quad \alpha = 1, 2, \tag{13.17}$$

with $H^{(\alpha)}$ given by (12.93) and $\tilde{g}^{(\alpha)}$ given by (12.94). The structure of this equation and the boundary condition lead to the symmetry of the perturbations with respect to both coordinates x_1 and x_2:

$$\begin{aligned} &\Phi^{(\alpha)}(x_1, x_2, \theta, c_p) = \Phi^{(\alpha)}(-x_1, x_2, \pi - \theta, c_p) \\ &= \Phi^{(\alpha)}(x_1, -x_2, -\theta, c_p) = \Phi^{(\alpha)}(-x_1, -x_2, \pi + \theta, c_p). \end{aligned} \tag{13.18}$$

The moments $\tilde{u}$ and $\tilde{q}$ are expressed in terms of the perturbations as

$$\begin{aligned} \begin{bmatrix} \tilde{u}(x_1, x_2) \\ \tilde{q}(x_1, x_2) \end{bmatrix} = \frac{1}{\pi} \int_0^{2\pi} \int_0^{\infty} &\left\{ \begin{bmatrix} 1 \\ c_p^2 - 1 \end{bmatrix} \Phi^{(1)}(x_1, x_2, \theta, c_p) \right. \\ &\left. + \begin{bmatrix} 0 \\ 1 \end{bmatrix} \Phi^{(2)}(x_1, x_2, \theta, c_p) \right\} e^{-c_p^2} c_p\, \mathrm{d}c_p\, \mathrm{d}\theta. \end{aligned} \tag{13.19}$$

Because of the symmetry (13.18), only the quarter-moments can be calculated as

$$\begin{aligned} \begin{bmatrix} \tilde{u}^{(1/4)}(x_1, x_2) \\ \tilde{q}^{(1/4)}(x_1, x_2) \end{bmatrix} = \frac{1}{\pi} \int_0^{\pi/2} \int_0^{\infty} &\left\{ \begin{bmatrix} 1 \\ c_p^2 - 1 \end{bmatrix} \Phi^{(1)}(x_1, x_2, \theta, c_p) \right. \\ &\left. + \begin{bmatrix} 0 \\ 1 \end{bmatrix} \Phi^{(2)}(x_1, x_2, \theta, c_p) \right\} e^{-c_p^2} c_p\, \mathrm{d}c_p\, \mathrm{d}\theta. \end{aligned} \tag{13.20}$$

Then the full moments are obtained

$$\begin{aligned} \tilde{u}(x_1, x_2) = \tilde{u}^{(1/4)}(x_1, x_2) + \tilde{u}^{(1/4)}(-x_1, x_2) \\ + \tilde{u}^{(1/4)}(x_1, -x_2) + \tilde{u}^{(1/4)}(-x_1, -x_2), \end{aligned} \tag{13.21}$$

$$\begin{aligned} \tilde{q}(x_1, x_2) = \tilde{q}^{(1/4)}(x_1, x_2) + \tilde{q}^{(1/4)}(-x_1, x_2) \\ + \tilde{q}^{(1/4)}(x_1, -x_2) + \tilde{q}^{(1/4)}(-x_1, -x_2). \end{aligned} \tag{13.22}$$

The boundary conditions for the new perturbations are written down as

$$\Phi^{(\alpha)} = 0 \quad \text{at} \quad x_1 = \pm\frac{1}{2}, \quad \text{for} \quad c_1 \lessgtr 0, \tag{13.23}$$

and

$$\Phi^{(\alpha)} = 0 \quad \text{at} \quad x_2 = \pm\frac{\beta}{2}, \quad \text{for} \quad c_2 \lessgtr 0. \tag{13.24}$$

13.1.4
Free-Molecular Regime

In the free-molecular regime, the kinetic equation takes the form (12.101) and has the solution (12.102) with a different expression for s_0. The details of such a cumbersome solution are given in the paper [150] where the flow rate G_P^* has been obtained in the following form:

$$G_P^* = \frac{1}{\sqrt{\pi}}\left[\beta \ln\left(\frac{1}{\beta} + \sqrt{1+\frac{1}{\beta^2}}\right) + \ln(\beta + \sqrt{1+\beta^2}) - \frac{\beta}{3}\left(\frac{1}{\beta+\sqrt{1+\beta^2}} + \frac{1}{1+\sqrt{1+\beta^2}}\right)\right]. \tag{13.25}$$

The other flow rates are related to the above-mentioned flow rate as

$$G_T^* = Q_P^* = \frac{1}{2}G_P^*, \quad Q_T^* = \frac{9}{4}G_P^*. \tag{13.26}$$

13.1.5
Numerical Scheme

Let us introduce the regularly distributed nodes in the computational domain $-1/2 \leq x_1 \leq 1/2$ and $-\beta/2 \leq x_2 \leq \beta/2$ as

$$x_{k_1} = \frac{k_1}{2N_x}, \quad -N_x \leq k_1 \leq N_x, \quad x_{k_2} = \frac{k_2}{2N_x}, \quad -\beta N_x \leq k_2 \leq \beta N_x, \tag{13.27}$$

so that the value βN_x should be integer. The angle nodes θ_j are also distributed uniformly over the interval $[0, \pi/2]$,

$$\theta_j = \frac{\pi j}{2N_\theta}, \quad 0 \leq j \leq N_\theta. \tag{13.28}$$

The perturbations Φ, functions H, and source terms $\tilde{g}$ are discretized as

$$\Phi_{k_1,k_2,j,i} = \Phi(x_{k_1}, x_{k_2}, \theta_j, c_i), \quad H_{k_1,k_2,i} = H(x_{k_1}, x_{k_2}, c_i), \quad \tilde{g}_i = \tilde{g}(c_i). \tag{13.29}$$

Then the centered scheme (9.50) takes the form

$$\Phi_{k_1,k_2} = [(C_1 - C_2 - \delta)\Phi_{k_1-1,k_2} + (C_2 - C_1 - \delta)\Phi_{k_1,k_2-1} + (C_1 + C_2 - \delta)\Phi_{k_1-1,k_2-1} + \delta(H_{k_1,k_2} + H_{k_1-1,k_2} + H_{k_1,k_2-1} + H_{k_1-1,k_2-1}) + 4\tilde{g}](C_1 + C_2 + \delta)^{-1}, \tag{13.30}$$

where C_1 and C_2 are defined as

$$C_1 = 4N_x c_i \cos\theta_j, \quad C_2 = 4N_x c_i \sin\theta_j. \tag{13.31}$$

The subscripts i and j and superscripts 1 and 2 are omitted in the scheme (13.30) because it is the same for all these superscipts and subscripts. The nodes x_{k_1} and x_{k_2} are swept always in the positive direction using the boundary conditions (13.23) and (13.24) for the nodes $k_1 = 0$ and $k_2 = 0$.

To calculate the quarter-moments (13.20), the following quadrature is used

$$\begin{bmatrix} \tilde{u}^{(1/4)}_{k_1,k_2} \\ \tilde{q}^{(1/4)}_{k_1,k_2} \end{bmatrix} = \sum_{j=0}^{N_\theta} \sum_{i=1}^{N_c} \left\{ \begin{bmatrix} 1 \\ c_i^2 - 1 \end{bmatrix} \Phi^{(1)}_{k_1,k_2,ji} + \begin{bmatrix} 0 \\ 1 \end{bmatrix} \Phi^{(2)}_{k_1,k_2,ji} \right\} W_i^c W_j^\theta, \tag{13.32}$$

where the weights W_j^θ correspond to Simpson's rule

$$W_j^\theta = \frac{\pi}{2N_\theta} W_j^S \tag{13.33}$$

with W_j^S given by (9.14). The nodes c_i and weights W_i^c for the speed are the same as those used for the axisymmetrical flow, that is, the expressions (11.131) and (12.115) are applied for a small rarefaction $\delta < 1$ and the values given in Section B.1.2 are used for $\delta \geq 1$. Considering the symmetry (13.21) and (13.22), the full moment are calculated

$$\begin{aligned} \tilde{u}_{k_1,k_2} &= \tilde{u}^{(1/4)}_{k_1,k_2} + \tilde{u}^{(1/4)}_{-k_1,k_2} + \tilde{u}^{(1/4)}_{k_1,-k_2} + \tilde{u}^{(1/4)}_{-k_1,-k_2}, \\ \tilde{q}_{k_1,k_2} &= \tilde{q}^{(1/4)}_{k_1,k_2} + \tilde{q}^{(1/4)}_{-k_1,k_2} + \tilde{q}^{(1/4)}_{k_1,-k_2} + \tilde{q}^{(1/4)}_{-k_1,-k_2}. \end{aligned} \tag{13.34}$$

Once the velocity $\tilde{u}$ and heat flow $\tilde{q}$ are calculated, the flow rates are obtained according to the definitions (13.5)–(13.8)

$$G_\mathrm{P}^* = -\sum_{k_2=0}^{\beta N_x} \sum_{k_1=0}^{N_x} \tilde{u}^{(\mathrm{P})}_{k_1,k_2} W_{k_1} W_{k_2}, \quad G_\mathrm{T}^* = \sum_{k_2=0}^{\beta N_x} \sum_{k_1=0}^{N_x} \tilde{u}^{(\mathrm{T})}_{k_1,k_2} W_{k_1} W_{k_2}, \tag{13.35}$$

$$Q_\mathrm{P}^* = \sum_{k_2=0}^{\beta N_x} \sum_{k_1=0}^{N_x} \tilde{q}^{(\mathrm{P})}_{k_1,k_2} W_{k_1} W_{k_2}, \quad Q_\mathrm{T}^* = -\sum_{k_2=0}^{\beta N_x} \sum_{k_1=0}^{N_x} \tilde{q}^{(\mathrm{T})}_{k_1,k_2} W_{k_1} W_{k_2}, \tag{13.36}$$

where the weights are given by Simpson's rule

$$W_{k_1} = \frac{2}{\beta N_x} W_{k_1}^S, \quad W_{k_2} = \frac{1}{N_x} W_{k_2}^S, \tag{13.37}$$

with $W_{k_1}^S$ and $W_{k_2}^S$ defined by (9.14).

The numerical codes "poiseuille_creep_chan.for" written in FORTRAN and an example with output data are given in Section B.8.

13.1.6
Numerical Results

The scheme parameters and results obtained by the code "poiseuille_creep_chan.for" with the numerical error within 0.5% are given in Table 13.2. Note that the reciprocal relation (11.110) is fulfilled within the numerical accuracy.

Table 13.2 Numerical scheme parameters and coefficients G_P^*, G_T^*, Q_P^*, and Q_T^* at $\beta = 2$ for gas flow through rectangular channel versus rarefaction δ based on the program listing in Section B.8.

δ	N_x	N_c	N_θ	Iters.	G_P^*	$G_T^* = Q_P^*$	Q_T^*
0^a					1.152	0.5762	2.593
0.01	40	40	60	7	1.137	0.5567	2.536
0.1	40	40	40	12	1.081	0.4823	2.256
1	40	8	40	37	1.059	0.3139	1.396
2	40	4	40	67	1.136	0.2457	1.025
5	40	4	40	185	1.432	0.1527	0.5736
10	40	4	40	493	1.973	0.0936	0.3283
20	40	4	40	1515	3.091	0.0524	0.1761
20^b					3.049	0.0587	0.1875

a) Equations (13.25) and (13.26).
b) G_P^* – (13.11), G_T^* – (11.87), Q_T^* – (11.91).

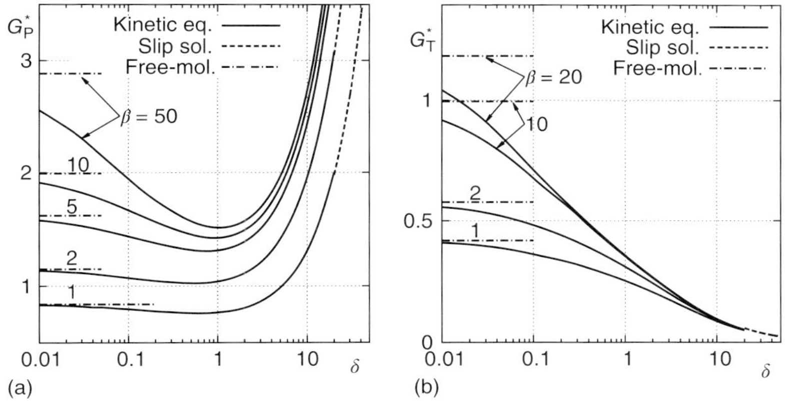

Figure 13.2 Poiseuille G_P^* (a) and thermal creep G_T^* (b) coefficients for rectangular channel versus rarefaction parameter δ and aspect ratio β: solid line – [148, 151]; dashed line – Eqs. (13.11) and (12.86); point-dashed line – Eqs. (13.25) and (13.26). Reproduced from Ref. [136] with permission.

More detailed information about the coefficients G_P^* and G_T^* are reported in Refs [136, 148, 151]. Their numerical values are plotted in Figure 13.2. For all values of the aspect ratio β, the coefficient G_P^* has the Knudsen minimum near the point $\delta \approx 1$. For the square channel ($\beta = 1$), the minimum is rather shallow, while for large values of the aspect ratio, $\beta \geq 10$, the Knudsen minimum is deep.

The slip solution, G_P^* given by (13.11) and G_T^* given by (12.86), presented in Figure 13.2 by the dashed lines works well in the range $\delta \geq 20$. For the small values of the aspect ratio $\beta \leq 5$, the numerical solution at $\delta = 0.01$ is close to the corresponding free molecular value of G_P^* given by Eq. (13.25), while for the large values of $\beta > 5$ the numerical solution is still far from the free molecular value even at $\delta = 0.01$. The thermal creep coefficient G_T^* vanishes in the viscous regime ($\delta \to \infty$) in accordance with Eq. (12.86), and it tends to a constant value given by Eqs. (13.25) and (13.26) in the free-molecular regime.

Additional information on the flow through a rectangular channel can be found in Refs [149, 150, 152, 153]. Data on the Poiseuille coefficient and thermal creep through an elliptical tube are reported in Refs [152, 154, 155]. A numerical solution for such a cross section is quite similar to that considered here, but it requires a nonregular mesh in the physical space. Flows of gaseous mixtures through long channels of rectangular cross sections are considered in Ref. [156, 157].

13.2
Flows Through Slits and Short Channels

Here, we again consider gas flows through a planar channel. In contrast to previous section, the channel width is infinite, but its length is finite and varies in a wide range. Owing to the simplicity of the present flow configuration, the slit and channel flows were included into the list of main benchmark problems of rarefied gas dynamics [158].

13.2.1
Formulation of the Problem

Consider two large reservoirs containing the same gas and connected by a channel as is shown in Figure 13.3. The gas in the left reservoir is maintained at a pressure p_0 and temperature T_0 far from the channel entrances, while the gas in the right reservoir is kept at a pressure p_1 and temperature T_1 far from the channel. The reservoir walls temperatures are constant and are equal to the corresponding gas

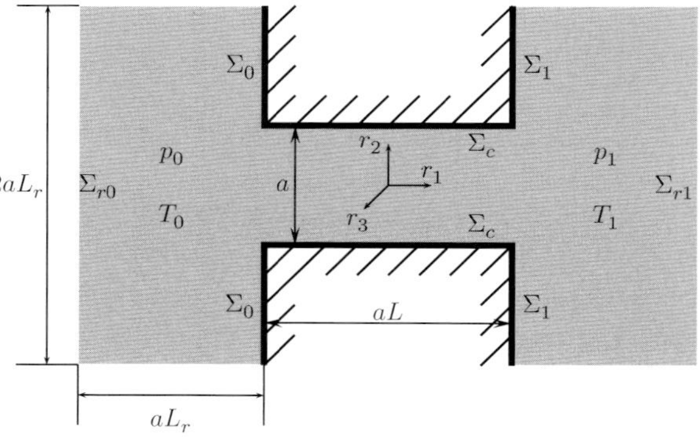

Figure 13.3 Scheme and computational domain for gas flow through slit and short channel.

temperatures T_0 and T_1. The temperature T_w of the channel wall can be arbitrary, that is, T_w is a given function of the longitudinal coordinate r_1.

The channel height is equal to a, while its length is equal to aL, that is, the quantity L represents the length-to-height ratio. The channel is degenerated into a slit at $L = 0$ and the opposite limit is the infinite plates $L \to \infty$. It is assumed that the channel width is so large that the gas flow is invariant with respect to the variable r_3, hence, all quantities, namely, number density n, bulk velocity $\boldsymbol{u}$, temperature T, and heat flow $\boldsymbol{q}$, are functions only of the coordinates r_1 and r_2. Moreover, the vectors $\boldsymbol{u}$ and $\boldsymbol{q}$ are two-dimensional

$$\boldsymbol{u} = [u_1, u_2, 0], \quad \boldsymbol{q} = [q_1, q_2, 0]. \tag{13.38}$$

The quantity of our interest is the mass flow rate $\dot{M}$ through the channel defined as

$$\dot{M} := m \int_{-a/2}^{a/2} n(r_1, r_2) u_1(r_1, r_2)\, \mathrm{d}r_2, \quad -\frac{aL}{2} \le r_1 \le \frac{aL}{2}. \tag{13.39}$$

Note that this definition means the flow rate per width unity so that $\dot{M}$ is measured in kg/m s. The mass flow rate $\dot{M}$ does not vary along the channel because of the mass conservation law.

The solution of the problem is determined by the following parameters: the length-to-height ratio L, pressure ratio p_1/p_0, temperature ratio T_1/T_0, temperature distribution $T_w(r_1)$, and rarefaction parameter δ defined by (7.46), that is, the channel height a is the characteristic size of the gas flow considered here.

13.2.2
Free-Molecular Regime

The limit case of the free-molecular flow ($\delta \to 0$) through a slit ($L \to 0$) into vacuum ($p_1/p_0 \to 0$) is called molecular effusion. In this case, the VDF in the slit section, $r_1 = 0$, reads

$$f(\boldsymbol{r}, \boldsymbol{v}) = \begin{cases} f_0^{\mathrm{M}} & v_1 > 0, \\ 0 & v_1 < 0, \end{cases} \tag{13.40}$$

where the Maxwellian f_0^{M} corresponds to the gas state in the left reservoir and is given by (5.1). Then the particle flow in the slit is calculated from the definition (2.18)

$$J_N = \int_{-\infty}^{\infty} \int_{-\infty}^{\infty} \int_{0}^{\infty} v_1 f_0^{\mathrm{M}}\, \mathrm{d}v_1\, \mathrm{d}v_2\, \mathrm{d}v_3 = \frac{n_0 v_0}{2\sqrt{\pi}}, \tag{13.41}$$

where v_0 is given by (5.2). Usually, this quantity is expressed via the mean speed $\langle v \rangle$ as

$$J_N = \frac{1}{4} n_0 \langle v \rangle, \tag{13.42}$$

where $\langle v \rangle$ is calculated by (2.44) with the temperature T_0. The mass flow rate is equal to J_N multiplied by the slit size a and by the molecule mass m

$$\dot{M}_0 := amJ_N = \frac{ap_0}{\sqrt{\pi} v_0}, \quad \text{for} \quad L \to 0 \quad \text{and} \quad \frac{p_1}{p_0} \to 0, \tag{13.43}$$

where the state equation (1.9) has been used. When the pressure p_1 in the right reservoir is not negligible, then the net flow rate is calculated as a difference of the two opposite flows

$$\dot{M} = \dot{M}_0 \left(1 - \frac{p_1}{p_0} \sqrt{\frac{T_0}{T_1}} \right), \quad \text{for} \quad L \to 0. \tag{13.44}$$

Thus, if the pressures p_0 and p_1 are equal to each other and $T_0 > T_1$, then $\dot{M}$ is negative, that is, the gas flows from the right (cold) reservoir to the left (hot) reservoir.

To calculate the mass flow rate into vacuum through a channel of finite length L, the quantity $\dot{M}_0$ is multiplied by the so-called transmission probability W_0,

$$\dot{M} = \dot{M}_0 W_0, \quad \text{for} \quad \frac{p_1}{p_0} \to 0. \tag{13.45}$$

The physical meaning of W_0 is the average probability that a molecule entering to the channel from the left reservoir will leave it into the right reservoir. The transmission probability W_0 can be calculated via Clausing's equation [159], which can also be found in the book by Cercignani [18] (Chapter V, Section 8) or in that by Saksaganskii [22]. The test particle Monte Carlo method, see the book by Bird [17], can be also applied to calculate W_0. This method is most appropriate for a nondiffuse gas–surface interaction. The following expression obtained by Berman [160] for the diffuse scattering

$$W_0 = \frac{1}{2}[1 + (1 + L^2)^{1/2} - L] - \frac{\frac{3}{2}\{L - \ln[L + (L^2 + 1)^{1/2}]\}^2}{L^3 + 3L^2 + 4 - (L^2 + 4)(1 + L^2)^{1/2}} \tag{13.46}$$

provides a good interpolation of numerical data. It is useful to have the asymptotic expressions for small value of L

$$W_0 = 1 - \frac{L}{2} + O(L^2), \quad \text{for} \quad L \ll 1 \tag{13.47}$$

and for its large value

$$W_0 = \frac{1}{L} \left[\ln \left(\frac{L}{2} \right) - \frac{1}{2} \right] + O(L^{-2}), \quad \text{for} \quad L \gg 1. \tag{13.48}$$

It should be emphasized that the transmission probability does not depend on the channel temperature T_w under the diffuse–specular scattering. However, the quantity W_0 is affected by T_w if the CL scattering kernel is used. If the down-flow pressure p_1 is not small, the expression (13.45) takes a more general form:

$$\dot{M} = \dot{M}_0 W_0 \left(1 - \frac{p_1}{p_0} \sqrt{\frac{T_0}{T_1}} \right). \tag{13.49}$$

13.2.3 Small Pressure and Temperature Drops

13.2.3.1 Definitions

It is worth to consider separately the special case when the pressure and temperature differences are small,

$$\xi_P = \frac{p_1 - p_0}{p_0}, \quad |\xi_P| \ll 1, \quad \xi_T = \frac{T_1 - T_0}{T_0}, \quad |\xi_T| \ll 1. \tag{13.50}$$

Since two small parameters ξ_P and ξ_T are used for the linearization, each moment is decomposed into two terms then Eqs. (5.12)–(5.14) and (5.16) take the form

$$n = n_0(1 + \varrho^{(P)}\xi_P + \varrho^{(T)}\xi_T), \tag{13.51}$$

$$\boldsymbol{u} = v_0(\tilde{\boldsymbol{u}}^{(P)}\xi_P + \tilde{\boldsymbol{u}}^{(T)}\xi_T), \tag{13.52}$$

$$T = T_0(1 + \tau^{(P)}\xi_P + \tau^{(T)}\xi_T), \tag{13.53}$$

$$\boldsymbol{q} = p_0 v_0(\tilde{\boldsymbol{q}}^{(P)}\xi_P + \tilde{\boldsymbol{q}}^{(T)}\xi_T). \tag{13.54}$$

The mass flow rate can be represented as

$$\dot{M} = \frac{a p_0}{L v_0}(-G_P\xi_P + G_T\xi_T). \tag{13.55}$$

Substituting (13.51) and (13.52) into the flow rate definition (13.39) and comparing it with (13.55), one can check that the Poiseuille and thermal creep coefficients are defined via the dimensionless bulk velocity

$$G_P := -2L\int_{-1/2}^{1/2} \tilde{u}_1^{(P)}(x_1, x_2)\,\mathrm{d}x_2, \quad G_T := 2L\int_{-1/2}^{1/2} \tilde{u}_1^{(T)}(x_1, x_2)\,\mathrm{d}x_2, \tag{13.56}$$

where the terms of the order ξ_P^2 and ξ_T^2 have been omitted and the coordinate x_1 can be any value being the range $[-L/2, L/2]$. The dimensionless coordinates (x_1, x_2) are related to r_1, r_2 via the channel height a by (7.43). Similar to G_P^* and G_T^* defined by (11.80), the coefficients G_P and G_T defined here are determined by the rarefaction parameter δ, but in addition, they are determined by the length-to-height ratio L. In the limit of infinite plates, the coefficients G_P and G_T tend to G_P^* and G_T^*

$$\lim_{L\to\infty} G_P(\delta, L) = G_P^*(\delta), \quad \lim_{L\to\infty} G_T(\delta, L) = G_T^*(\delta). \tag{13.57}$$

The second limit is valid if the temperature distribution $T_w(r_1)$ is linear along the channel. Thus, the quantities G_P and G_T will also be called as Poiseuille and thermal creep coefficients, respectively.

However, the representation (13.55) is not appropriate for the limit of slit ($L = 0$) because both G_P and G_T are just zero in this case. To avoid such a nonsense, other dimensionless flow rates are introduced as

$$\dot{M} = \dot{M}_0\left(-\mathcal{G}_P\xi_P + \frac{1}{2}\mathcal{G}_T\xi_T\right), \tag{13.58}$$

where $\dot{M}_0$ is given by (13.43). The new introduced coefficients are related to previous coefficients as

$$\mathcal{G}_\mathrm{P} = \frac{\sqrt{\pi}}{L} G_\mathrm{P}, \quad \mathcal{G}_\mathrm{T} = \frac{2\sqrt{\pi}}{L} G_\mathrm{T} \tag{13.59}$$

and can also be used for a finite length L, but they vanish in the limit of infinite plates $L \to \infty$. Thus, both representations (13.55) and (13.58) will be used in this section. Using the relations (13.59), it is verified that the new coefficients are defined in terms of the bulk velocities as

$$\mathcal{G}_\mathrm{P} := -2\sqrt{\pi} \int_{-1/2}^{1/2} \tilde{u}_1^{(\mathrm{P})}(x_1, x_2)\, \mathrm{d}x_2, \quad \mathcal{G}_\mathrm{T} := 4\sqrt{\pi} \int_{-1/2}^{1/2} \tilde{u}_1^{(\mathrm{T})}(x_1, x_2)\, \mathrm{d}x_2, \tag{13.60}$$

where the coordinate x_1 varies in the interval $[-L/2, L/2]$. Their free-molecular value is equal to W_0

$$\mathcal{G}_\mathrm{P} = \mathcal{G}_\mathrm{T} = W_0, \quad \text{for} \quad \delta \to 0 \tag{13.61}$$

that follows from (13.44).

13.2.3.2 Kinetic Equation

In the considered problem, the linearization is performed near the global Maxwellian f_0^M using ξ_P and ξ_T given by (13.50) as the small parameters,

$$f(\boldsymbol{x}, \boldsymbol{c}) = f_0^\mathrm{M}[1 + h^{(\mathrm{P})}(\boldsymbol{x}, \boldsymbol{c})\xi_\mathrm{P} + h^{(\mathrm{T})}(\boldsymbol{x}, \boldsymbol{c})\xi_\mathrm{T}]. \tag{13.62}$$

Both perturbation functions $h^{(\mathrm{P})}$ and $h^{(\mathrm{T})}$ obey the same kinetic equation. If we choose the S-model , that is, Eq. (7.47) with $A_1 = 1$, $A_2 = 0$, and $A_3 = 1$, then the kinetic equation takes the form (13.14) with the difference that $\tilde{g} = 0$. The function H is given by (7.48) and contains the six moments: ϱ, $\tilde{u}_1$, $\tilde{u_2}$, τ, $\tilde{q}_1$, and $\tilde{q}_2$, which are calculated via the corresponding perturbations $h^{(\mathrm{P})}$ and $h^{(\mathrm{T})}$ by (5.17).

The equations for $h^{(\mathrm{P})}$ and $h^{(\mathrm{T})}$ are exactly the same, but they satisfy different boundary conditions. In this problem, five different boundaries can be distinguished. Let us denote as Σ_0 the wall surface of the left reservoir, Σ_1 the wall surface of the right one, Σ_c the wall surface of the channel, Σ_{r0} the imaginary surface restricting the left reservoir, and Σ_{r1} is also the imaginary surface restricting the right reservoir. All surfaces are shown in Figure 13.3. The surfaces Σ_0, Σ_1, and Σ_c are impermeable, then under the diffuse reflection assumption the condition (5.70) is applied that takes the form

$$h^{(\mathrm{P})} = \varrho_r^{(\mathrm{P})} \quad \text{for} \quad c_n > 0, \tag{13.63}$$

and

$$h^{(\mathrm{T})} = \varrho_r^{(\mathrm{T})} + \tau_w(c^2 - 2) \quad \text{for} \quad c_n > 0, \tag{13.64}$$

in case of the flow due to the pressure ξ_P and temperature ξ_T differences, respectively. Here, $c_n > 0$ means that the perturbations correspond to reflected particles and the quantity τ_w is defined for each surface as

$$\tau_w = \begin{cases} 0 & \text{on } \Sigma_0, \\ 1 & \text{on } \Sigma_1, \\ (T_w - T_0)/(\xi_T T_0) & \text{on } \Sigma_c. \end{cases} \tag{13.65}$$

The quantities $\varrho^{(P)}$ and $\varrho^{(T)}$ are calculated by (5.71) via the corresponding perturbations of incident particles. The surfaces Σ_{r0} and Σ_{r1} are permeable so that the perturbations of particles entering into the computational domain have the form

$$h^{(P)} = \begin{cases} 0 & \text{on } \Sigma_{r0}, \\ 1 & \text{on } \Sigma_{r1}, \end{cases} \qquad h^{(T)} = \begin{cases} 0 & \text{on } \Sigma_{r0}, \\ \left(c^2 - \frac{5}{2}\right) & \text{on } \Sigma_{r1} \end{cases} \tag{13.66}$$

and do not depend on that of particles leaving the domain.

The numerical scheme to solve this problem is practically the same as that to calculate the flow rate through a long rectangular channel (see Section 13.1.5). The main difference is the computational domain that contains significant parts of the reservoir. The computational domains in the reservoirs could be rectangular as is shown in Figure 13.3. The size L_r must be sufficiently large so that its further increase would not change the results within an assumed accuracy. As is pointed out in the paper [161], the size L_r varies from 40 for $\delta \leq 1$ to 10 for $\delta = 10$. The nodes in the velocity space can be exactly the same as those described in Section 13.1.5, while the nodes for position coordinates x_1 and x_2 must be carefully chosen because of the domain size. These nodes should be denser in the channel and nearer its entrances, while they should be rarer far from the channel. Such a nonregular nodes distribution is justified by a sharp variation of the flow characteristics inside the channel and by their smooth variation far from the channel.

Since we are interested only in the mass flow rate, the computational effort to obtain G_P and G_T can be reduced using the reciprocal relation (5.83). The expressions of the kinetic coefficients satisfying this relation were obtained in Refs [38, 45]. As a consequence, the thermal creep coefficient G_T is related to the heat flow vector $\boldsymbol{q}^{(P)}$ in the gas flow due to the pressure difference ξ_P, namely,

$$G_T = \int_{-1/2}^{1/2} q_1^{(P)}\left(\frac{L}{2}, x_2\right) \mathrm{d}x_2 + 2\int_{-L/2}^{L/2} \tau_w(x_1) q_2^{(P)}\left(x_1, \frac{1}{2}\right) \mathrm{d}x_1, \tag{13.67}$$

where τ_w is given by (13.65) on Σ_c. In case of the slit flow ($L = 0$), this relation in terms of $\mathcal{G}_T$ takes the form

$$\mathcal{G}_T = 4\sqrt{\pi} \int_{-1/2}^{1/2} \tilde{q}_1^{(P)}(0, x_2)\, \mathrm{d}x_2. \tag{13.68}$$

These two relations allow us to calculate both coefficients G_P and G_T (or $\mathcal{G}_P$ and $\mathcal{G}_T$) solving only the kinetic equation for $h^{(P)}$.

13.2.3.3 Hydrodynamic Solution

In the hydrodynamic limit ($\delta \to \infty$), the equation similar to (13.9) should be solved. First, the mass conservation law (3.26) is written down as

$$\frac{\partial u_1}{\partial r_1} + \frac{\partial u_2}{\partial r_2} = 0, \tag{13.69}$$

where the nonstationary $\partial \rho/\partial t$ and nonlinear $\boldsymbol{u} \cdot \nabla \rho$ terms have been omitted. Then a combination of the conservation law (3.27) and Newton's law (6.2) for two-dimensional flow leads to the following two equations

$$\mu \left(\frac{\partial^2 u_i}{\partial r_1^2} + \frac{\partial^2 u_i}{\partial r_2^2} \right) = \frac{\partial p}{\partial r_i}, \quad i = 1, 2, \tag{13.70}$$

where again the nonstationary $\partial \boldsymbol{u}/\partial t$ and nonlinear $\boldsymbol{u} \cdot \nabla \boldsymbol{u}$ terms have been omitted.

In case of the slit flow ($L = 0$), these equations are solved analytically [162] under the nonslip boundary condition. Substituting this solution into the definition (13.60), the dimensionless flow rate $\mathcal{G}_\mathrm{P}$ is obtained as [161]

$$\mathcal{G}_\mathrm{P} = \frac{\pi^{3/2}}{16}\delta \quad \text{for} \quad L \to 0 \quad \text{and} \quad \delta \to \infty. \tag{13.71}$$

In case of channel ($L > 0$), a numerical solution of the system (13.69) and (13.70) is reported in the work [163]. However, an approximate analytical solution can also be obtained using the electrical resistance analogy. The resistance, or inverse flow rate G_P^{-1}, of a channel with a finite length L can be considered as a sum of resistance of infinite channel $(\delta/6)^{-1}$, see Eq. (11.87), and that of a slit $(L\mathcal{G}_\mathrm{P}/\sqrt{\pi})^{-1}$, where $\mathcal{G}_\mathrm{P}$ is given by (13.71). As a result, the flow rate takes the form

$$G_\mathrm{P} = \frac{\delta L}{6L + 16/\pi} \quad \text{for} \quad \delta \to \infty. \tag{13.72}$$

This formula smoothly connects the two limits: $L \to 0$ given by (13.71) with (13.59) and $L \to \infty$ given by (11.87). The numerical solution obtained in Ref. [163] is in a good agreement with the analytical expression (13.72).

13.2.3.4 Numerical Results

In case of the slit flow ($L = 0$), the kinetic equations for $h^{(\mathrm{P})}$ and $h^{(\mathrm{T})}$ were solved in the work [164]. Since both solutions $h^{(\mathrm{P})}$ and $h^{(\mathrm{T})}$ were obtained, the reciprocal relation (13.68) was used as an additional criterion of the numerical accuracy of 1%. The numerical data reported in Ref. [164] are briefly reproduced in Table 13.3, which shows that the Poiseuille coefficient $\mathcal{G}_\mathrm{P}$ monotonically increases, while the thermal creep coefficient $\mathcal{G}_\mathrm{T}$ monotonically decreases. The coefficient $\mathcal{G}_\mathrm{P}$ was also calculated on the basis of the BGK equation in the work [161] where the reported

Table 13.3 Coefficients $\mathcal{G}_P$ and $\mathcal{G}_T$ for gas flow through slit ($L = 0$) versus rarefaction δ, Ref. [164].

δ	$\mathcal{G}_P$	$\mathcal{G}_T$
0.0	1.000	1.000
0.04	1.036	0.9968
0.1	1.074	0.9841
1	1.462	0.7728
2	1.831	0.6535
10	4.590	0.4621
20	7.988	0.3384

values are very close to those based on the S-model given in Table 13.3. The following analytical formula

$$\mathcal{G}_P = \begin{cases} 1 - (0.2439 \lg \delta - 0.3833)\delta - (0.0338 \lg \delta - 0.055)\delta^2 & \text{at} \quad \delta \le 8, \\ 0.3480\delta - (4.449 - 25.17 \lg \delta)/\delta + (138.7 - 238.3 \lg \delta)/\delta^2 & \text{at} \quad \delta \ge 8 \end{cases} \tag{13.73}$$

proposed in Ref. [161] interpolates the numerical data within 1%. The papers [161, 164] provided numerical values of $\mathcal{G}_P$ and $\mathcal{G}_T$ for the nondiffuse gas–surface interaction too. These data point out that the Poiseuille coefficient $\mathcal{G}_P$ is weakly affected by the gas–surface interaction law, while the thermal creep $\mathcal{G}_T$ is sensitive to this factor.

Numerical data on the coefficients G_P and G_T for a short channel ($L > 0$) are reported in the papers [163, 165, 166] applying the integro-moment method. Two temperature distributions were used to calculate G_T: linear

$$\tau_w(x_1) = \frac{1}{2} + \frac{x_1}{L} \tag{13.74}$$

and stepwise

$$\tau_w(x_1) = \begin{cases} 0 & \text{for} \quad -L/2 \le x_1 \le 0, \\ 1 & \text{for} \quad 0 \le x_1 \le L/2. \end{cases} \tag{13.75}$$

The papers [167–169] report numerical data on G_P and G_T obtained by the DVM for the linear temperature distribution (13.74). Partially, these data are reproduced in Tables 13.4 and 13.5. The Poiseuille coefficient G_P is also plotted on Figure 13.4. First, we should note that when the length L is finite, both coefficients G_P and G_T are finite in the free-molecular limit ($\delta \to 0$). In other words, if the parallel plates confining a gas are restricted in their width, see Eq. (13.25), or in their length, see Eqs. (13.59), (13.61), and (13.46), the singularity (11.135) at $\delta \to 0$ disappears. Since in practice channel width and length are always restricted, the flow rates G_P and G_T are never infinite in the free-molecular limit, but their values significantly

Table 13.4 Poiseuille coefficient G_P for gas flow through short channel versus length-to-height ratio L and rarefaction δ: Refs [167, 168].

	G_P				
δ	$L = 1$	5	10	100	1000
0.0[a)]	0.386	1.01	1.36	2.67	4.00
0.02	0.391	1.00	1.35	2.34	2.64
0.2	0.422	1.02	1.29	1.74	1.80
1	0.520	1.09	1.29	1.52	1.54
2	0.622	1.21	1.39	1.58	1.60
4	0.810	1.46	1.65		
10	1.345	2.28	2.50		
20			3.98	4.35	4.40
200			28.0	33.4	34.1
200[b)]	18.0	28.5	30.7	33.1	33.3

a) Equations (13.59), (13.61), and (13.46).
b) Equation (13.72)

Table 13.5 Thermal creep coefficient G_T for gas flow through short channel versus length-to-height ratio L and rarefaction δ: $L = 1$ and 5 – Refs [166, 170], $L = 10$ and 100 – Ref. [167].

	G_T					
	$L = 1$		5		10	100
δ	τ_w (13.74)	(13.75)	(13.74)	(13.75)	(13.74)	(13.74)
0.0[a)]	0.193	0.193	0.503	0.503	0.679	1.33
0.02	0.192	0.191	0.491	0.491	0.6465	1.024
0.2	0.180	0.180	0.416	0.417	0.5000	0.5953
2	0.118	0.118	0.217	0.211	0.2420	0.2697
20					0.04902	0.05349
200					0.00487	0.00578

a) Equations (13.59), (13.61), and (13.46).

depends on the length-to-height L or/and width-to-height β ratios. Even for relatively small values of the rarefaction parameter, say $\delta = 0.02$, the flow rates G_P and G_T are still far from their free-molecular values when the length L is large. According to Figure 13.4, the Poiseuille coefficient G_P does not have the Knudsen minimum for the small values of L, namely, for $L \leq 5$. The thermal creep coefficient always decreases when the flow regime varies from the free-molecular to hydrodynamic one. Comparing the values of thermal creep G_T from Table 13.5 for the two different temperature distributions (13.74) and (13.75), we see that the coefficient G_T is weakly sensitive to this distribution.

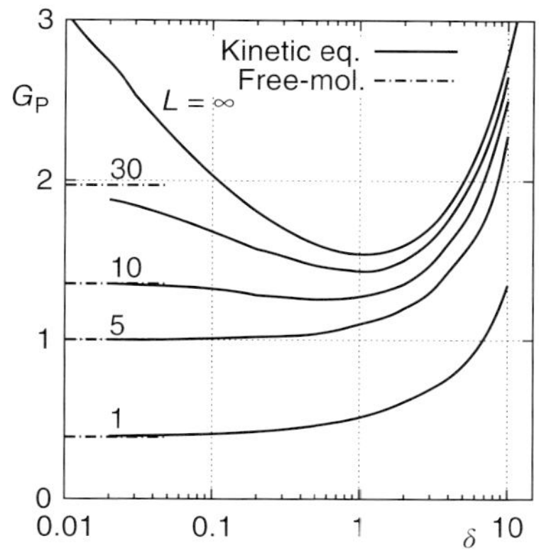

Figure 13.4 Poiseuille coefficient G_P for gas flow through short channel versus rarefaction δ: solid line – numerical solution [166]; point-dashed line – free-molecular solution, Eqs. (13.25) and (13.26).

13.2.4
Arbitrary Pressure Drop

13.2.4.1 Definition

Now let us consider a gas flow through slits and channels under an arbitrary pressure drop. Numerical data for such flows due to an arbitrary temperature drops are very poor so that we assume $T_1 = T_0$ in this section. For our purpose, one more dimensionless flow rate W is defined as

$$W := \frac{\dot{M}}{\dot{M}_0}, \tag{13.76}$$

where $\dot{M}_0$ is given by (13.43). The flow rate defined by such a way can be used for any finite channel length L. In particular case of the free-molecular limit ($\delta \to 0$), this flow rate is related to the transition probability (13.45) as

$$W = W_0 \left(1 - \frac{p_1}{p_0}\right) \quad \text{for} \quad \delta \to 0. \tag{13.77}$$

This is one of few problems of rarefied gas dynamics solved by many methods and in wide ranges of the determined parameters. It began to draw more attention after its inclusion into the list of benchmark problems [158]. First, this problem was solved in case of the gas flow into vacuum ($p_1/p_0 \to 0$) [171–173] applying the DSMC technique. The same situation was considered in the work [174] solving the BGK and S-model without linearization. The gas flow under an arbitrary pressure ratio p_1/p_0 was modeled by the DSMC in Refs [175–177]. The BGK and the S-model were applied by the authors of papers [168, 178, 179]. Finally, this problem was solved also applying the full BE [180–183].

13.2.4.2 Kinetic Equation

If we want to apply a model equation to a two-dimensional planar problem, the kinetic equation (7.44) is reduced to

$$c_1 \frac{\partial f}{\partial x_1} + c_2 \frac{\partial f}{\partial x_2} = \delta A_1 \frac{p\mu_0}{p_0\mu}(F - f), \tag{13.78}$$

where F is given by (7.45) and the rarefaction parameter δ is calculated by (7.46) via the parameters in the left container. Most of the papers considering the problem adopted the assumption that $\mu/\mu_0 = \sqrt{T/T_0}$ corresponding to the HS molecular model. The assumption of the diffuse reflection on the reservoirs and channel walls is also adopted in most of the papers so that the reflected VDF is given by (4.11). The VDF tends to the Maxwellian (2.37) in the reservoirs far from the channel entrances. In the left reservoir, the Maxwellian corresponds to the density $n_0 = p_0/k_B T_0$ and temperature T_0, while in the right reservoir, the Maxwellian has the same temperature but the density is equal to $n_1 = n_0 p_1/p_0$.

As is noted in Section 9.4.5, the numerical methods to solve the nonlinearized kinetic equations such as (13.78) are quite similar to those applied to the linearized ones so that the details are omitted here, but they can be found in the papers [174, 178].

13.2.4.3 Numerical Results

A comparison of numerical data on the dimensionless flow rate W based on the DSMC technique with those obtained from the BGK is performed by Graur *et al.* [174, 178] in case of the slit flow W. The largest discrepancy between two approaches is observed in the transitional regime ($\delta \approx 1$) and reaches 5% that is enough for practical calculations. A comparison of the DSMC results [172, 175] with those obtained from the BE by Aristov *et al.* [180] are in a good agreement within the numerical accuracy 1% so that the approaches based on the DSMC and BE are most reliable, but in contrast to the model equations they require significant computational effort. Moreover, the DSMC techniques are not appropriate in case of small pressure drop, when the ratio p_1/p_0 is close to unity because the statistical scattering in this case becomes very large. In such situations, the model equations are the most optimal tool in practice. The numerical results on the flow rate W for the slit flow ($L = 0$) obtained in Refs [175] and [178] are briefly given in Table 13.6. The analogous data for the

Table 13.6 Dimensionless flow rate W for gas flow through slit ($L = 0$) versus pressure ratio p_1/p_0 and rarefaction δ: $0 \leq p_1/p_0 \leq 0.7$ – DSMC, Ref. [175]; $p_1/p_0 = 0.9$ and 0.99 – BGK, Ref. [178].

	W					
δ	$\frac{p_1}{p_0} = 0$	**0.1**	**0.5**	**0.7**	**0.9**	**0.99**
0	1.000	0.900	0.500	0.3	0.1	0.01
0.1	1.026	0.923	0.520	0.315	0.1062	0.01065
1	1.148	1.060	0.640	0.397	0.1419	0.01437
10	1.479	1.467	1.237	0.940	0.4058	0.04298
20	1.541	1.531	1.344	1.098	0.5910	0.06891
100	1.568	1.561	1.383	1.147	0.7293	0.1891

Table 13.7 Dimensionless flow rate W for gas flow through short channel versus length-to-height ratio L, pressure ratio p_1/p_0, and rarefaction δ.

δ	W			
	$\frac{p_1}{p_0} = 0$	0.1	0.5	0.9
		$L = 1$[b)]		
0[a)]	0.684	0.616	0.342	
0.1	0.698	0.630	0.354	
1	0.767	0.706	0.419	
10	1.04	1.03	0.832	
20	1.15	1.15	1.03	
100	1.36	1.36	1.30	
		$L = 5$[c),d)]		
0[a)]	0.357	0.321	0.178	0.0356
0.1	0.357	0.321	0.181	0.0354
1	0.358	0.325	0.186	0.0385
10	0.490	0.477	0.332	0.0776
20	0.626	0.621	0.489	0.121
100	1.02	1.02	0.947	0.369
		$L = 10$[e)]		
0[a)]	0.241	0.217	0.120	0.0241
0.1	0.236	0.230	0.112	0.0206
1	0.220	0.215	0.111	0.0220
10	0.294	0.283	0.187	0.0437
20	0.388	0.393	0.278	0.0660
100	0.796	0.800	0.705	0.223

a) Equations (13.46) and (13.77).
b) DSMC, [177].
c) $p_1/p_0 = 0$, 0.1, and 0.5 – DSMC, [177].
d) $p_1/p_0 = 0.9$ – BGK, [182].
e) BGK, [182].

channel flow ($L = 1, 5$, and 10) obtained in Refs [177] and [182] are summarized in Table 13.7. In order to analyze the behavior of the flow rate W, it is plotted in Figure 13.5. In both cases $L = 0$ and $L = 5$, the qualitative dependency of W on δ is similar. More specifically, the reduced flow rate tends to its theoretical value (13.77) in the free molecular limit ($\delta \to 0$). The flow rate W increases very slowly by varying the parameter δ from 0 to 1, then the quantity W increases sharply in the range $1 \leq \delta \leq 20$.

For the large pressure drop, when the ratio $p_1/p_0 \leq 0.7$, the flow rate W through a slit ($L = 0$) tends to a constant value. The same behavior should be observed in the other situations, namely, for large length L and small pressure drop ($p_1/p_0 \to 1$), but for significantly larger rarefaction parameter δ, say $\delta > 1000$. Since the

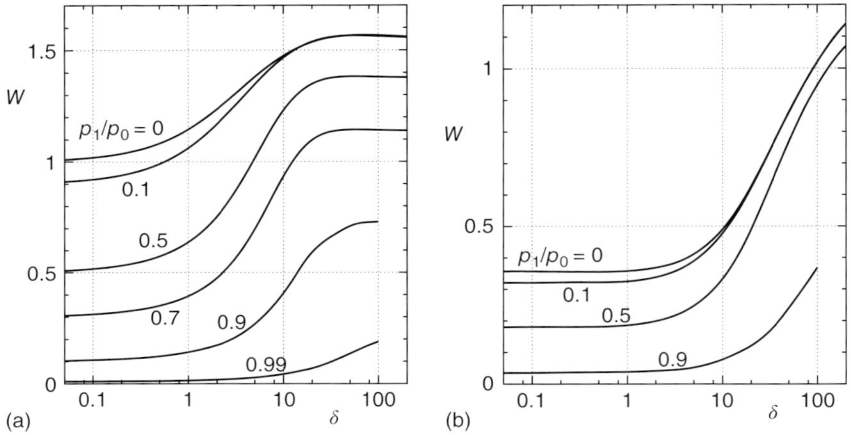

Figure 13.5 Dimensionless flow rate W for gas flow through a slit $L = 0$ (a) and short channel $L = 5$ (b) versus rarefaction δ: $0 \leq p_1/p_0 \leq 0.7$ – Ref. [175]; $p_1/p_0 = 0.9$ and 0.99 – Ref. [178].

numerical solving for so large values of δ requires significant effort, no numerical results are available to confirm such a behavior. It is curious that the results for $p_1/p_0 = 0$ and 0.1 are almost identical when $\delta > 10$. This is a typical choked flow because the mass flow rate does not depend on the down-flow pressure p_1.

The numerical results [175] based on the DSMC method are well interpolated by the following formula:

$$W = \left(1 - \frac{p_1}{p_0}\right)\left(1 + \frac{A\delta - B\delta \ln(\delta) + C\delta^2}{1 + D\delta + E\delta^2}\right), \tag{13.79}$$

where the interpolating coefficients are given in Table 13.8. The formula (13.79) provides the numerical results with the numerical uncertainty 1% practically for the whole range of the gas rarefaction.

The numerical solution of the nonlinearized model equation for the pressure ratio p_1/p_0 close to unity allows us to check the range of applicability of the linear theory described in Section 13.2.3. For this purpose, we should compare the quantity $\mathcal{G}_\mathrm{P}$ defined by (12.5) with the quantity

$$W' = W\frac{p_0}{p_0 - p_1}. \tag{13.80}$$

Table 13.8 Coefficients for interpolating formula (13.79).

p_1/p_0	A	B	C	D	E
0	0.182	0.006	0.0011	0.257	0.0020
0.1	0.218	0.006	0.0052	0.235	0.0072
0.5	0.276	0.026	0.0241	0.076	0.0137
0.7	0.297	0.049	0.0253	0.017	0.0090

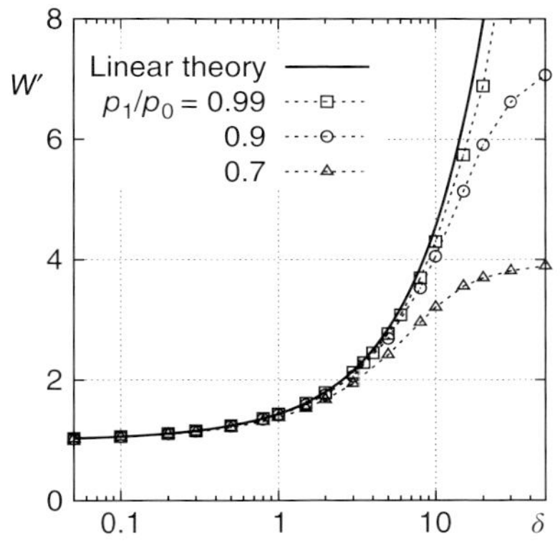

Figure 13.6 Comparison between linear [161] and nonlinear [178] theories for slit flow ($L = 0$).

It is easy to check that $W' \to \mathcal{G}_P$ when $p_1/p_0 \to 1$. The quantity $\mathcal{G}_P$ corresponding to the linearized BGK model is plotted in Figure 13.6 by the solid line. The quantity W' obtained from the nonlinearized BGK model is presented by the square, circle, and triangle symbols for $p_1/p_0 = 0.99$, 0.9, and 0.7, respectively. It can be seen that for the small values of the rarefaction ($\delta < 1$), the results based on the linearized equation are in a good agreement with the solution of the nonlinearized BGK equation even for the pressure ratio $p_1/p_0 = 0.7$ that corresponds to 30% of the relative pressure drop. But when the rarefaction parameter increases ($\delta > 1$), the results based on the linearized BGK equation diverge from those based on the nonlinear BGK even for the pressure ratio close to unity $p_1/p_0 = 0.99$ that corresponds to 1% of the relative pressure difference. For the rarefaction interval considered here, $0.01 \le \delta \le 100$, it is possible to state that the relative difference of the flow rate W' obtained from the linearized BGK model [161] and that obtained from the full BGK model does not exceed the quantity $\delta(1 - p_1/p_0)$. Thus, if the rarefaction parameter is small, the linear theory provides reasonable results even if the pressure drop $(p_0 - p_1)/p_0$ is large. However, for large values of δ, the linear theory fails even for a small pressure drop.

The distributions of the density n/n_0, temperature T/T_0, and longitudinal component of the bulk velocity u_1/v_0 along the symmetry axis obtained by the DSMC method in the paper [175] are presented in Figure 13.7 for the rarefaction parameter $\delta = 1000$. The density n, temperature T, and velocity u_1 have a sharp variation in the points $x_1 = 5.7$, 10.8, and 16.7 at the pressure ratio $p_1/p_0 = 0.1$. Such a phenomenon for the two-dimensional flow is called Mach belts. The intensity of the Mach belt depends on the rarefaction parameter. The density, temperature, and bulk velocity distributions for the pressure ratios $p_1/p_0 = 0.5$ and 0.7 have a sharp variation just past the slit. No Mach belts are observed for the large values of the rarefaction parameter when $p_1/p_0 = 0.5$ or 0.7. For smaller values of the parameter δ, the Mach belts become weaker and disappear in the transitional and free-molecular regimes.

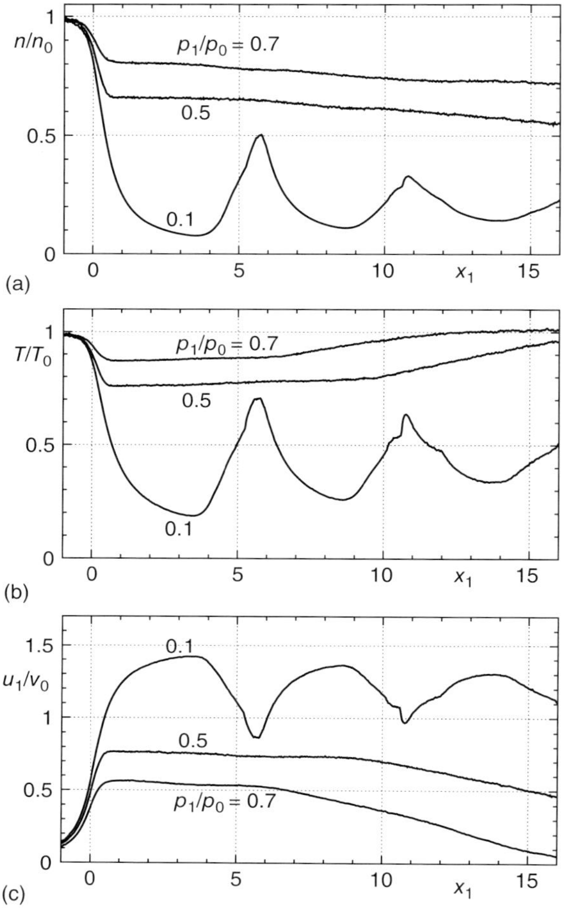

Figure 13.7 Distributions of density n/n_0, temperature T/T_0, and bulk velocity u_1/v_0 along the slit axis at $\delta = 1000$. Reproduced from Ref. [175] with permission.

13.3 End Correction for Channel

13.3.1 Definitions

A calculation of gas flow through a channel of finite length is a nontrivial task and it becomes harder by increasing the channel length. That is why it is attractive to apply the infinite channel results to situations when the length is not so large. It is possible when the EFP ℓ is significantly larger than the channel length aL. This condition can be written down as $\delta L \gg 1$. Under this condition, the gas flow in the middle section of channel is similar to that in the infinite channel so that it is possible to introduce the concept of end correction, which does not depend on the channel length. The physical meaning of such a correction is that the gas flow in reservoirs connected by a channel and in the channel inlet/outlets creates an additional resistance. That is why, the coefficient G_P determined by δ and L for a finite length is always smaller than that for the infinite plates, $G_P(\delta, L) < G_P^*(\delta)$. The additional resistance can be treated as an additional length ΔL so that the value $G_P^*(\delta)$ can be corrected in order to calculate $G_P(\delta, L)$ as

$$G_P(\delta, L) = \frac{L}{L + 2\,\Delta L} G_P^*(\delta), \tag{13.81}$$

where the end correction ΔL depends only on the rarefaction δ, but it is independent of the length L. The factor 2 appears because of the corrections on both inlet and outlet, which are the same for small pressure difference. In general, the representation (13.81) is valid under the condition $\delta L \gg 1$, but more specific condition can be found calculating the gas flow in the reservoir and in that part of the channel adjacent to the reservoir. Note that Eq. (13.72) is a particular case of the representation (13.81) in the hydrodynamic limit ($\delta \to \infty$).

The end correction ΔL was initially calculated for slow viscous flow through a tube by Weissberg [184]. Then, the same idea was used for rarefied gas flows in the papers [163, 169, 185, 186]. This section provides a definition of this concept, the method of its calculation, and the manner of its application.

The scheme and coordinates for the two-dimensional flow considered to calculate the end correction is displayed in Figure 13.8, where the region $r_1 < 0$ represents the reservoir, while the region $r_1 > 0$ corresponds to the channel of height a. The gas flows into the long channel in the r_1 direction from the infinitely large reservoir, where it is maintained at pressure p_0 and temperature T_0 far from the channel inlet. Inside the channel ($r_1 > 0$), a small pressure gradient ξ_P is maintained far from the channel entrance

$$\xi_P = \frac{a}{p}\frac{\partial p}{\partial r_1} \quad \text{at} \quad r_1 \to \infty, \quad |\xi_P| \ll 1. \tag{13.82}$$

The channel height a is assumed to be the unit length so that the rarefaction parameter is determined by (7.46) and dimensionless coordinates by (7.43).

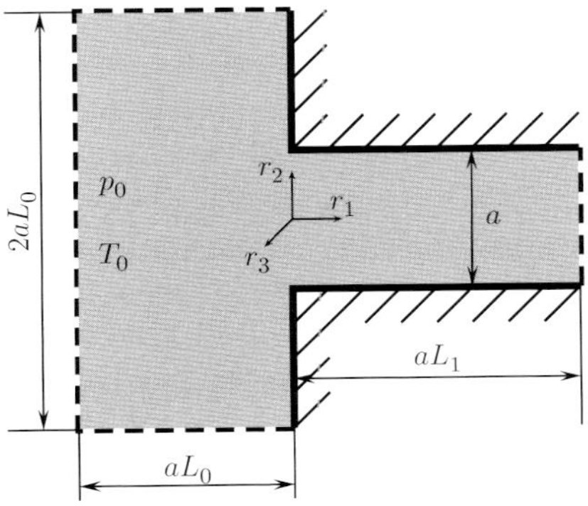

Figure 13.8 Scheme of end correction for channel.

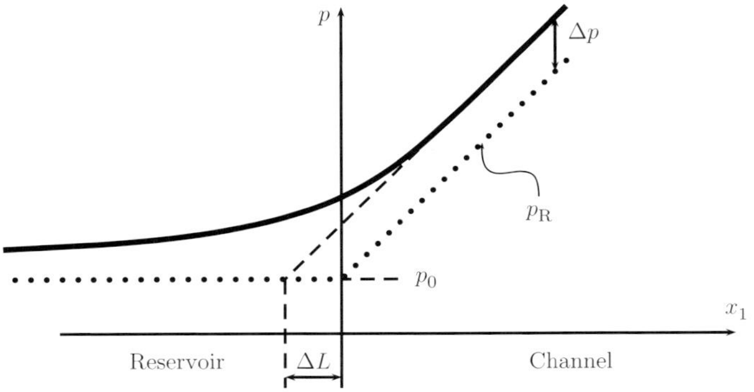

Figure 13.9 Pressure distribution along the symmetry axis at the inlet part.

The expected pressure variation along the symmetry axis ($x_2 = 0$) is depicted on Figure 13.9, that is, it smoothly varies from its constant value p_0 to the linear dependence on the longitudinal coordinate x_1. However, at $x_1 \to \infty$, the pressure does not tend to the linear function $x_1\xi_P$, but it has a jump Δp so that its asymptotic behavior reads

$$p(x_1) \to p_0(1 + x_1\xi_P) + \Delta p, \quad \text{at} \quad x_1 \to -\infty. \tag{13.83}$$

If we extrapolate the linear function (13.83) by a straight line, it will cut the horizontal line $p = p_0$ in the point $x_1 = -\Delta L$ as is shown on Figure 13.9. Using the geometrical reasonings, we can relate the pressure jump Δp to the end correction ΔL as

$$\Delta L = \frac{\Delta p}{\xi_P p_0}. \tag{13.84}$$

Note that the correction ΔL is always positive because when $\xi_P < 0$ then $\Delta p < 0$. Thus, if the pressure jump Δp is known, the end correction ΔL is calculated immediately.

The above-formulated problem of pressure jump is similar to the phenomena of velocity slit and temperature jump described in Chapter 10.

13.3.2
Kinetic Equation

The linearization of the kinetic equation is performed using the pressure gradient ξ_P as the small parameter. In order to approach the linear trend (13.83) of the pressure at large distances from the channel inlet and constant pressure in the reservoir, the reference moments are defined as

$$\varrho_R = \begin{cases} 0 & \text{at} \quad x_1 < 0 \\ x_1 & \text{at} \quad x_1 > 0 \end{cases}, \quad \tau_R = 0, \quad \boldsymbol{u}_R = 0. \tag{13.85}$$

The reference pressure $p_R = n_R k_B T_0 = n_0(1 + \varrho_R\xi_P)$ is plotted by the dotted line on Figure 13.9.

Since we are not interested in the heat flow, the BGK model is appropriate for the considered problem, that is, Eq. (7.47) is applied with $A_1 = 1, A_2 = 0$, and $A_3 = 0$. After its linearization, the kinetic equation takes the same form as that for the two-dimensional flow through a rectangular channel (13.14) with the function H given by (7.48) with the four moments: ϱ, $\tilde{u}_1$, $\tilde{u}_2$, and τ. The source term $\tilde{g}$ is calculated by substituting (13.85) into (5.30) and then into (7.49)

$$\tilde{g} = \begin{cases} 0 & \text{at} \quad x_1 < 0, \\ -c_1 & \text{at} \quad x_1 > 0. \end{cases} \tag{13.86}$$

The discontinuous source function is the main difference of the present problem from all considered previously.

To formulate the boundary conditions, we should distinguish the solid boundaries and imaginary boundaries shown on Figure 13.8 by the solid and dashed lines, respectively. Similar to the previous problem, the boundary condition has the form (13.63) on the solid boundaries representing the channel and reservoir walls. The perturbation function is zero on the imaginary surface restricting the reservoir. The perturbation function of the imaginary surface representing the cross section at $x_1 = L_1$ should be given by the solution of the kinetic equation corresponding to the infinite plates (11.92).

Once the kinetic equation (13.14) with the source term (13.86) subject to the boundary conditions described earlier is solved in the domain depicted on Figure 13.8, the moments ϱ, $\tilde{u}_1$, $\tilde{u}_2$, and τ are known. Then the length increment ΔL is calculated from the density asymptotic value far from the entrance inside the channel,

$$\Delta L = \lim_{x_1 \to \infty} \varrho(x_1, x_2), \tag{13.87}$$

which follows from Eqs. (13.83) and (13.84), from the state equation $p = nk_BT$, and from the fact that $T \to T_0$ at $x_1 \to \infty$.

The parameters of the numerical scheme for this problem is described in detail in the paper by Pantazis *et al.* [187].

13.3.3 Numerical Results

The density perturbation ϱ along the symmetry axis ($x_2 = 0$) is shown in Figure 13.10 for δ =0.2, 1, and 10. It is seen that as $x_1 \to \infty$, the density perturbation reaches a constant value. The perturbation ϱ reaches its asymptotic value faster for the larger values of rarefaction. It can be said that when $x_1 > 1/\delta$ (or $r_1 > \ell$), the quantity ϱ does not vary anymore, that is, similar to the velocity slip problem, here we also deal with the Knudsen layer having the size of the EFP.

The values of ΔL for the rarefaction parameter δ varying in the range from 0.2 to 10 and the computational domain sizes L_0 and L_1 are given in Table 13.9. It is observed that the quantity ΔL decreases by increasing the rarefaction parameter δ and tends to the value $4\pi/3$ extracted from (13.72). The data of Table 13.9 should be used together with the Poiseuille coefficient G_P^* for infinite channel given in

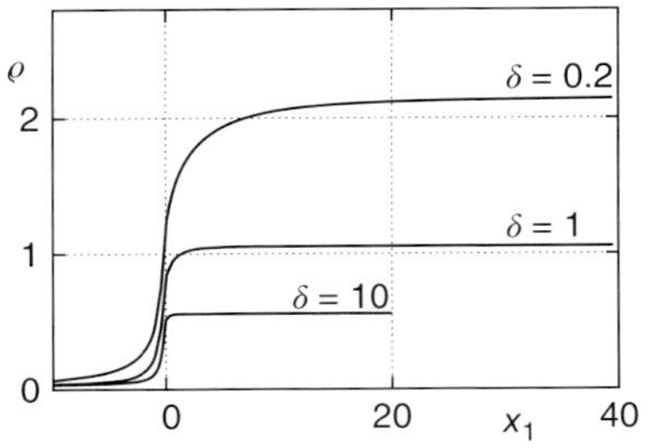

Figure 13.10 Density deviation ϱ in the end-correction problem versus longitudinal coordinate x_1 at $x_2 = 0$. Reproduced from Ref. [187] with permission.

Table 13.9 Poiseuille coefficient G_P^*, end-correction ΔL, and domain sizes L_0, L_1 for channel versus rarefaction δ, Ref. [187].

δ	G_P^* [188]	ΔL	L_0	L_1
0.2	1.8079	2.15	10	60
0.4	1.6408	1.55	10	50
1	1.5389	1.05	12	40
2	1.5942	0.827	12	30
4	1.8440	0.654	15	30
10	2.7638	0.556	15	20
∞	$\delta/6$[a]	0.424[b]		

a) Asymptote (11.87).
b) Equation (13.72).

Table 11.3 or with those obtained by Cercignani and Pagani [188] from the BGK model and reproduced in Table 13.9. A comparison of the Poiseuille coefficient $G_P(\delta, L)$ calculated by (13.81) with the corresponding values of $G_P(\delta, L)$ calculated directly, see Table 13.4, is performed on Figure 13.11. This comparison shows that the end-correction idea works well even for the short channel $L = 1$, but it fails when $\delta L < 1$.

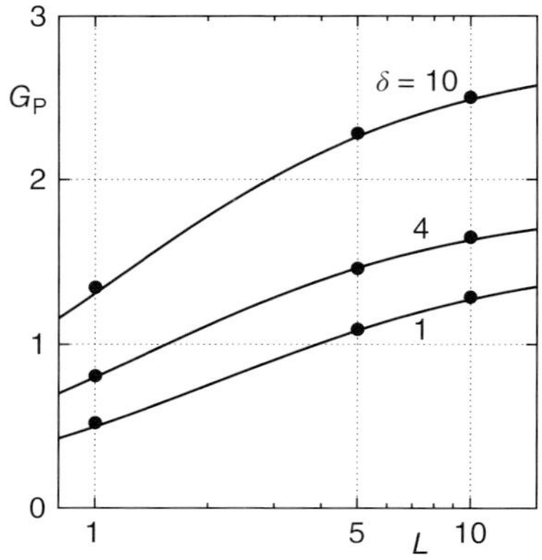

Figure 13.11 Comparison of the end-correction formula for channel (13.81) supported by the data of Table 13.9 (solid lines) with numerical values (circles) of $G_P(\delta, L)$ from Table 13.4.

Exercises

13.1 Consider the helium gas flowing through a rectangular cross-sectional channel of height $a = 1$ mm, length-to-height ratio $L = 100$, and width-to-height ratio $\beta = 2$. Calculate the mass flow rate $\dot{M}$ under the following conditions: the temperature is constant and equal to 300 K, the average pressure is $p = 44$ Pa, and the pressure drop $\Delta p = 0.5$ Pa.
Solution: Since $\delta = 2$ and $L \gg 1$, Eq. (13.4) is used with $G_P = 1.136$ and $\xi_P = \Delta p/pL$. Then $\dot{M} = 1.0 \times 10^{-8}$ g/s.

13.2 Consider the helium gas flowing through a rectangular cross-sectional channel of height $a = 1$ mm, length-to-height ratio $L = 5$, and width-to-height ratio $\beta = 100$. Calculate the mass flow rate $\dot{M}$ under the same conditions as in the previous exercise.
Solution: Since $\delta = 2$ and $\beta \gg 1$, Eq. (13.55) is used with $G_P = 1.21$ and $\xi_P = \Delta p/p$. Moreover, the result is multiplied by βa. Then $\dot{M} = 1.1 \times 10^{-5}$ g/s.

13.3 Solve the previous exercise using the end-correction concept, Eq. (13.81).
Solution: Using Eq. (13.81) and Table 13.9, we have $G_P = 1.20$. Then $\dot{M} = 1.1 \times 10^{-5}$ g/s.

14
Two-Dimensional Axisymmetrical Flows

This chapter is dedicated to gas flows that have an axial symmetry and at the same time their characteristics vary along the symmetrical axis. Such flows are most hard for numerical calculations among those considered in this book.

14.1
Flows Through Orifices and Short Tubes

14.1.1
Formulation of the Problem

A flow through a short circular tube is very similar to that through a short channel. Most of the equations describing the short tube flows are the same as those for the channel given in Section 13.2 so that only different expressions will be given in this section.

Consider two large reservoirs connected by a circular tube as is shown in Figure 14.1. The reservoirs contain the same gas maintained at different pressures far from the tube, namely, p_0 and p_1 in the left and right reservoirs, respectively. Data on nonisothermal flows in this geometrical configuration are very poor and not presented here. Thus, it is assumed that all reservoirs and tube walls are maintained at the same temperature T_0. The tube radius is denoted as a and its length is equal to aL so that L is the length-to-radius ratio. The limit $L = 0$ corresponds to the orifice flow, while the opposite limit $L \to \infty$ represents a long tube considered in Section 12.3.

Because of the axial symmetry, the kinetic equation is written down in the cylindrical coordinates, that is, the longitudinal coordinate r_1 is kept, while the coordinates r_2 and r_3 are replaced by r and ϕ as

$$\boldsymbol{r} = [r_1, r, \phi], \quad r = \sqrt{r_2^2 + r_3^2}, \quad \tan\phi = \frac{r_3}{r_2}. \tag{14.1}$$

The components u_2 and u_3 of the bulk velocity vector u are substituted by the radial u_r and azimuthal u_ϕ ones

$$u_r = u_2 \cos\phi + u_3 \sin\phi, \quad u_\phi = -u_2 \sin\phi + u_3 \cos\phi. \tag{14.2}$$

Rarefied Gas Dynamics: Fundamentals for Research and Practice, First Edition. Felix Sharipov.

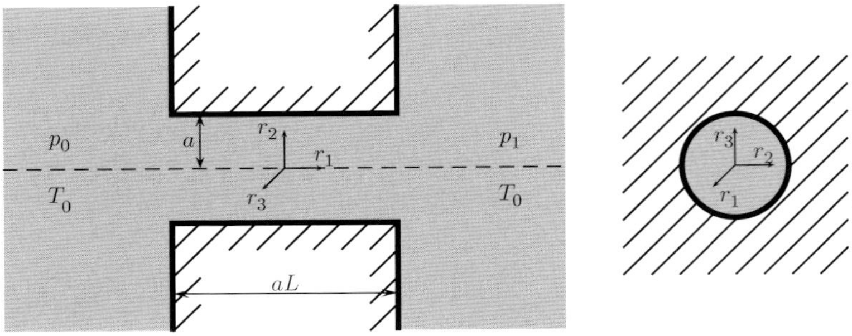

(a) Longitudinal section (b) Cross section

Figure 14.1 Scheme for flow through a short tube.

In case of axial symmetry, the azimuthal coordinate disappears $u_\phi = 0$ and the vector u becomes two-dimensional

$$\boldsymbol{u} = [u_1, u_r]. \tag{14.3}$$

All moments are the functions of the two coordinates

$$n = n(r_1, r), \quad u_1 = u_1(r_1, r), \quad u_r = u_r(r_1, r), \quad T = T(r_1, r). \tag{14.4}$$

As in the previous section, we are interested in the mass flow rate calculated as

$$\dot{M} := 2\pi m \int_0^a n(r_1, r)u_1(r_1, r)r \,\mathrm{d}r, \quad -\frac{aL}{2} \leq r_1 \leq \frac{aL}{2}, \tag{14.5}$$

which is independent of the longitudinal coordinate r_1. The parameters determining the solution are the length-to-radius ratio L, pressure ratio p_1/p_0, and the rarefaction parameter δ defined by (7.46) using the radius a as the characteristic length.

14.1.2
Free-Molecular Flow

The mass flow rate through an orifice $L \to 0$ into vacuum $p_1/p_0 \to 0$ in the free-molecular regime $\delta \to 0$ is calculated by simple multiplication of the particle flow expression (13.41) by the molecular mass m and orifice area πa^2

$$\dot{M}_0 = \pi a^2 m J_N = \frac{\sqrt{\pi} a^2 p_0}{v_0}, \quad \text{for} \quad L \to 0, \quad \text{and} \quad \frac{p_1}{p_0} \to 0. \tag{14.6}$$

If the pressure p_1 is not negligible, then the mass flow rate reads

$$\dot{M} = \dot{M}_0 \left(1 - \frac{p_1}{p_0}\right), \quad \text{for} \quad L \to 0 \tag{14.7}$$

with $\dot{M}_0$ defined by (14.6). If the tube length is not negligible, the mass flow rate (14.7) is multiplied by the transmission probability W_0 defined generally by (13.45),

$$\dot{M} = \dot{M}_0 W_0 \left(1 - \frac{p_1}{p_0}\right). \tag{14.8}$$

The numerical values of W_0 as a function of the length-to-radius ratio L are well known and can be found in the book by Saksaganskii [22] or in the review [166]. Berman [160] proposed the following expression valid in case of the diffuse scattering

$$W_0 = 1 + \frac{L^2}{4} - \frac{L}{4}(L^2+4)^{1/2} - \frac{[(8-L^2)(L^2+4)^{1/2} + L^3 - 16]^2}{72L(L^2+4)^{1/2} - 288\ln[L/2 + (L^2/4+1)^{1/2}]}. \tag{14.9}$$

The asymptotic behaviors of the transmission probability in the limits of short and long tubes are given as

$$W_0 = 1 - \frac{L}{2} + O(L^2) \quad \text{for} \quad L \ll 1, \tag{14.10}$$

and

$$W_0 = \frac{8}{3L} + O\left(\frac{1}{L^2}\right) \quad \text{for} \quad L \gg 1. \tag{14.11}$$

14.1.3 Small Pressure Drop

14.1.3.1 Definitions

The limit of the small pressure drop is considered separately. In this case, the relative pressure difference

$$\xi_P = \frac{p_1 - p_0}{p_0}, \quad |\xi_P| \ll 1 \tag{14.12}$$

is used as the small parameter of the linearization. Since we consider only flows caused by the pressure drop ξ_P, it is not necessary to distinguish two types of the linearized moments, but their expressions are given as

$$n = n_0(1 + \varrho\xi_P), \quad u_1 = v_0\tilde{u}_1\xi_P, \quad u_r = v_0\tilde{u}_x\xi_P, \quad T = T_0(1 + \tau\xi_P). \tag{14.13}$$

The Poiseuille flow coefficient G_P is defined as

$$\dot{M} = -\frac{\pi a^2 p_0}{L v_0} G_P \xi_P, \tag{14.14}$$

which tends to G_P^* defined by (12.79) in the limit of long tube $L \to \infty$, but is not appropriate for the orifice flow ($L \to 0$). In this limit, another coefficient is defined as

$$\dot{M} = -\dot{M}_0 \mathcal{G}_P \xi_P = -\frac{\sqrt{\pi} a^2 p_0}{v_0} \mathcal{G}_P \xi_P, \tag{14.15}$$

where the expression (14.6) for $\dot{M}_0$ has been used. This coefficient can be used for short tube ($L > 0$) too, but not for an infinitely long tube. Substituting the expressions (14.13) into (14.5) and comparing it with (14.14) and (14.15), we check that the dimensionless flow rates are defined as

$$G_{\mathrm{P}} := -4L \int_0^1 \tilde{u}_1(x_1, x)x \, \mathrm{d}x, \quad \mathcal{G}_{\mathrm{P}} := -4\sqrt{\pi} \int_0^1 \tilde{u}_1(x_1, x)x \, \mathrm{d}x, \tag{14.16}$$

where the two-dimensional vector $\boldsymbol{x} = [x_1, x]$ is related to $\boldsymbol{r} = [r_1, r]$ by (7.43) via the tube radius a.

The relation of G_{P} to $\mathcal{G}_{\mathrm{P}}$ is the same as in the planar two-dimensional flow (13.59). In the free-molecular limit ($\delta \to 0$), the coefficient $\mathcal{G}_{\mathrm{P}}$ is equal to the transmission probability W_0, while the coefficient G_{P} is proportional to W_0

$$\mathcal{G}_{\mathrm{P}} = W_0, \quad G_{\mathrm{P}} = \frac{L}{\sqrt{\pi}} W_0 \quad \text{for} \quad \delta \to 0. \tag{14.17}$$

14.1.3.2 Kinetic Equations

The linearization is performed near the global Maxwellian representing the distribution function as (5.6) using the pressure drop ξ_{P} as the small parameter. In a general case, the perturbation function h depends on the six variables, $h = h(x_1, x, \phi, c_1, c_x, c_\phi)$, where the velocity components c_2 and c_3 are replaced by the radial c_x and azimuthal c_ϕ component. Then the derivatives $c_2 \partial h/\partial x_2 + c_3 \partial h/\partial x_3$ are substituted by the right-hand side of (9.60). In order to take into account the axial symmetry and to reduce the number of variables, these two components are replaced by the variables c_p and θ related by (9.62). Then the variable ϕ is eliminated because of (9.61), and the expression (9.60) is reduced to the right-hand side of Eq. (9.63). Thus, for the considered problem, the perturbation function depends on the five variables $h = h(x_1, x, \theta, c_1, c_p)$. Keeping the initial form for the derivative with respect to x_1 and using the form (9.63) for the variables x_2 and x_3, the kinetic equation (7.47) takes the form

$$c_1 \frac{\partial h}{\partial x_1} + c_p \cos\theta \frac{\partial h}{\partial x} - \frac{c_p \sin\theta}{x} \frac{\partial h}{\partial \theta} = \delta(H - h). \tag{14.18}$$

Here, the BGK model has been already assumed, that is, $A_1 = 1, A_2 = A_3 = 0$. The function H in this case reads

$$H = \varrho + 2(c_1\tilde{u}_1 + c_p\tilde{u}_x \cos\theta) + \tau\left(c^3 - \frac{3}{2}\right). \tag{14.19}$$

The moments are calculated by (5.17) replacing $\mathrm{d}c_2 \, \mathrm{d}c_3$ by $c_p \, \mathrm{d}c_p \, \mathrm{d}\theta$

$$\begin{bmatrix} \varrho \\ \tilde{u}_1 \\ \tilde{u}_x \\ \tau \end{bmatrix} = \frac{2}{\pi^{3/2}} \int_0^{\pi} \int_0^{\infty} \int_{-\infty}^{\infty} \begin{bmatrix} 1 \\ c_1 \\ c_p \cos\theta \\ \frac{2}{3}c^2 - 1 \end{bmatrix} e^{-c^2} c_p \, \mathrm{d}c_1 \, \mathrm{d}c_p \, \mathrm{d}\theta, \tag{14.20}$$

where the integration with respect to θ in done in the range $[0, \pi]$ instead of $[0, 2\pi]$ multiplying the expression by the factor 2.

The boundary conditions for the perturbation h are the same as those for $h^{(\mathrm{P})}$ in Section 13.2.3, namely, the condition (13.63) is used on the reservoirs and tube walls, and the first expression of (13.66) is valid in the reservoirs far from the tube entrances.

14.1.3.3 Hydrodynamic Solution

In the hydrodynamic limit ($\delta \to \infty$), the system of Navier–Stokes equations composed from Eqs. (3.26), (3.27), and (6.2) written down in the cylindrical coordinates are solved. In case of the orifice flow ($L \to 0$), the system was solved analytically by Roscoe [162]. According to this solution, the flow rate in terms of the coefficient $\mathcal{G}_P$ reads

$$\mathcal{G}_P = \frac{2}{3\sqrt{\pi}}\delta. \tag{14.21}$$

Following the same idea of Section 13.2.3, an approximate analytical solution can be obtained using the electrical resistance analogy. The resistance of a tube of finite length L is calculated as a sum of resistance of infinite tube inversely proportional to G_P given by Eq. (12.86), and that of an orifice related to (14.21). Then the flow rates are obtained as

$$\mathcal{G}_P = \frac{\delta\sqrt{\pi}}{4L + 3\pi/2}, \quad G_P = \frac{\delta L}{4L + 3\pi/2}. \tag{14.22}$$

14.1.3.4 Numerical Results

The numerical scheme to solve Eq. (14.18) is a combination of those described in Section 9.4.2 for one-dimensional flows and in Section 9.4.4 for axisymmetrical flows. The details of the scheme and numerical values of the coefficient $\mathcal{G}_P$ can be seen in the paper by Pantazis and Valougeorgis [189]. Additionally, numerical data in terms of the Poiseuille coefficient G_P are reported by Titarev [190] where the linearized kinetic equation was solved in its three-dimensional form in terms of the Cartesian coordinates.

The results for orifice flow ($L = 0$) and for short tube, namely, $L = 0.5$ and 1, reported in Ref. [189] are partially shown in Table 14.1. The numerical data reported in the papers [190] for longer tubes $10 \leq L \leq 50$ are presented in Table 14.2 in terms of the Poiseuille coefficient G_P. The numerical results of both papers [189] and [190] are plotted in Figure 14.2 in terms of the Poiseuille coefficient G_P and compared with the data for infinite tube calculated by the numerical code described in Section 12.3. Tables 14.1 and 14.2 and Figure 14.2 show that the flow rate G_P through a tube of finite length does not have the Knudsen minimum. The value of G_P practically reaches its free-molecular value at $\delta = 0.1$. The coefficient G_P increases by increasing the rarefaction parameter δ for all values of the length L. At $\delta = 50$ and larger, the hydrodynamic limit expression (14.22) provide a good approximation.

It is useful to obtain an interpolating formula, at least in particular case of the orifice flow. The expression

$$\mathcal{G}_P = 1 + \delta\frac{19.2 - 1.22\ln\delta + 0.376\,\delta}{50 + \delta} \quad \text{for} \quad L \to 0 \tag{14.23}$$

interpolates within 1% the numerical data given in the second column of Table 14.1. Moreover, it provides the exact values in the free-molecular limit, $\mathcal{G}_P = 1$ and in the hydrodynamic solution (14.21).

Table 14.1 Dimensionless flow rate $\mathcal{G}_P$ for gas flow through short tube versus δ, Ref. [189].

δ	$\mathcal{G}_P$		
	$L = 0$	0.5	1
0[a)]	1.00	0.801	0.672
0.1	1.04	0.833	0.696
0.5	1.19	0.947	0.786
1	1.37	1.08	0.892
2	1.72	1.35	1.10
5	2.77	2.13	1.70
10	4.35	3.32	2.63
10[b)]	3.76	2.64	2.03

a) Equations (14.17) and (14.9).
b) Equation (14.22).

Table 14.2 Dimensionless flow rate G_P for gas flow through tube of finite length versus rarefaction δ, Ref. [190].

δ	G_P		
	$L = 10$	20	50
0[a)]	1.08	1.23	1.37
0.1	1.086	1.224	1.338
0.5	1.142	1.257	1.338
1	1.232	1.341	1.416
5	2.059	2.197	2.289
10	3.144	3.340	3.450
20	5.338	5.659	5.871
50	11.88	12.61	13.09
50[b)]	11.2	11.8	12.2

a) Equations (14.17) and (14.9).
b) Equation (14.22).

14.1.4 Arbitrary Pressure Drop

Similar to the gas flow through a finite channel, one more flow rate is introduced to describe the gas flow through an orifice and short tube under an arbitrary pressure drop

$$W := \frac{\dot{M}}{\dot{M}_0}, \tag{14.24}$$

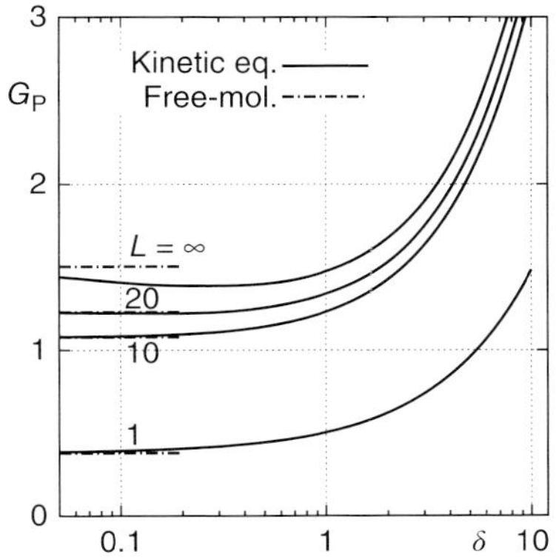

Figure 14.2 Poiseuille coefficient G_P for gas flow through short tube versus rarefaction δ, Refs [189, 190]: solid lines – kinetic equation, point-dashed lines – free-molecular limit.

where $\dot{M}_0$ is defined by (14.6). This flow rate is valid for any length-to-radius ratio L except infinite. In the limit of free-molecular flow ($\delta \to 0$), this quantity is calculated by (13.77) with W_0 given by (14.9).

Numerical data on this problem based on the kinetic equation are not so rich as those for the flow through a short channel. Misdanitis *et al.* [179] reported numerical results on the orifice flow $L \to 0$ based on the nonlinearized BGK model that for axisymmetrical flow reads

$$c_1 \frac{\partial f}{\partial x_1} + c_p \cos\theta \frac{\partial f}{\partial x} - \frac{c_p \sin\theta}{x} \frac{\partial f}{\partial \theta} = \delta \frac{p\mu_0}{p_0\mu}(F - f), \tag{14.25}$$

where F is given by (7.45) with $A_2 = A_3 = 0$. Titarev and Shakhov [191] solved the three-dimensional nonlinearized S-model , that is, when $A_2 = 0$ and $A_3 = 1$ in the expression (7.45). They considered a gas flow through a short tube into vacuum. The same situation was considered by Aristov *et al.* [192] applying the nonlinearized BE. The DSMC technique was used to calculate the gas flow through an orifice in Refs [193–195] and through a short tube in Refs [196, 197]. The works [192–194, 196, 197] are based on the HS molecular model. A comparison of the results [192] based on the BE with those [196] obtained by the DSMC method demonstrates an excellent agreement between two completely different techniques. To validate the results based on the HS potential, the orifice flow was also modeled by the DSMC based on the AI potential [195]. Partially, these results are reproduced in Table 14.3. The main conclusion of Ref. [195] is that the orifice flow is very weakly affected by the intermolecular potential. If an uncertainty of 2% or larger is enough in some practical calculations, then numerical results on the orifice and short tube flows obtained on the bases of the HS molecular model can be successfully applied.

The orifice and short tube flows are also used to validate results based on the model equation. For this purpose, it is interesting to compare the results reported in Ref. [195] obtained by the DSMC method and those reported in Ref. [179] from the BGK model. The discrepancy between these results reaches its maximum value of 5% in the transitional regime $\delta \sim 1$. These data give us an idea about the

Table 14.3 Dimensionless flow rate W for gas flow through orifice versus pressure ratio p_1/p_0 and rarefaction δ based on the AI potential for gas argon at $T = 300$ K, [195].

	W			
δ	$\frac{p_1}{p_0} = 0.1$	0.3	0.5	0.7
0[a)]	0.9	0.7	0.5	0.3
0.1	0.9141	0.7140	0.5117	0.3077
1	1.0384	0.8371	0.6179	0.3824
5	1.3269	1.1999	0.9978	0.6920
10	1.4357	1.3504	1.1935	0.9188
20	1.4946	1.4282	1.2907	1.0431
100	1.5260	1.4730	1.3492	1.1168

a) Free-mol. solution, Eqs. (14.7), (14.24).

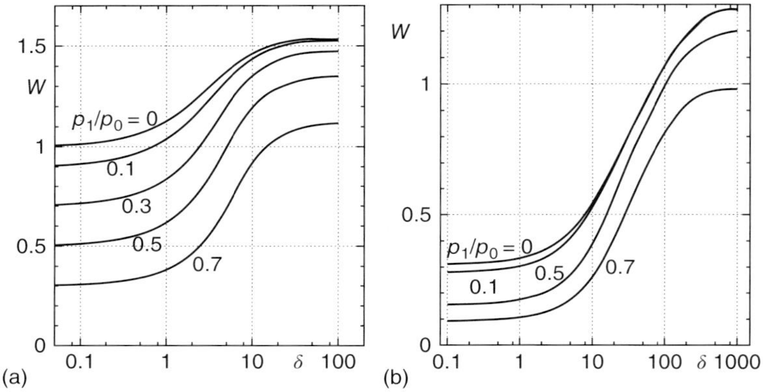

Figure 14.3 Dimensionless flow rate W for gas flow through orifice $L = 0$ (a) and short tube $L = 5$ (b) versus δ [194, 195, 197].

uncertainty of results based on the model equation for strongly nonequilibrium flows.

Numerical results on the gas flow through a short tube obtained in Refs [196, 197] by the DSMC method based on the HS molecular model are partially reproduced in Table 14.4. To visualize the behavior of the flow rate W, it is plotted against the rarefaction δ on Figure 14.3. The qualitative behavior of W is the same for the orifice flow $L = 0$ and for the short tube flow $L = 5$. In case of the orifice, the flow W undergoes a sharp increase in the range $1 \leq \delta \leq 20$ and tends to a constant value in the hydrodynamic limit $\delta \to \infty$. The flow W for the short tube $L = 5$ is different just quantitatively. It undergoes the same increase in the range $1 \leq \delta \leq 200$, and the variation of W is quite larger than that for the orifice.

In order to check the applicability of the results based on the linearized kinetic equation given in Table 14.1, let us compare them with those for the pressure ratios

Table 14.4 Dimensionless flow rate W for gas flow through short tube versus pressure ratio p_1/p_0 and rarefaction δ obtained by the DSMC method based on HS molecular model, Refs [196, 197].

	W			
δ	$\frac{p_1}{p_0} = 0$	0.1	0.5	0.7
		$L = 0.5$		
0.0[a)]	0.801	0.856	0.399	0.241
0.1	0.812	0.869	0.409	0.246
1	0.902	0.984	0.488	0.300
10	1.220	1.38	1.01	0.762
20	1.302	1.45	1.15	0.937
100	1.435	1.51	1.35	1.15
		$L = 1$		
0.0[a)]	0.672	0.605	0.336	0.201
0.1	0.680	0.613	0.343	0.205
1	0.754	0.689	0.405	0.249
10	1.062	1.05	0.866	0.640
20	1.163	1.16	1.04	0.831
100	1.358	1.35	1.29	1.10
		$L = 5$		
0.0[a)]	0.311	0.279	0.155	0.093
0.1	0.312	0.281	0.156	0.093
1	0.334	0.304	0.175	0.106
10	0.543	0.529	0.388	0.263
20	0.695	0.689	0.571	0.411
100	1.068	1.07	0.993	0.814

a) Equations (13.77) (14.9).

$p_1/p_0 = 0.9$ and 0.7 obtained by the DSMC method in the papers [194] and [195], respectively. More specifically, the interpolating formula (14.23) should be compared with the flow rate W' defined by (13.80). The flow rate $\mathcal{G}_P$ given by (14.23) is plotted by the solid line on Figure 14.4. The values of W' calculated by (13.80) using the data of W reported in Ref. [194] for $p_1/p_0 = 0.9$ are plotted by the circle symbols, and those calculated using the data from Ref. [195] for $p_1/p_0 = 0.7$ are depicted by the square symbols. As in case of the slit flow, see Figure 13.6, the results based on the linearized equation are in a good agreement with the nonlinear data even for the pressure ratio $p_1/p_0 = 0.7$ when the rarefaction is small ($\delta \leq 0$). When the pressure ratio is closer to unity, $p_1/p_0 = 0.9$, the range of the applicability of the linear theory becomes larger, that is, $0 \leq \delta \leq 10$. It is expected that considering the pressure ratio p_1/p_0 quite close to unity, say 0.99, the applicability range of the linear theory will be larger, but it is always restricted. At the moments, it is difficult to propose a general criterion of such an applicability because of the lack of the corresponding data. The criterion also can be obtained using the second approximation h_2 in the expansion (5.72).

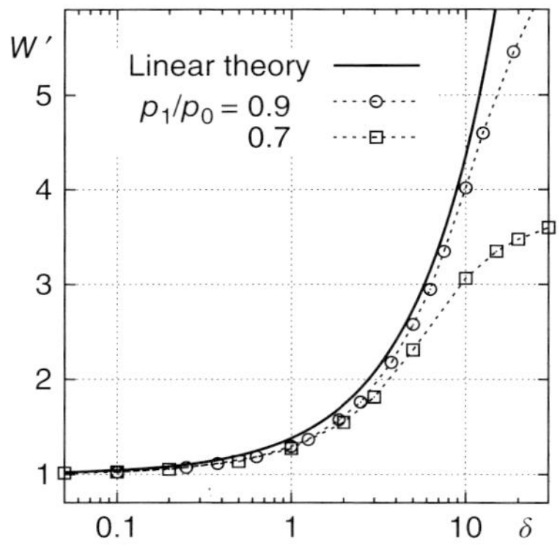

Figure 14.4 Comparison between linear Eq. (14.23) and nonlinear [194, 195] theories for orifice flow ($L = 0$).

The distributions of the density n/n_0, temperature T/T_0, and longitudinal component of the bulk velocity u_1/v_0 along the symmetry axis obtained by the DSMC method in the paper [195] are presented in Figure 14.5 for the rarefaction parameter $\delta = 100$. These flow fields are similar to those shown on Figure 13.7. The density n, temperature T, and velocity u_1 have a sharp variation in the point $x_1 \approx 4$ at the pressure ratio $p_1/p_0 = 0.1$. A similar variation is observed in the point $x_1 \approx 2$ for the ratio $p_1/p_0 = 0.3$. These variations represent the so-called Mach disks, when shock waves in the downflow reservoir occur. The density, temperature, and bulk velocity distributions for the pressure ratios $p_1/p_0 = 0.5$ and 0.7 vary sharply just past the orifice, and then all quantities change smoothly. Thus, when the rarefaction parameter δ is large, the flow field changes qualitatively by varying the pressure ratio p_1/p_0.

14.2
End Correction for Tube

14.2.1
Definitions

A numerical solution of the kinetic equation (14.18) is a hard task so that we should employ all ideas to avoid a numerical solution of the time-consuming problems. One of such ideas is to use the concept of the end correction described in Section 13.3. In the present section, we will introduce the same concept to the axisymmetrical flow through a tube of finite length.

The end correction consists of an increment of the tube length ΔL relating the Poiseuille coefficient G_P for a finite length tube to that for an infinite tube G_P^*. The relation for the axisymmetrical flow is the same as that for the planar one (13.81). To calculate the correction ΔL, consider a gas flow at the outlet of tube and in a reservoir adjacent to the tube as is shown in Figure 14.6. Thus, the region $r_1 > 0$ represents the tube, while the region $r_1 \le 0$ corresponds to the reservoir. The gas

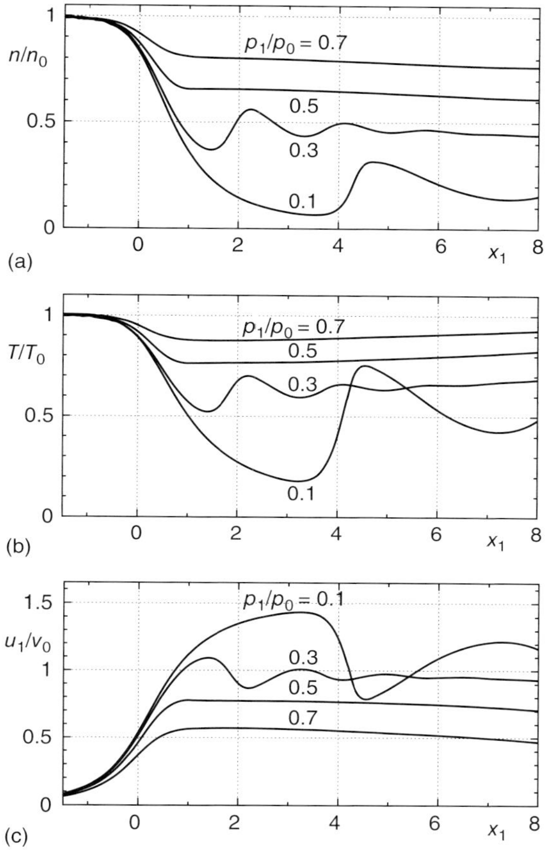

Figure 14.5 Distributions of density n/n_0, temperature T/T_0, and bulk velocity u_1/v_0 along the orifice axis at $\delta = 100$. Reproduced from Ref. [195] with permission.

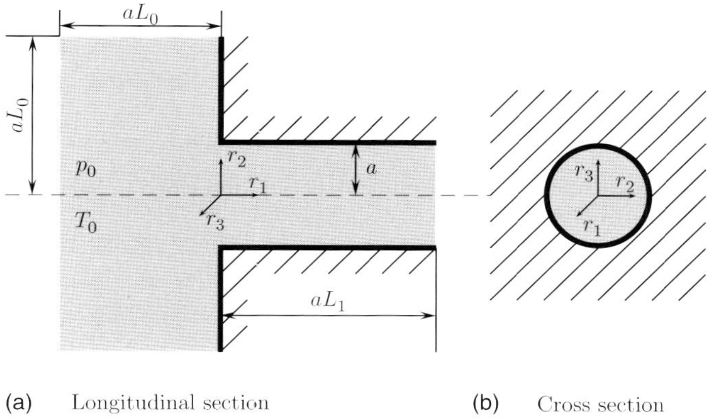

Figure 14.6 Scheme of end correction for tube.

flows from the long tube into the reservoir where it is maintained at pressure p_0 and temperature T_0 far from the tube outlet. Thus, the geometrical flow configuration differs from that considered in Section 13.3 only by the axial symmetry so that the left-hand side of Eq. (13.14) is substituted by the left-hand side of Eq. (14.18)

$$c_1 \frac{\partial h}{\partial x_1} + c_p \cos\theta \frac{\partial h}{\partial x} - \frac{c_p \sin\theta}{x} \frac{\partial h}{\partial \theta} = \delta(H - h) + \tilde{g}. \tag{14.26}$$

The function H has the form (14.19) with the moments given by (14.20). The source term $\tilde{g}$ is determined by (13.86). The boundary conditions imply the impermeability at all solid walls and zero perturbation in the reservoir far from the tube. At the cross section $x_1 = L_1$, the perturbation function is equal to that obtained for an infinite tube from Eq. (12.87).

14.2.2
Numerical Results

The kinetic equation (14.26) was solved numerically by Pantazis *et al.* [198]. The values of the end correction ΔL and the computational domain sizes L_0, L_1 are presented in Table 14.5 for various values of the rarefaction parameter δ covering a wide range of the gas rarefaction. It is observed that the values of ΔL monotonically decrease by increasing δ. Such a behavior indicates that the end effect is more significant for highly rarefied flows. The domain size L_0, see Figure 14.6, slightly varies with the rarefaction δ, while the size L_1 decreases by increasing the rarefaction. These results are qualitatively similar to those obtained previously for the plane geometry in Section 13.3. In the hydrodynamic limit ($\delta \to \infty$), the increment ΔL tends to the value 0.681 obtained from the Navier–Stokes equation in Ref. [184]. In the opposite limit, $\delta \to 0$, the length increment ΔL tends to infinity.

Table 14.5 Poiseuille coefficient G_P^*, end correction ΔL, and domain sizes L_0, L_1 for tube versus rarefaction δ, Ref. [198].

δ	G_P^* [199]	ΔL	L_0	L_1
0.1	1.404	1.52	10	60
0.2	1.382	1.33	10	60
0.4	1.380	1.16	10	50
1	1.459	0.964	12	40
2	1.661	0.841	12	30
4	2.119	0.735	15	30
10	3.582	0.682	15	20
∞	$\delta/4$[a]	0.681[b]		

a) Asymptote (12.86).
b) Reference [184].

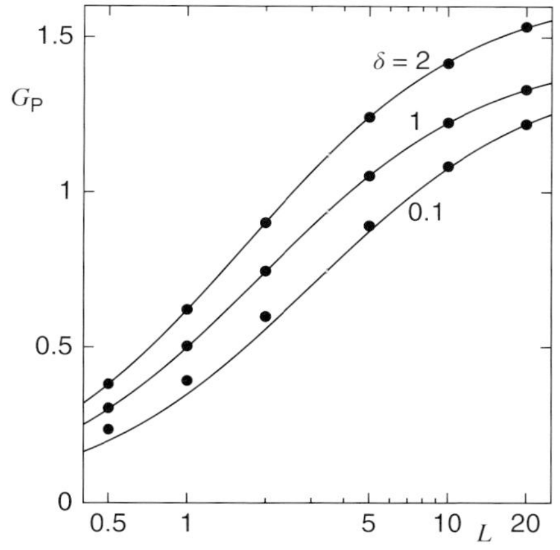

Figure 14.7 Comparison of end-correction formula for tube (13.81) supported by data in Table 14.5 (solid lines) with numerical values (circles) of $G_P(\delta, L)$ from Ref. [189].

The numerical data for the small values of δ indicate the following asymptotic behavior of the increment

$$\Delta L = 0.9 - 0.27 \ln \delta \quad \text{at} \quad \delta \to 0. \tag{14.27}$$

To make easier the use of these results, the values of the Poiseuille coefficient G_P^* for infinite tube ($L = \infty$) obtained from the BGK model by Cercignani and Sernagiotto [199] are also given in Table 14.5.

The Poiseuille coefficient $G_P(\delta, L)$ calculated by (13.81) with the data given in Table 14.5 is shown in Figure 14.7. This comparison shows that the end-correction formula works well even for the short tube such as $L = 0.5$ when the rarefaction parameter is rather large $\delta \geq 1$. For the small rarefaction $\delta = 0.1$, the expression (13.81) provides a good approximation for long tubes, $L \geq 5$. Thus, the criterion $\delta L > 0.5$ can be adopted for the validity of the end-correction formula for short tubes.

14.3 Transient Flow Through a Tube

In the previous sections, stationary flows have been considered when the macroscopic quantities are fully established and independent of the time. However, a knowledge of transient flows is also important in modeling of gas dynamics in diverse technological systems. In the present section, flows through orifices and tubes with characteristics varying in time will be considered. Thus, we consider the same geometrical configuration as shown in Figure 14.1 of Section 14.1 so that the same notations are adopted in this section. Let us assume that the inlet ($x_1 = -L/2$) of the tube is closed by a membrane, while the tube outlet is open so that the tube contains the gas at the pressure of the right recipient p_1. In the time instant $t = 0$, the membrane is removed instantaneously and the gas begins to flow through the tube from the left recipient to the right recipient. In this situation, we are interested in the mass flow rate $\dot{M}$ through the tube and flow field as a function of the time t. Numerical results of such flows based on the DSMC methods are reported in the papers [200, 201]. This problem was also solved on the basis of the S-model by Ho and Graur [202].

Just after the membrane opening, the distribution functions of the opposite flows in the inlet section are not affected by the intermolecular collisions. Thus, at any rarefaction parameter δ, the total flow rate through this section is equal to its free-molecular value for the corresponding pressure ratio p_1/p_0. Since the gas fills the tube with the pressure p_1 in the beginning, the flow rate through the outlet cross section will be zero at $t = 0$. Finally, at any rarefaction parameter δ and at any length L, the initial flow rate W $(t = 0)$ reads

$$W = \begin{cases} 1 - \frac{p_1}{p_0}, & \text{at} \quad x_1 = -L/2, \\ 0, & \text{at} \quad x_1 = L/2, \end{cases} \tag{14.28}$$

where W is given by (14.24). In order to describe the temporal evolution of W and other quantities, it is convenient to introduce the dimensional time as

$$\tilde{t} = \frac{v_0}{a} t. \tag{14.29}$$

The results on the flow rates W reported in the paper [201] for two values of the length-to-radius ratios $L = 1$ and 5 are shown in Figure 14.8, respectively. All flow rates begin from their initial values (14.28) and tend to their steady values given in Table 14.4. For the small and intermediate values of the rarefaction parameter, $\delta = 0.1$ and 1, the inlet flow rate (W at $x_1 = -L/2$) smoothly decreases from its initial value $(1 - p_1/p_0)$ to its stationary value for the all length-to-radius ratios L and pressure ratios p_1/p_0 considered here. Under the same conditions, the outlet flow rate (W at $x_1 = L/2$) smoothly increases from zero to its stationary value. Such a behavior is expected for small vales of δ because no shock wave appears.

The behavior of the flow rates for the large values of the rarefaction $\delta = 10$ and 100 qualitatively differs dependent on the values L and pressure ratio p_1/p_0. According to Figure 14.8 at $L = 1$ and $p_0/p_1 = 0.1$, the inlet flow rate W first increases exceeding its stationary value and then it approaches to the stationary value from above. For the higher pressure ratio, $p_1/p_0 = 0.5$, at the value of $L = 1$, both inlet and outlet flow rates smoothly increase up to their stationary value. However, to reach before the stationary value, the outlet flow rate, W at $x_1 = L/2$, exceeds than in the inlet. For the longer tube $L = 5$, the behaviors of the flow rates at $\delta = 10$ and 100 shown in Figure 14.8 are quite unexpected. At the small pressure ratio $p_1/p_0 = 0.1$, the inlet flow rate $(x_1 = -L/2)$ first sharply increases and then smoothly decreases tending to its stationary value. The outlet flow rate $(x_1 = L/2)$ increases from zero up to its stationary flow. At the larger pressure ratio $p_1/p_0 = 0.5$, the flow rate behaviors are most curious especially in the hydrodynamic regime $(\delta = 100)$. The inlet flow rate $(x_1 = -L/2)$ first increases, then

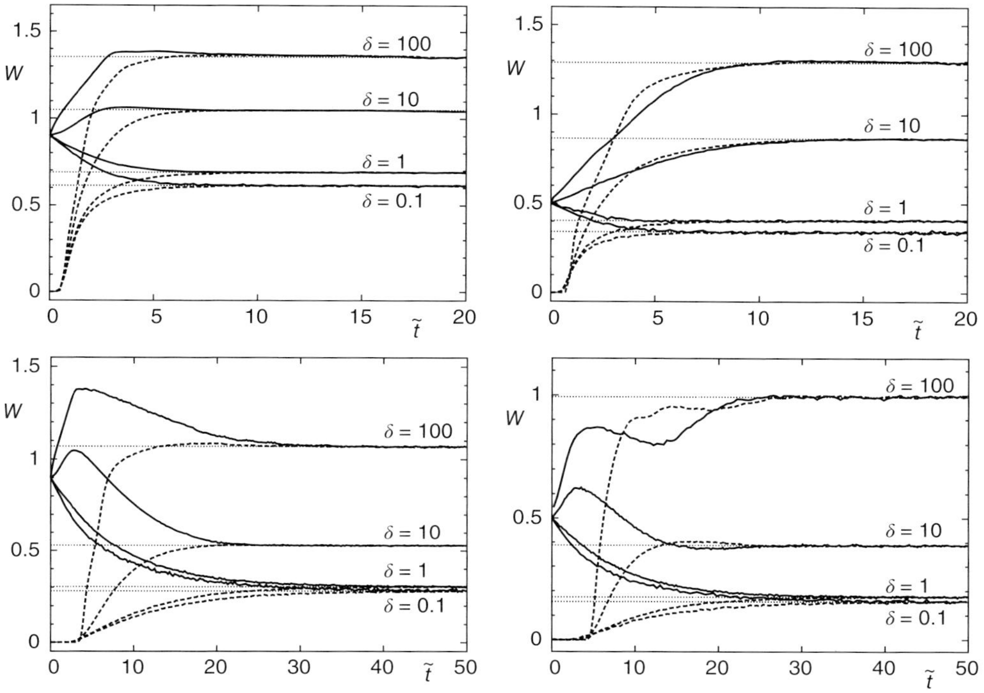

Figure 14.8 Flow rates W through short tube versus time $\tilde{t}$: up – $L = 1$, down – $L = 5$, left – $p_1/p_0 = 0.1$, right – $p_1/p_0 = 0.5$; solid lines – inlet, dashed lines – outlet, pointed lines – stationary flow, Table 14.4. Reproduced from Ref. [201] with permission.

decreases, and then increases again tending to its stationary value. The outlet flow rate (W at $x_1 = L/2$) sharply increases and slowly approaches to its stationary value. Such behavior can be explained by a shock wave moving to the downflow direction and by a rarefaction wave going to the upflow container.

The distributions along the symmetry axis of the density n/n_0, temperature T/T_0, and bulk velocity u_2/v_0 obtained in Ref. [201] at the pressure ratio $p_1/p_0 = 0.1$, rarefaction $\delta = 100$, and length-to-radius ratio $L = 5$ are shown in Figure 14.9. It can be clearly seen that a shock wave going from the inlet cross section to the outlet one and a rarefaction wave going in the opposite direction into the upflow container appear. It is curious that the temperature inside the tube first increases and then it drops. The velocity distribution always has a maximum, which moves to the downflow container. The flow fields in the whole domain considered here are established within the dimensionless time $\tilde{t}$ about 30. Naturally, it takes less time to establish the stationary flow in the upflow container than that in the downflow container.

In practice, it is interesting to know how long does it take to establish the stationary flow rates in both inlet and outlet sections. We may define the time $\tilde{t}_s$

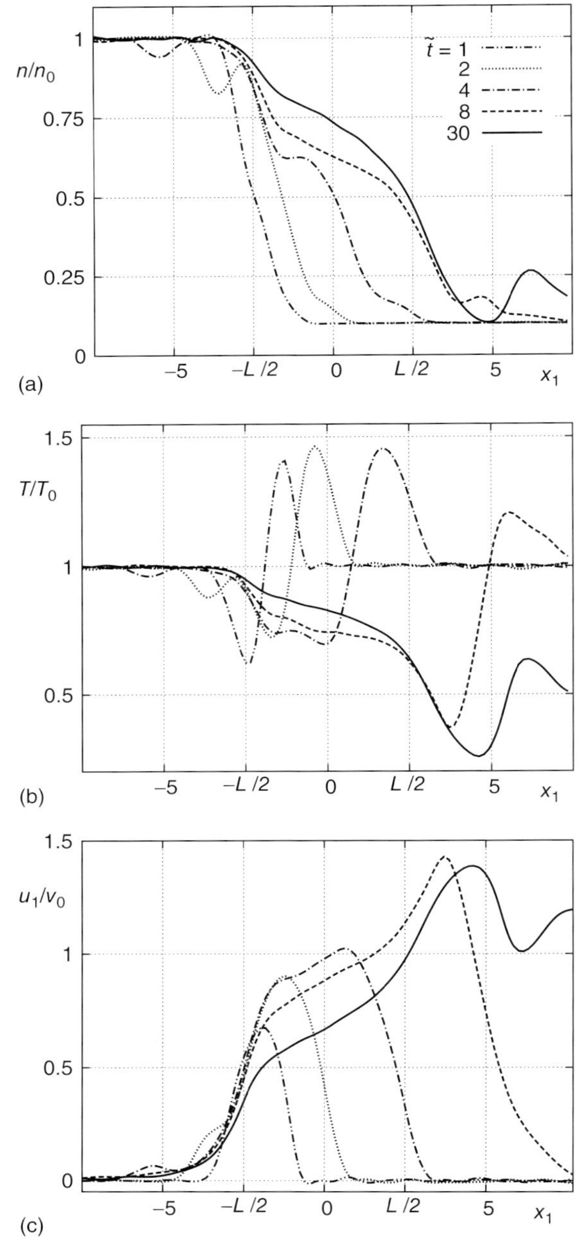

Figure 14.9 Distributions of the density n/n_0, temperature T/T_0, and longitudinal bulk velocity u_1/v_0 along tube axis at $p_1/p_0 = 0.1$, $\delta = 100$, and $L = 5$. Reproduced from Ref. [201] with permission.

Table 14.6 Approximate time $\tilde{t}_s$ needed to establish steady flow in tube. Reproduced from Ref. [201] with permission.

L	p_1/p_0	$\tilde{t}_s$			
		$\delta = 0.1$	1	10	100
0	0.1	—	6.4	7.3	4.4
	0.5	—	8.7	9.1	14
1	0.1	7	7	6.5	10
	0.5	6.5	7	13	12
5	0.1	45	40	23	30
	0.5	40	35	25	30

corresponding to the moment when both inlet and outlet flow rates differ from their stationary values for 1%. The deviation from the stationary value should vanish at $\tilde{t} \to \infty$ being always within 1% for any subsequent moment $\tilde{t} > \tilde{t}_s$. The values of $\tilde{t}_s$ calculated in the papers [200, 201] are given in Table 14.6. Note that these values are not very accurate because of the statistical scattering, but they give us an idea about the steady flow establishment. The magnitude of $\tilde{t}_s$ for orifice ($L = 0$) and short tube ($L = 1$) varies from 6.5 to 13. For the longer tube ($L = 5$), the time $\tilde{t}_s$ is naturally larger than that for the short tube and varies from 23 to 45. It is curious that for the long tube ($L = 5$), the smallest time $\tilde{t}_s$ is observed at $\delta = 10$. The values of $\tilde{t}_s$ for the orifice flow reported in work [202] based on the S-model are similar to those given in Table 14.6.

Exercises

14.1 Consider the helium gas flowing through a thin orifice of radius $a = 1$ mm due to the relative pressure drop $\Delta p/p_0 = 10^{-2}$ at the temperature $T_0 = 300$ K. Calculate the mass flow rate $\dot{M}$ for the upflow pressure (a) $p_0 = 0.23$ Pa, (b) $p_0 = 23$ Pa.
Solution: (a) Since $\delta = 0.01$, Eq. (14.7) is used. Then $\dot{M} = 3.7 \times 10^{-9}$ g/s; (b) Since $\delta = 1.0$, Eq. (14.15) with $\mathcal{G}_P = 1.37$ (see Table 14.1) and $\xi_P = 0.01$. Then $\dot{M} = 5.0 \times 10^{-7}$ g/s.

14.2 Repeat the previous exercise considering the pressure ratio $p_1/p_0 = 0.1$.
Solution: (a) Since $\delta = 0.01$, Eq. (14.7) is used. Then $\dot{M} = 3.3 \times 10^{-7}$ g/s; (b) Since $\delta = 1.0$, Eq. (14.24) with $W = 1.04$ (see Table 14.3). Then $\dot{M} = 3.8 \times 10^{-5}$ g/s. Note that the value of W for helium differs from that for argon, given in Table 14.3 for 2%.

14.3 Consider the helium gas flowing through a tube of radius $a = 1$ mm at the length-to-radius ratio $L = 1$ due to the relative pressure drop $\Delta p/p_0 = 10^{-2}$ at the temperature $T_0 = 300$ K. Calculate the mass flow rate $\dot{M}$ using Eq. (14.15) and then Eqs. (14.14) and (13.81) with Table 14.5 for the upflow pressure (a) $p_0 = 2.3$ Pa, (b) $p_0 = 230$ Pa.

Solution: (a) Using (14.15) with $\mathcal{G}_P = 0.696$ (see Table 14.1) and $\xi_P = 0.01$, we obtain $\dot{M} = 2.5 \times 10^{-8}$ g/s. Using Eqs. (14.14) and Eq. (13.81), $\dot{M} = 2.2 \times 10^{-8}$ g/s. (b) Using (14.15) with $\mathcal{G}_P = 2.63$ (see Table 14.1) and $\xi_P = 0.01$, we obtain $\dot{M} = 9.6 \times 10^{-6}$ g/s. Using Eqs. (14.14) and Eq. (13.81), $\dot{M} = 9.8 \times 10^{-6}$ g/s.

14.4 Consider the helium gas flowing through a tube of radius $a = 1$ mm and the length-to-radius ratio $L = 1$ at the pressure ratio $p_1/p_0 = 0.5$ and temperature $T_0 = 300$ K. Calculate the mass flow rate $\dot{M}$ for the upflow pressure (a) $p_0 = 0.23$ Pa, (b) $p_0 = 23$ Pa.
Solution: (a) Since $\delta = 0.01$, Eqs. (14.8) and (14.9) are used. Then $\dot{M} = 1.2 \times 10^{-7}$ g/s; (b) Since $\delta = 1.0$, Eq. (14.24) with $W = 0.405$ (see Table 14.4) is used. Then $\dot{M} = 1.5 \times 10^{-5}$ g/s.

14.5 Calculate the time t_s to establish the stationary flow under the conditions of the previous exercise. Consider only $p_0 = 23$ Pa.
Solution: (a) $t_s = a\tilde{t}_s/v_0$. According to Table 14.6, $\tilde{t}_s = 7$. Then $t_s = 6.3 \times 10^{-6}$ s.

15
Flows Through Long Pipes Under Arbitrary Pressure and Temperature Drops

15.1
Stationary Flows

15.1.1
Main Equations

In Sections 11.3, 12.3, and 13.1, the flow rates G_P^* and G_T^* through long channels and tubes were calculated as functions of the local rarefaction parameter δ. Once these coefficients are known, the total mass flow rate is calculated by Eqs. (11.80), (12.79), and (13.4) via the gradients of pressure ξ_P and temperature ξ_T. The temperature distribution and consequently its local gradient are determined by thermal properties of the pipe and can be measured or calculated independent of the gas flow. Here, we will use the word "pipe" meaning channel or tube with any kind of the cross section. However, the pressure distribution along a pipe and its gradient are not known and determined by the gas flow. In practice, only the pressures on the pipe ends are measured. Numerical calculations of the mass flow rate and flow field as functions of these pressures, see Sections 13.2 and 14.1, are difficult and become harder by increasing the pipe length. However, when the length is sufficiently long, the flow rate through such a pipe caused by large pressure and temperature drops can be easily calculated using the data on the coefficient G_P^* and G_T^* in spite of the fact that these coefficients are obtained from the linearized kinetic equation. In this chapter, a methodology to calculate flow rates through long pipes caused by arbitrary pressure and temperature drops is described.

Consider two chambers containing a gas and connected by a pipe as is depicted in Figure 15.1. The cross section of the pipe can be arbitrary with a characteristic size equal to a. The length of the pipe aL is significantly larger than its transversal size, $L \gg 1$. The gas in the left reservoir is maintained at a pressure p_0 and temperature T_0, while the pressure is p_1 and the temperature is T_1 in the right chamber. The origin of the longitudinal coordinate r is fixed at the left entrance of the pipe. Since the flow considered here is assumed to be one dimensional, the coordinate subscript is omitted. The temperature distribution along the pipe is denoted as $T_w(r)$.

Rarefied Gas Dynamics: Fundamentals for Research and Practice, First Edition. Felix Sharipov.

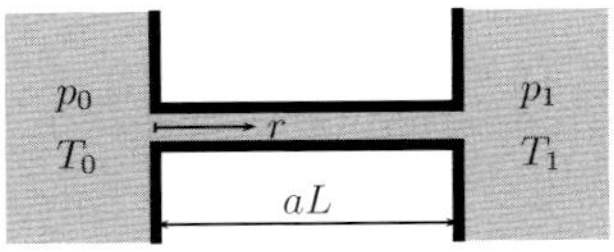

Figure 15.1 Scheme of stationary flow through long pipe.

To calculate the flow rate $\dot{M}$ between the chambers as the function of pressures p_0 and p_1 and temperatures T_0 and T_1, the two rarefaction parameters δ_0 and δ_1 are introduced as

$$\delta_0 = \frac{ap_0}{\mu_0 v_0}, \quad \delta_1 = \frac{ap_1}{\mu_1 v_1}. \tag{15.1}$$

The first parameter δ_0 is exactly the same defined by (7.46), while the second parameter δ_1 is defined via the viscosity μ_1 and MPS v_1 corresponding to the temperature T_1

$$\mu_1 = \mu(T_1), \quad v_1 = \left(\frac{2k_B T_1}{m}\right)^{1/2}. \tag{15.2}$$

For the problem under question, it is convenient to introduce the following dimensionless flow rate as

$$G := \dot{M}\frac{Lv_0}{Ap_0}, \tag{15.3}$$

where A is the cross-sectional area, for example, $A = a$, $A = ab$, and $A = \pi a^2$ for infinite plates, rectangular channel, and circular tube, respectively. The coefficient G can be expressed as

$$G = \frac{L}{\sqrt{\pi}}W, \tag{15.4}$$

where W is defined by Eqs. (13.76) and (14.24) for short channel and tube, respectively.

The pressure and temperature gradients in a long tube are estimated as

$$\xi_P = \frac{a}{p}\frac{dp}{dr} \sim \frac{p_0 - p_1}{p_0 L}, \quad \xi_T = \frac{a}{T}\frac{dT}{dr} \sim \frac{T_0 - T_1}{T_0 L}. \tag{15.5}$$

Since $L \gg 1$, the gradients ξ_P and ξ_T are always small even if the pressure and temperature drops are large. Then Eqs. (11.80), (12.79), and (13.4) are valid locally in each cross section of pipe. These equations can be written in a unique form as

$$\dot{M} = \frac{Ap}{v_m}\left[-G_P^*(\delta)\frac{a}{p}\frac{dp}{dr} + G_T^*(\delta)\frac{a}{T}\frac{dT}{dr}\right], \quad \delta = \frac{ap}{\mu v_m}, \tag{15.6}$$

where the pressure p and temperature T are local. The viscosity μ and MPS v_m are also referred to the local temperature T. Instead of the dimensionless coordinate x introduced by (7.43), here we will use the following:

$$x = \frac{r}{aL}, \tag{15.7}$$

which allows us to write the equation without the quantity L. Substituting the expression (15.6) into (15.3) and considering that the gas temperature in each section is equal to the pipe wall temperature T_w, the differential equation for the local pressure is obtained as

$$G = \frac{1}{p_0}\sqrt{\frac{T_0}{T_w}}\left[-G_{\mathrm{P}}^*(\delta)\frac{\mathrm{d}p}{\mathrm{d}x} + G_{\mathrm{T}}^*(\delta)\frac{p}{T_w}\frac{\mathrm{d}T_w}{\mathrm{d}x}\right]. \tag{15.8}$$

In general, this equation is solved numerically by a finite difference method.

If the flow rate G and the pressure p_0 are known, the integration of Eq. (15.8) is realized from $x = 0$ to $x = 1$ with the boundary condition $p(0) = p_0$. As a result of the integration, the pressure p_1 is obtained. If the pressures p_0 and p_1 are known, the quantity G is fitted to satisfy the boundary conditions $p(0) = p_0$ and $p(1) = p_1$. In the subsequent sections, some particular examples of application of Eq. (15.8) are given.

15.1.2 Isothermal Flows

First, let us consider an isothermal flow, $T_w = T_0 = T_1$. Then the integral equation (15.8) is simplified to

$$G\ \mathrm{d}x = -\frac{1}{\delta_0}G_{\mathrm{P}}^*(\delta)\,\mathrm{d}\delta, \tag{15.9}$$

where the relation $p/p_0 = \delta/\delta_0$ has been used. Integrating the left-hand side of this equality with respect to x from 0 to 1 and the right-hand side with respect to δ from δ_0 to δ_1, we obtain the expression for G as

$$G = \frac{1}{\delta_0}\int_{\delta_1}^{\delta_0} G_{\mathrm{P}}^*(\delta)\,\mathrm{d}\delta. \tag{15.10}$$

Once the function $G_{\mathrm{P}}^* = G_{\mathrm{P}}^*(\delta)$ is known, the integration (15.10) is easily performed. The interpolating formulas (11.136) and (12.119) can be used for the integration in case of planar channel and circular tube, respectively. Numerical values of G for tube flow are given in Table (15.1) for some values of δ_0 and pressure ratio p_1/p_0, which is equal to the parameter ratio δ_1/δ_0. The integration was performed using the interpolating formula (12.119). Analyzing the second and third columns of this table, we note that they are practically the same. It means that the flow rate does not depend on the downflow pressure p_1 when it is significantly smaller than p_0.

Additional details on the isothermal flow through a long tube under arbitrary pressure drop are given in Ref. [203]. Some examples of the coefficient G calculated via the integration (15.10) can be found in Ref. [148] for a rectangular channels and in Ref. [154] for an elliptical tube. An analysis of these numerical data shows that in case of a square channel $\beta = 1$ and a cylindrical tube, the approximation

$$G \approx \frac{\delta_0 - \delta_1}{\delta_0}G_{\mathrm{P}}^*\left(\frac{\delta_0 + \delta_1}{2}\right) \tag{15.11}$$

Table 15.1 Dimensionless flow rate G for tube calculated by (15.10) with (12.119).

	G				
δ_0	$\frac{p_1}{p_0} = 0$	0.01	0.1	0.5	0.9
0.1	1.438	1.424	1.290	0.7099	0.1411
1	1.420	1.405	1.276	0.7190	0.1470
10	2.385	2.371	2.243	1.481	0.3447
100	13.54	13.53	13.30	9.887	2.477

provides a good accuracy, that is, the disagreement between the exact integration (15.10) and approximate formula (15.11) does not exceed 2%. For a rectangular channel with a large aspect ratio, say $\beta = 100$, Eq. (15.11) provides an accuracy of about 6%.

When the flow rate G is known, the pressure distribution along a pipe can be calculated integrating Eq. (15.9) with respect to x from 0 to any value smaller than 1 and with respect to δ from δ_0 to any value larger than δ_1

$$x = \frac{1}{G\delta_0} \int_{\delta}^{\delta_0} G_{\mathrm{p}}^*(\delta)\, \mathrm{d}\delta, \tag{15.12}$$

where $\delta_1 \leq \delta \leq \delta_0$. This equation provides the function $x = x(\delta)$, which is inverted into $\delta = \delta(x)$. Since $p(x)/p_0 = \delta(x)/\delta_0$, the pressure distribution $p(x)$ is known. Some typical distributions for two values of the pressure ratio $p_1/p_2 = 0$ (obtained in Ref. [136]) and $p_1/p_2 = 0.5$ are shown in Figure 15.2. For the small value of the rarefaction $\delta_0 = 0.1$, the density distribution is linear. For the large values of rarefaction, $\delta = 10$ and 100, the distributions are slightly different from the linear form when $p_1/p_0 = 0.5$. However, in case of the gas flow into vacuum ($p_1/p_0 = 0$), the pressure linearly depends on the coordinate x near the inlet of the tube, and then it sharply decreases up to zero near the tube outlet.

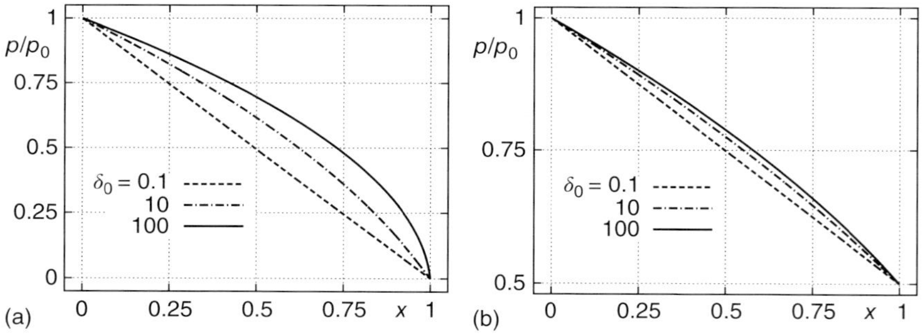

Figure 15.2 Pressure distribution along circular tube: - (a) $p_1/p_0 = 0$, reproduced from Ref. [136] with permission; -(b) $p_1/p_0 = 0.5$.

The expression (15.10) can be corrected taking into account the end effects calculated in Sections 13.3 and 14.2. Following the methodology described in Refs [187, 198], the integration of the left-hand side of (15.9) with respect to x in the range from $-\Delta L_0$ to $L + \Delta L_1$ leads to the expression

$$G = \frac{L}{\delta_0 \left(L + \Delta L_0 + \Delta L_1\right)} \int_{\delta_1}^{\delta_0} G_{\mathrm{P}}^*(\delta)\,\mathrm{d}\delta, \tag{15.13}$$

where the end corrections ΔL_0 and ΔL_1 correspond to δ_0 and δ_1, respectively. Physically, the expression (15.13) is treated as an increment of the pipe length for ΔL_0 and ΔL_1 at the left and right ends, respectively.

15.1.3
Nonisothermal Flows

If the temperatures T_0 and T_1 are different, Eq. (15.8) is solved numerically. To calculate the local rarefaction parameter $\delta(x)$, its dependence on the temperature should be known, which is determined by the viscosity $\mu = \mu(T)$. For the sake of simplicity, let us use the theoretical expression of the viscosity (6.28) corresponding to the HS potential. In this case, the local rarefaction parameter is related to δ_0 as

$$\delta(x) = \delta_0 \frac{T_0}{p_0} \frac{p(x)}{T_w(x)}. \tag{15.14}$$

To construct the finite difference scheme, the pressure gradient is obtained from (15.8)

$$\frac{\mathrm{d}p}{\mathrm{d}x} = \frac{1}{G_{\mathrm{P}}^*(\delta)} \left(G_{\mathrm{T}}^*(\delta) \frac{p}{T_w} \frac{\mathrm{d}T_w}{\mathrm{d}x} - \sqrt{\frac{T_w}{T_0}} p_0 G \right). \tag{15.15}$$

To approximate it, the regularly distributed nodes are introduced

$$x_k = \frac{k}{N_x}, \quad 0 \le k \le N_x, \tag{15.16}$$

where N_x is integer. Since the calculations are fast, Eq. (15.15) can be approximated by the one-sided scheme that is coarse but simple. Then the pressure in the node $k+1$ is expressed via the parameters in the previous one k

$$p\left(x_{k+1}\right) = p\left(x_k\right) + \frac{1}{G_{\mathrm{P}k}^*} \left[G_{\mathrm{T}k}^* p\left(x_k\right) \left(\frac{T_w(x_{k+1})}{T_w(x_k)} - 1 \right) \right.$$
$$\left. - \sqrt{\frac{T_w(x_k)}{T_0} \frac{p_0}{N_x}} G \right], \tag{15.17}$$

where the index k varies from 0 to $N_x - 1$, the quantities $G_{\mathrm{P}k}^*$ and $G_{\mathrm{P}k}^*$ are calculated via the local rarefaction (15.14) as

$$G_{\mathrm{P}k}^* = G_{\mathrm{P}}^* \left(\delta_0 \frac{T_0}{p_0} \frac{p(x_k)}{T_w(x_k)} \right), \quad G_{\mathrm{T}k}^* = G_{\mathrm{T}}^* \left(\delta_0 \frac{T_0}{p_0} \frac{p(x_k)}{T_w(x_k)} \right). \tag{15.18}$$

Table 15.2 Dimensionless flow rate G for tube calculated by (15.17) with (12.119) and (12.120), $T_1/T_0 = 0.263$.

	G				
δ_0	$\frac{p_1}{p_0} = 0$	0.01	0.1	0.5	0.9
0.1	1.466	1.438	1.192	0.1699	−0.7970
1	1.577	1.553	1.358	0.5446	−0.2939
10	3.604	3.592	3.469	2.400	0.5034
100	25.06	25.05	24.74	18.58	4.668

If the pressures $p_0 = p(0)$ and $p_1 = p(1)$ are known, then the quantity G is fitted to satisfy the boundary conditions for the pressures.

Usually, a reservoir with a lower pressure ($p_1 < p_0$) has also a lower temperature, $T_1 < T_0$. The value $T_1/T_0 = 0.263$ corresponds to an experimental situation when $T_1 = 77$ K (boiling temperature of nitrogen) and $T_0 = 293$ K (room temperature). Some values of the flow rate G for this temperature ratio, for the linear temperature distribution, $T_w(x) = T_0 + (T_1 - T_0)x$, for the same values of the pressure ratio p_1/p_0 and rarefaction δ chosen for the isothermal flow, see Table 15.1, are given in Table 15.2. Comparing the isothermal and nonisothermal flows, we see that the temperature variation can increase, decrease, and even change the sign of the flow rate G depending on the rarefaction δ and pressure ratio p_1/p_0. To understand such a behavior, let us return to the expression (15.8) containing the two terms in the right-hand side. Both pressure and temperature gradients are negative in our example so that the first term is positive, while the second term in negative. For the large values of the rarefaction $\delta = 100$, the lower temperature T_1 leads to an increase in the flow rate G because the rarefaction parameter given by (15.14) becomes larger at lower temperature. As a result, the Poiseuille coefficient G_P^* given by (12.86) for $\delta \gg 1$ increases, but the thermal creep coefficient G_P^* in this case is negligible. For the small value of the rarefaction $\delta_0 = 0.1$, both terms in Eq. (15.6) are comparable so that the thermal creep reduces the total flow rate G. When the pressure drop is small, $p_1/p_0 = 0.9$, the thermal creep dominates and the total flow rate G becomes negative, that is, the gas flows from the colder reservoir to the hotter reservoir.

Additional data on the nonisothermal flow through long pipes of various shapes of the cross can be found in Refs [142, 155, 204, 205].

15.2 Pipes with Variable Cross Section

Equation (15.8) can be generalized for a pipe of variable cross section following the methodology proposed in Ref. [206]. For the sake of simplicity, only a flow through a circular tube having a variable radius $a = a(x)$ is considered here. It is assumed that the tube radius smoothly depends on the r coordinate, that is, the derivative

$$\left|\frac{da(r)}{dr}\right| \ll 1, \tag{15.19}$$

is sufficiently small. In this case, we should specify the cross-sectional area in the definitions (15.3) and (15.7)

$$G := \dot{M}\frac{Lv_0}{\pi a_0^2 p_0}, \quad x = \frac{r}{a_0 L}, \quad a_0 = a(0). \tag{15.20}$$

Then the relation (15.8) is generalized as

$$G = \frac{1}{p_0}\sqrt{\frac{T_0}{T_w}}\left[\frac{a(x)}{a_0}\right]^3 \left[-G_P^*(\delta)\frac{dp}{dx} + G_T^*(\delta)\frac{p}{T_w}\frac{dT_w}{dx}\right]. \tag{15.21}$$

This differential equation is solved by the same method described in the previous section.

Some numerical examples of the isothermal flow, $T_w = T_0 = T_1$, calculated in Ref. [206] on the basis of Eq. (15.21) are given in Table (15.3), where a conical tube is considered,

$$a(x) = a_0 + (a_1 - a_0)x, \tag{15.22}$$

where $a_1 = a(1)$ is the tube radius at the tube outlet. The pressure ratio p_1/p_0 and rarefaction are the same as those chosen for the isothermal flow through a tube of constant radius (see Table 15.1). Evidently, the flow rate through the cone tube with the increasing radius is larger than that through the tube of the constant radius.

An analytical integration of (15.21) can be performed in the free molecular ($\delta_1, \delta_2 \ll 1$) and viscous regimes ($\delta_1, \delta_2 \gg 1$) where the Poiseuille coefficient G_P^* has a simple expression. Using Eq. (12.105), the flow rate G takes the form

$$G = \frac{16}{3\sqrt{\pi}}\frac{(a_1/a_0)^2}{1 + a_1/a_0}\left(1 - \frac{p_1}{p_0}\right), \quad \text{for} \quad \delta_0, \delta_1 \ll 1. \tag{15.23}$$

Table 15.3 Dimensionless flow rate G for the tube of variable radius calculated by (12.119) and (12.120), $a_1/a_0 = 10$, T =constant, Ref. [206].

	G			
δ_0	$\frac{p_1}{p_0} = 0$	**0.1**	**0.5**	**0.9**
0.1	25.95	23.21	12.78	2.547
1	25.99	23.57	13.80	2.910
10	52.29	50.66	35.56	8.531
100	354.5	350.1	263.1	66.21

Considering only the main term in Eq. (12.86), the integration of (15.21) leads to

$$G = \frac{3\delta_0}{8} \frac{(a_1/a_0)^3}{1 + a_1/a_0 + (a_1/a_0)^2} \left[1 - \left(\frac{p_1}{p_0}\right)^2\right], \quad \text{for} \quad \delta_0, \delta_1 \gg 1. \qquad (15.24)$$

These expressions give an idea about the influence of the radius variation.

The paper [206] contains some examples of the nonisothermal flow through a conical tube, which show that the influence of the temperature variation has the same tendency as that for a tube of constant radius, that is, the flow rate G increases by decreasing T_1 in the hydrodynamic regime. However, the flow rate decreases and even changes its own sign by decreasing T_1 near the free-molecular regime.

15.3 Transient Flows

15.3.1 Main Equations

In this section, the methodology to apply the coefficients G_{P}^* and G_{T}^* to gas flows through long capillaries under arbitrary pressure and temperature drops is generalized to the transient flows when the pressure distribution and mass flow rate vary with time. Such a generalization was proposed in Ref. [207]. In this case, we should specify not only pressures p_0, p_1 and temperatures T_0, T_1 on the capillary ends, but also the reservoirs of volumes V_0 and V_1 connected by a capillary as is shown in Figure 15.3. In general, all these quantities could be functions of time.

As was shown in Ref. [208], the characteristic time to establish a steady flow over a cross section of capillary has the order of a/v_m. According to the numerical results shown in Section 14.3, the time to establish a steady flow over the whole capillary increases by increasing its length. Thus, under the assumption $L \gg 1$, the time to establish a steady flow in a cross section is significantly smaller than that to establish a steady flow in the whole capillary. It allows us to apply a steady solution for each cross section, that is, the mass flow rate $\dot{M}(t, r)$ is now a function of the time t and position r, and it is written as

$$\dot{M}(t, r) = \frac{Ap}{v_m} \left[-G_{\mathrm{P}}^*(\delta) \frac{a}{p} \frac{\partial p}{\partial r} + G_{\mathrm{T}}^*(\delta) \frac{a}{T} \frac{\partial T}{\partial r}\right]. \qquad (15.25)$$

The mass balance in a cross section can be expressed in terms of the local mass density $\rho(t, r)$ as

$$A \frac{\partial \rho}{\partial t} = -\frac{\partial \dot{M}}{\partial r}. \qquad (15.26)$$

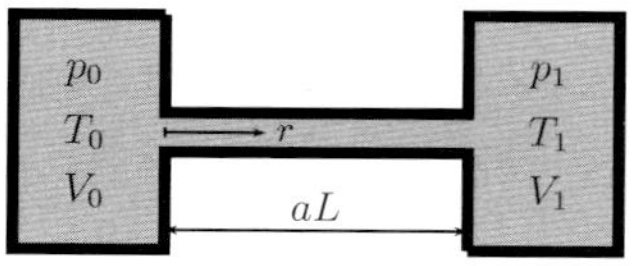

Figure 15.3 Scheme of transient flow through long pipe.

Combining this equation with Eq. (15.25) and with the state equation (1.9), the following equation for the evolution of the pressure $p(t, r)$ inside the capillary is obtained

$$\frac{\partial p}{\partial t} = \frac{a v_m^2}{2} \frac{\partial}{\partial r} \frac{p}{v_m} \left[G_{\mathrm{P}}^*(\delta) \frac{\partial \ln p}{\partial r} - G_{\mathrm{T}}^*(\delta) \frac{\partial \ln T}{\partial r} \right] + \frac{p}{T} \frac{\partial T}{\partial t}. \tag{15.27}$$

Again, we assume that the temperature distribution of the gas along the capillary is the same as that of the capillary wall and given as a function of the time and position, $T = T_w(t, r)$. If we assume that the containers do not have any other gas inflow or outflow apart from those through the capillary, the pressures $p_0(t)$ and $p_1(t)$ in the containers are determined by their volumes V_0 and V_1, temperatures T_0 and T_1, and mass flow rate $\dot{M}$ through the capillary ends. Thus, the dependence of the pressures $p_0(t)$ and $p_1(t)$ on time is unknown *a priori*.

To solve Eq. (15.27), an initial condition, $p(0, r)$, and two boundary conditions,

$$p(t, r) = p_0(t)|_{r=C}, \quad p(t, r) = p_1(t)|_{r=aL}, \tag{15.28}$$

are needed. The initial pressure distribution $p(0, r)$ is usually known, while the pressures $p_0(t)$ and $p_1(t)$ are obtained from the mass balance in the containers. Starting from the state equation in each container

$$M_0 = \frac{m p_0 V_0}{k_{\mathrm{B}} T_0} = 2 \frac{p_0 V_0}{v_0^2}, \quad M_1 = \frac{m p_1 V_1}{k_{\mathrm{B}} T_1} = 2 \frac{p_1 V_1}{v_1^2}, \tag{15.29}$$

where M_0 and M_1 are the total masses of gas in the containers, and considering that

$$\frac{\mathrm{d}M_0}{\mathrm{d}t} = -\dot{M}|_{r=C}, \quad \frac{\mathrm{d}M_1}{\mathrm{d}t} = \dot{M}|_{r=aL}, \tag{15.30}$$

the pressure variations are obtained as

$$\frac{\mathrm{d} \ln p_0}{\mathrm{d}t} = \frac{v_0 V_p}{2 V_0} \left(G_{\mathrm{P}}^* \frac{\partial \ln p}{\partial r} - G_{\mathrm{T}}^* \frac{\partial \ln T}{\partial r} \right) \Bigg|_{r=0} - \frac{\mathrm{d}}{\mathrm{d}t} \ln \left(\frac{V_0}{T_0} \right), \tag{15.31}$$

$$\frac{\mathrm{d} \ln p_1}{\mathrm{d}t} = -\frac{v_1 V_p}{2 V_1} \left(G_{\mathrm{P}}^* \frac{\partial \ln p}{\partial r} - G_{\mathrm{T}}^* \frac{\partial \ln T}{\partial r} \right) \Bigg|_{r=aL} - \frac{\mathrm{d}}{\mathrm{d}t} \ln \left(\frac{V_1}{T_1} \right), \tag{15.32}$$

where $V_p = aA$ is the pipe volume. Thus, Eq. (15.27) with the boundary conditions (15.31) and (15.32) completely determines the solution of the problem on nonsteady flow of rarefied gas through a long capillary. Once the pressure distribution $p(t, r)$ is known, the mass flow rate through any cross section is given by Eq. (15.25).

15.3.2 Approaching to Equilibrium

As an example of application of the above-described technique, consider a nonsteady isothermal flow of rarefied gas through a circular tube of radius a. The temperatures in both containers and that of the tube are maintained constant, $T_w = T_0 = T_1$. The pressures in the tube and in right container are initially so small that they can be assumed to be zero, $p(0,r) = 0$ and $p_1(0) = 0$. We are going to calculate the evolution of both pressures $p_0(t)$ and $p_1(t)$ assuming $V_0 = V_1 = V_c$. The solution is determined by the initial rarefaction parameter in the left container denoted as

$$\delta_{00} = \frac{a p_{00}}{\mu_0 v_0}, \quad p_{00} = p_0(t)|_{t=0}. \tag{15.33}$$

Under the isothermal conditions, Eqs. (15.27), (15.31), and (15.32) are simplified to

$$\frac{\partial p}{\partial \tilde{t}} = \left(\frac{\partial G_{\mathrm{P}}^*}{\partial x}\frac{\partial p}{\partial x} + G_{\mathrm{P}}^* \frac{\partial^2 p}{\partial x^2} \right), \tag{15.34}$$

$$\frac{\mathrm{d}p_0}{\mathrm{d}\tilde{t}} = G_{\mathrm{P}}^* \left.\frac{\partial p}{\partial x}\right|_{x=0}, \quad \frac{\mathrm{d}p_1}{\mathrm{d}\tilde{t}} = -G_{\mathrm{P}}^* \left.\frac{\partial p}{\partial x}\right|_{x=1}, \tag{15.35}$$

where the dimensionless time has been introduced as

$$\tilde{t} = \frac{v_0}{2aL^2} t. \tag{15.36}$$

When the pressure $p(\tilde{t}, x)$ is known, the local dimensionless flow rate is calculated as

$$G(\tilde{t}, x) = -G_{\mathrm{P}}^*(\delta)\frac{\partial p}{\partial x}. \tag{15.37}$$

The finite difference scheme to solve Eq. (15.34) is based on the nodes (15.16) for the coordinate x. The time is advanced by the step $\Delta\tilde{t}$,

$$\tilde{t}_{i+1} = \tilde{t}_i + \Delta\tilde{t}, \quad \tilde{t}_0 = 0. \tag{15.38}$$

Then, Eq. (15.34) is approximated as

$$\begin{aligned} p_{i+1,k} = p_{ik} + \Big\{ \frac{1}{4} \left[G_{\mathrm{P}}^*\left(\delta_{i,k+1}\right) - G_{\mathrm{P}}^*\left(\delta_{i,k-1}\right)\right] \left(p_{i,k+1} - p_{i,k-1}\right) \\ + G_{\mathrm{P}}^*(\delta_{ik})(p_{i,k+1} - 2p_{ik} + p_{i,k-1})\Big\} N_x^2 \Delta\tilde{t}, \end{aligned} \tag{15.39}$$

where

$$p_{ik} = p(\tilde{t}_i, x_k), \quad \delta_{ik} = \delta_{00}\frac{p_{ik}}{p_{00}}. \tag{15.40}$$

Equations (15.35) are approximated as

$$p_{i+1,0} = p_{i,0} + G_{\mathrm{P}}^*\left(\delta_{i,0}\right)\left(p_{i,1} - p_{i,0}\right) N_x \Delta\tilde{t}, \tag{15.41}$$

$$p_{i+1,N_x} = p_{i,N_x} - G_{\mathrm{P}}^*\left(\delta_{i,N_x}\right)\left(p_{i,N_x} - p_{i,N_x-1}\right) N_x \Delta\tilde{t}. \tag{15.42}$$

The numerical calculations are carried out for the scheme parameters $N_x = 100$ and $\Delta\tilde{t} = 10^{-6}$, which provide the numerical error of the flow rate G and pressures p_0 and p_1 less than 1%.

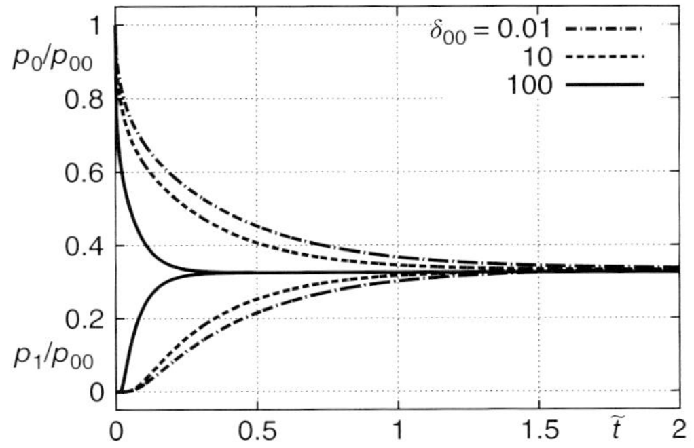

Figure 15.4 Pressures p_0/p_{00} and p_1/p_{00} for transient flow through long tube versus dimensionless time $\tilde{t}$. Reproduced from Ref. [207] with permission.

The evolution of the pressure p_0 and p_1 for three values of the rarefaction parameter $\delta_{00} = 0.01$, 10, and 100 are plotted on Figure 15.4. An analysis of the data corresponding to $\delta_0 = 0.1$ and $\delta_0 = 1$ showed that the functions $p_0(t)$ and $p_1(t)$ are practically the same as those in the free-molecular regime $\delta_0 = 0.01$. From Figure 15.4 it can be seen that in all cases the pressures p_0 and p_1 tend to the value $p_{00}/3$ because initially the gas fills only one container, while in the equilibrium state ($\tilde{t} \to \infty$) the same quantity of the gas fills both containers and the capillary in each of them having the same volumes. The time to establish the equilibrium pressures decreases by increasing the rarefaction parameter δ_{00}.

The dependence of the dimensionless flow rates G on the dimensionless time $\tilde{t}$ in the inlet and outlet is plotted in Figure 15.5 for the limit values of the rarefaction parameter $\delta_{00} = 0.01$ and $\delta_{00} = 100$. It shows that the inlet flow rate $G(\tilde{t}, 0)$ has a high value in the beginning because of the high pressure gradient at $x = 0$. The outlet flow rates $G(\tilde{t}, 1)$ increase from zero and then vanish in the equilibrium

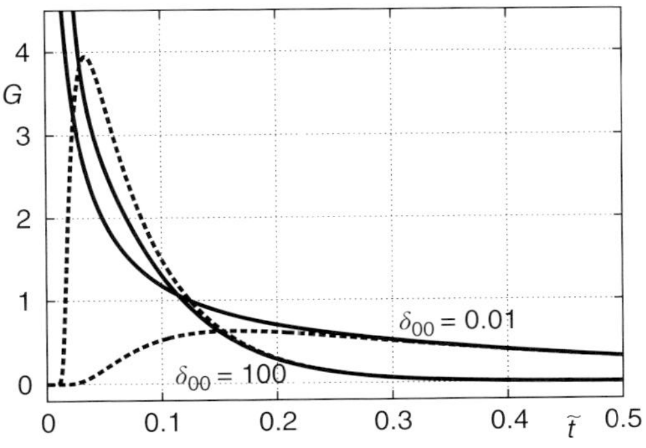

Figure 15.5 Dimensionless flow rate G given by Eq. (15.37) for transient flow through long tube versus dimensionless time $\tilde{t}$: solid line – inlet $x = 0$, dashed line – outlet $x = 1$. Reproduced from Ref. [207] with permission.

Table 15.4 Dimensionless time $\tilde{t}_{eq}$ needed to establish the equilibrium in tube versus δ_{00}, Ref. [207].

δ_{00}	0.01	0.1	1	10	100
$\tilde{t}_{eq}$	1.84	1.88	1.94	1.46	0.41

state ($\tilde{t} \to \infty$). The maximum values and time to reach this value depends on the rarefaction parameter δ_{00}. In the hydrodynamic regime ($\delta_{00} = 100$), the maximum of G is equal to 3.9 reached at $\tilde{t} = 0.034$, while in the free-molecular regime ($\delta_{00} = 0.01$), the maximum is quite smaller, $G_{max} = 0.63$, which is reached quite later, at $\tilde{t} = 0.16$.

It is interesting to know the time $\tilde{t}_{eq}$ needed to reach the equilibrium when the pressures p_0 and p_1 are the same and the flow rate G is zero. Since the flow rate for the problem in question never vanishes but always approaches to zero, the time to reach the equilibrium could be defined as that when both flow rates in the inlet and outlet decrease up to the value 0.01. The values of $\tilde{t}_{eq}$ are given in Table 15.4. It is curious that the time t_{eq} reaches its maximum in the transitional regime, at $\delta_{00} = 1$. This time tends to a constant value in the free-molecular regime ($\delta_{00} = 0.01$) and decreases in the hydrodynamic regime, at $\delta_{00} \to \infty$.

Exercises

15.1 Consider the helium gas flowing through a circular tube of radius $a = 1$ mm and length-to-radius ratio of $L = 100$ at the temperature $T_0 = 300$ K. Calculate the mass flow rate $\dot{M}$ under the following conditions: (a) $p_0 = 23$ Pa, $p_1 = 0.5p_0$; (b) $p_0 = 2.3$ Pa, $p_1 = 0.1p_0$.
Solution: (a) Using Eq. (15.3) with $G = 0.7190$ (see Table 15.1), we obtain $\dot{M} = 4.7 \times 10^{-7}$ g/s; (b) Using Eq. (15.3) with $G = 1.290$ (see Table 15.1), we obtain $\dot{M} = 8.3 \times 10^{-8}$ g/s.

15.2 Consider the previous exercise with the difference that the tube temperature varies linearly from T_0 to $T_1 = 0.263T_0$.
Solution: (a) Using Eq. (15.3) with $G = 0.5446$ (see Table 15.2), we obtain $\dot{M} = 3.5 \times 10^{-7}$ g/s; (b) Using Eq. (15.3) with $G = 1.192$ (see Table 15.2), we obtain $\dot{M} = 7.7 \times 10^{-8}$ g/s.

15.3 Consider a circular tube of radius $a = 1$ mm and length-to-radius ratio of $L = 100$ connecting two containers of volumes $V_0 = V_1 = \pi a^3 L$, see Figure 15.3, at the temperature $T_0 = 300$ K. In the beginning, the left container is closed from the tube by a membrane at a pressure p_{00}, while the tube and right container are maintained at significantly lower pressure $p_1 \ll p_{00}$. Calculate the time t_{eq} to establish the equilibrium after the membrane removal. Consider (a) $p_{00} = 2.3$ kPa, (b) $p_{00} = 0.23$ Pa.
Solution: (a) Since $\delta_{00} = 100$, Eq. (15.36) is used with $\tilde{t}_{eq} = 0.41$ (see Table 15.4). Then $t_{eq} = 7.3 \times 10^{-3}$ s. (b) Since $\delta_{00} = 0.01$, Eq. (15.36) is used with $\tilde{t}_{eq} = 1.84$ (see Table 15.4). Then $t_{eq} = 0.033$ s.

16
Acoustics in Rarefied Gases

16.1
General Remarks

16.1.1
Description of Waves in Continuous Medium

Wave propagations are usually described using the mathematical formalism based on complex numbers, see Section A.9, which will be employed in this chapter. Let us consider a plane harmonic wave propagating in the direction r_1 through a continuous medium. According to the classical theory of sound propagation [23], all properties of the medium oscillate near their equilibrium values according to

$$\Psi(r_1, t) = A_{\Psi 0} \Re \exp[i(kr_1 - \omega t)], \tag{16.1}$$

where Ψ is a deviation of some quantity, $A_{\Psi 0}$ is the deviation amplitude, i is the imaginary unit, $\Re$ means the real part of complex number, k is the wave number, and ω is the angular frequency of oscillation. The wave number k is a complex number

$$k = k_{\mathrm{R}} + ik_{\mathrm{I}}, \tag{16.2}$$

so that its real part is related to the sound speed v_s as

$$k_{\mathrm{R}} = \frac{\omega}{v_s}. \tag{16.3}$$

The imaginary part k_{I} represents the so-called attenuation characterizing the rate of decay of the amplitude

$$A_{\Psi}(r_1) = A_{\Psi 0} e^{-k_{\mathrm{I}} r_1}. \tag{16.4}$$

Thus, the quantities $A_{\Psi 0}$ and ω characterize a wave generator, while the complex wave number k depends on the properties of the medium where the wave propagates.

The concept of continuous medium is applicable under two conditions. First of all, the characteristic size of the wave propagation should be larger than the EFP. Second, the wave frequency ω should be quite smaller than the frequency of intermolecular collisions having the order p/μ (see Eq. (7.15)). If at least one of

Rarefied Gas Dynamics: Fundamentals for Research and Practice, First Edition. Felix Sharipov.

these conditions is broken, the wave propagation is described on the basis of the kinetic equation.

Some authors, see example, Refs [209–214], analyzed a collisionless wave propagation and found that the variations of the gas properties cannot be described by the simple expression (16.1), but the wave number is not constant anymore. In this case, the properties can be represented in the following form:

$$\Psi(r_1, t) = A_\Psi(r_1)\exp[i(\varphi_\Psi(r_1) - \omega t)], \tag{16.5}$$

where A_Ψ and φ_Ψ are amplitudes and phases, respectively. These quantities are functions of the coordinate r_1 and are different for each gas property. This chapter is devoted to the calculation of the functions $A_\Psi(r_1)$ and $\varphi_\Psi(r_1)$ for transversal and longitudinal waves propagating between two plane plates.

16.1.2
Complex Perturbation Function

Let us imagine that some characteristics of a solid surface bounding a gas oscillate with an angular frequency ω and their intensities are characterized by some small parameter ξ. When such a source acts during a long time, the gas flow becomes established and its perturbation function also harmonically depends on the time. In the frame of this formalism, the complex perturbation function is introduced as

$$f(t, \boldsymbol{r}, \boldsymbol{v}) = f_0^{\mathrm{M}}(\boldsymbol{v})\{1 + \xi\ \Re[h(\boldsymbol{r}, \boldsymbol{v})e^{-i\omega t}]\}, \tag{16.6}$$

where $\Re$ denotes the real part of complex number in accordance with (A.39). Substituting (16.6) into the moment definitions (2.4), (2.16), (2.26), (2.28), and (2.32), they can be written down via their perturbation as

$$\begin{bmatrix} n(t,\boldsymbol{r})/n_0 - 1 \\ \boldsymbol{u}(t,\boldsymbol{r})/v_0 \\ T(t,\boldsymbol{r})/T_0 - 1 \\ \mathsf{P}(t,\boldsymbol{r})/p - \mathsf{I} \\ \boldsymbol{q}/p_0 v_0 \end{bmatrix} = \xi\Re \begin{bmatrix} \varrho(\boldsymbol{r}) \\ \tilde{\boldsymbol{u}}(\boldsymbol{r}) \\ \tau(\boldsymbol{r}) \\ \Pi(\boldsymbol{r}) \\ \tilde{\boldsymbol{q}}(\boldsymbol{r}) \end{bmatrix} e^{-i\omega t}. \tag{16.7}$$

The complex functions ϱ, $\tilde{\boldsymbol{u}}$, τ, Π, and $\tilde{\boldsymbol{q}}$ are related to the complex perturbation function by the same manner as defined in Section 5.2, more exactly, by the relation (5.17). According to Eq. (A.40), any complex number can be written in terms of the absolute value and argument. Physically, the absolute values are oscillation amplitudes A and the arguments are wave phases φ, so that the moments can be represented as

$$\begin{aligned} &\varrho = A_\varrho e^{i\varphi_\varrho}, \quad \tilde{\boldsymbol{u}} = \boldsymbol{A}_u e^{i\varphi_u}, \quad \tau = A_\tau e^{i\varphi_\tau}, \\ &\Pi = \mathsf{A}_\Pi e^{i\varphi_\Pi}, \quad \tilde{\boldsymbol{q}} = \boldsymbol{A}_q e^{i\varphi_q}. \end{aligned} \tag{16.8}$$

A substitution of the representation (16.6) into the full BE (3.1) leads to the following linearized BE:

$$-i\omega h + \boldsymbol{v}\cdot\nabla_r h = \hat{L}h, \tag{16.9}$$

instead of the form (5.20). To write down the model equation in the dimensionless form, the dimensionless velocity $\boldsymbol{c}$ defined by (5.2) will be used here. However, the dimensionless coordinates defined by (7.43) are not used for oscillating flows, but it is more convenient to define them via the oscillation frequency as

$$\boldsymbol{x} = \frac{\omega}{v_0}\boldsymbol{r}. \tag{16.10}$$

A gas flow described by the linearized BE (16.9) is also characterized by the rarefaction parameter δ defined by (7.46), but it does not include the oscillation frequency ω determining the flow as well. Thus, one more quantity determining such flows called oscillating speed parameter will be used here. It is defined via the equilibrium pressure p_0 and viscosity μ_0 at the equilibrium temperature as

$$\theta = \frac{p_0}{\mu_0\omega}. \tag{16.11}$$

Since the ratio p_0/μ_0 has the order of the intermolecular collisions, the parameter θ represents the ratio of the collisions frequency to the oscillating frequency ω. Thus, the limit $\theta \to 0$ corresponds to a flow with an extremely high frequency ω in comparison with the molecular collision frequency, while the opposite limit $\theta \to \infty$ represents oscillations with a low frequency so that the gas flow can be considered as quasi-stationary.

Dividing Eq. (16.9) by ω and considering the linearized model operators (7.8), (7.19), and (7.34), the linearized model equation can be written in the following form:

$$-ih + \boldsymbol{c}\cdot\nabla_x h = A_1\theta(H-h), \tag{16.12}$$

where the complex function H is expressed by (7.48) with the complex moments defined by (16.7). Formally, this equation is stationary so that it can be solved by the DVM described in Chapter 9.

If we assume that the gas–surface interaction happens instantaneously, that is, the time interaction is significantly smaller than $1/\omega$, the boundary condition for the complex perturbation is the same as that for the real one and also given by (5.50), where the surface source function h_w can be complex.

16.1.3 One-Dimensional Flows

Here, we will restrict ourselves by considering a gas confined between two infinite plates fixed at $r_1 = 0$ and $r_1 = a$. The left surface ($r_1 = 0$) has some characteristics oscillating in time, while in the right plate properties are fixed. In such a situation, the gas flow depends only on one dimensionless coordinate x_1 defined by (16.10) so that the subscript "1" at x_1 will be further omitted. Moreover, we introduce the dimensionless distance L, which is related to the two determining parameters θ and δ

$$L = \frac{\omega}{v_0}a = \frac{\delta}{\theta}. \tag{16.13}$$

In this particular case, Eq. (16.12) is reduced to

$$-(i - A_1\theta)h + c_1\frac{\partial h}{\partial x} = A_1\theta H, \tag{16.14}$$

where the coordinate x varies in the interval $[0, L]$. If we replace x by x' in this equation, multiply it by $\exp[-(i - A_1\theta)x'/c_1]$ and integrate it with respect to x' from 0 to x for $c_1 > 0$ and from L to x for $c_1 < 0$, then it can be written down in the integral form

$$h(x, \boldsymbol{c}) = h(0, \boldsymbol{c}) \exp\left[(i - A_1\theta)\frac{x}{c_1}\right] + \frac{A_1\theta}{c_1}\int_0^x H(x', \boldsymbol{c}) \times \exp\left[(i - A_1\theta)\frac{x - x'}{c_1}\right] \mathrm{d}x' \quad \text{at} \quad c_1 > 0, \tag{16.15}$$

$$h(x, \boldsymbol{c}) = h(L, \boldsymbol{c}) \exp\left[(i - A_1\theta)\frac{L - x}{|c_1|}\right] - \frac{A_1\theta}{c_1}\int_x^L H(x', \boldsymbol{c}) \times \exp\left[(i - A_1\theta)\frac{x' - x}{|c_1|}\right] \mathrm{d}x' \quad \text{at} \quad c_1 < 0, \tag{16.16}$$

where the perturbations $h(0, \boldsymbol{c})$ for $c_1 > 0$ and $h(L, \boldsymbol{c})$ for $c_1 < 0$ are given by the boundary conditions.

Some useful conclusions can be made from the expressions (16.15) and (16.16). First, the perturbation h is an oscillating function of c_1 with an increasing frequency when the velocity c_1 tends to zero. Such a behavior is due to the terms e^{iA/c_1} with $A > 0$. It means that the nodes of the molecular velocity c_1 should be carefully chosen taking into account this behavior. Second, the oscillation amplitude of the perturbation tends to zero because of the terms e^{-A/c_1}

$$\lim_{c_1 \to 0} h(x, \boldsymbol{c}) = 0 \tag{16.17}$$

when the parameters θ and δ are different from zero.

16.2 Oscillatory Couette Flow

16.2.1 Definitions

Here, we consider a gas flow similar to that described in Section 11.1, namely, a gas confined between two parallel plates separated by a distance a as is shown in Figure 16.1. The coordinate origin is fixed at the left plate, which oscillates in its own plane along the axis r_2, with a frequency ω so that the wall velocity $\boldsymbol{u}_w$ has only the second component depending on the time as

$$\boldsymbol{u}_w = [0, u_w, 0], u_w = u_m \cos(\omega t) = u_m \, \Re e^{-i\omega t}, \tag{16.18}$$

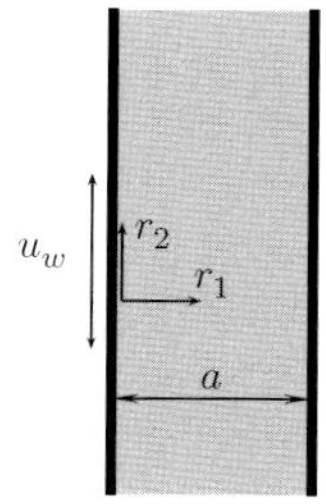

Figure 16.1 Scheme of oscillating Couette flow.

where u_m is the speed amplitude. We assume that the magnitude u_m is small compared to the most probable molecular speed v_0 so that the ratio

$$\xi = \frac{u_m}{v_0}, \quad \xi \ll 1 \tag{16.19}$$

is used as the small parameter for the linearization.

After a sufficiently long time, a transversal wave is established between the plates. In this case, only two quantities oscillate, namely, the bulk velocity u_2 and shear stress P_{12}, which are linearized as

$$u_2(t, r_1) = \xi v_0 \ \Re[\tilde{u}(x) e^{-i\omega t}], \quad P_{12}(t, r_1) = \xi p_0 \ \Re[\Pi(x) e^{-i\omega t}], \tag{16.20}$$

where subscripts "2" and "12" have been omitted at $\tilde{u}$ and Π, respectively. Moreover, the unique coordinate x is related to r_1 via (16.10). Once the complex functions $\tilde{u}(x)$ and $\Pi(x)$ are known, the amplitudes A_u, A_Π and phases φ_u, φ_Π are calculated via the representations (16.8).

16.2.2
Slip Regime

The Navier–Stokes equation can be applied to the stationary Couette flow, see Section 11.1, if the rarefaction parameter is large $\delta \gg 1$. In case of the nonstationary flow considered here, the applicability of the Navier–Stokes equation is subject to one more condition, namely, the oscillation parameter θ must also be large $\theta \gg 1$. Physically, this condition means that the frequency of the intermolecular collisions must be quite larger than the frequency of variation in macroscopic quantities. A combination of the momentum conservation law (3.27) and Newton's law (6.2) leads to the following equations:

$$-\rho \frac{\partial u_2}{\partial t} + \mu \frac{\partial^2 u_2}{\partial r_1^2} = 0, \quad P_{12} = -\mu \frac{\partial u_2}{\partial r_1}. \tag{16.21}$$

The slip boundary condition (10.2) is applied to both plates and reads as

$$u_2 = \begin{cases} u_w + \sigma_P \ell \frac{\partial u_2}{\partial r_1} & \text{at} \quad r_1 = 0, \\ -\sigma_P \ell \frac{\partial u_2}{\partial r_1} & \text{at} \quad r_1 = a. \end{cases} \tag{16.22}$$

Let us write down these equations in the dimensionless quantities as

$$2\theta i\tilde{u} + \frac{d^2\tilde{u}}{dx^2} = 0, \quad \Pi = -\frac{1}{\theta}\frac{d\tilde{u}}{dx}, \tag{16.23}$$

$$\tilde{u} = \begin{cases} 1 + \frac{\sigma_P}{\theta}\frac{d\tilde{u}}{dx} & \text{at} \quad x = 0, \\ -\frac{\sigma_P}{\theta}\frac{d\tilde{u}}{dx} & \text{at} \quad x = L. \end{cases} \tag{16.24}$$

Note that the complex functions $\tilde{u}$ and Π depend only on one variable x so that the partial derivatives in (16.21) and (16.22) are substituted by the ordinary derivatives.

The solution of Eqs. (16.23) and (16.24) was obtained in the paper [215] and reads

$$\tilde{u}(x) = \left[\sin\Theta + (1+i)\frac{\sigma_P}{\sqrt{\theta}}\cos\Theta\right]\mathcal{B}, \tag{16.25}$$

$$\Pi(x) = \frac{1}{\sqrt{\theta}}\left[(1+i)\cos\Theta - 2i\frac{\sigma_P}{\sqrt{\theta}}\sin\Theta\right]\mathcal{B}, \tag{16.26}$$

where

$$\mathcal{B} = \left[\left(1 - 2i\frac{\sigma_P^2}{\theta}\right)\sin\Theta_0 + 2(1+i)\frac{\sigma_P}{\sqrt{\theta}}\cos\Theta_0\right]^{-1}, \tag{16.27}$$

$$\Theta = (1+i)\frac{\delta}{\sqrt{\theta}}\left(1 - \frac{x}{L}\right), \quad \Theta_0 = (1+i)\frac{\delta}{\sqrt{\theta}}. \tag{16.28}$$

As expected, this solution is harmonic, that is, the phases φ_u and φ_Π are linear functions of the coordinate x. It is curious that the frequency parameter appears in this solution as $\sqrt{\theta}$, but not θ as one could expect.

It is interesting to consider two limit situations. First, when $\delta/\sqrt{\theta} \ll 1$ and the oscillation is very slow, then Eqs. (16.25) and (16.26) are expanded into the Taylor series with respect to the small parameter $\delta/\sqrt{\theta}$. Keeping just the main term in this expansion, we obtain

$$\tilde{u}(x) = \frac{\delta(1 - x/L) + \sigma_P}{\delta + 2\sigma_P}, \quad \Pi = \frac{1}{\delta + 2\sigma_P} \tag{16.29}$$

that coincide with the stationary Couette flow in the slip flow regime (see Eqs. (11.12) and (11.13)). The opposite situation $\delta/\sqrt{\theta} \to \infty$, the trigonometrical functions can be represented as

$$\sin\Theta = -\frac{i}{2}[\exp(i\Theta) - \exp(-i\Theta)] \to \frac{i}{2}\exp(-i\Theta), \tag{16.30}$$

$$\cos\Theta = \frac{1}{2}[\exp(i\Theta) + \exp(-i\Theta)] \to \frac{1}{2}\exp(-i\Theta). \tag{16.31}$$

The functions $\sin \Theta_0$ and $\cos \Theta_0$ are represented in the same way. Then the expressions (16.25) and (16.26) take the form

$$\tilde{u}(x) = \frac{\sqrt{\theta} \exp\left[(i-1)\sqrt{\theta} x\right]}{\sqrt{\theta} - (i-1)\sigma_{\mathrm{P}}}, \quad \Pi(x) = (1-i)\frac{\exp\left[(i-1)\sqrt{\theta} x\right]}{\sqrt{\theta} - (i-1)\sigma_{\mathrm{P}}}. \tag{16.32}$$

First, these expressions were obtained in Ref. [216]. In this case, the velocity $\tilde{u}$ and shear stress Π at the oscillating plate ($x = 0$) does not depend on the rarefaction parameter δ, that is, the position L of the fixed plate does not matter. The expression (16.32) can be written in the form (16.1) where the wave number takes the form

$$k = (1+i)\sqrt{\theta}\frac{\omega}{v_0} = (1+i)\sqrt{\omega \frac{n_0 m}{2\mu}}, \tag{16.33}$$

so that the real and imaginary parts are the same and depend on the frequency ω. It means that both wave speed and attenuation vary by varying its frequency ω.

16.2.3
Kinetic Equation

Similar to the stationary Couette flow, see Section 11.1, the oscillating one is solved on the basis of the BGK model, that is, Eq. (16.14) with $A_1 = 1, A_2 = A_3 = 0$. The function H contains the unique moment $\tilde{u}$ so that the kinetic equation reads

$$-ih + c_1 \frac{\partial h}{\partial x} = \theta(2c_2\tilde{u} - h), \tag{16.34}$$

which differs from (11.15) by the first term in the left-hand side. Moreover, the rarefaction parameter δ has been substituted by the frequency parameter θ considering its range of variation $[0, L]$ instead of $[0,1]$. The complex moments $\tilde{u}$ and Π are related to the complex perturbation in the same way as those for the stationary Couette flow (11.16).

The boundary conditions are obtained from the general form (5.50) with

$$h_w = 2c_2\tilde{u}_w, \quad \tilde{u}_w = \begin{cases} 1 & \text{at} \quad x = 0, \\ 0 & \text{at} \quad x = L. \end{cases} \tag{16.35}$$

If the diffuse scattering of gaseous particles on both plates is assumed, the boundary conditions take the form (5.70). Using the same argument given just after (11.17), we conclude that $\varrho_r = 0$ and obtain the boundary condition in the form

$$h = \begin{cases} 2c_2 & \text{at} \quad x = 0 \quad \text{and} \quad c_1 > 0, \\ 0 & \text{at} \quad x = L \quad \text{and} \quad c_1 < 0. \end{cases} \tag{16.36}$$

In order to eliminate the variables c_2 and c_3, a new perturbation function is introduced by (11.19), which obeys the equation

$$(\theta - i)\Phi + c_1 \frac{\partial \Phi}{\partial x} = \theta\tilde{u}. \tag{16.37}$$

The boundary conditions (16.36) are written in terms of the new function Φ as

$$\Phi = \begin{cases} 1 & \text{at} \quad x = 0 \quad \text{and} \quad c_1 > 0, \\ 0 & \text{at} \quad x = L \quad \text{and} \quad c_1 < 0. \end{cases} \tag{16.38}$$

The moments are expressed via the new perturbation function as

$$\begin{bmatrix} \tilde{u}(x) \\ \Pi(x) \end{bmatrix} = \frac{1}{\sqrt{\pi}} \int_{-\infty}^{\infty} \begin{bmatrix} 1 \\ 2c_1 \end{bmatrix} e^{-c_1^2} \Phi(x, c_1)\, \mathrm{d}c_1, \tag{16.39}$$

that follows from (11.16) and (11.19). For this particular case, Eqs. (16.15) and (16.16) are reduced to

$$\Phi(x, c_1) = e^{(i-\theta)x/c_1} + \frac{\theta}{c_1} \int_0^x \tilde{u}(x') e^{(i-\theta)(x-x')/c_1}\, \mathrm{d}x' \quad \text{at} \quad c_1 > 0, \tag{16.40}$$

$$\Phi(x, c_1) = -\frac{\theta}{c_1} \int_x^L \tilde{u}(x') e^{(i-\theta)(x'-x)/|c_1|}\, \mathrm{d}x' \quad \text{at} \quad c_1 < 0. \tag{16.41}$$

16.2.4
Free-Molecular Regime

Consider a situation when the rarefaction parameter δ tends to zero, while the oscillation speed parameter θ is arbitrary and fixed. It is possible when the pressure is fixed, but the distance a between the plates decreases. Since $x \leq L$, then x tends to zero at $L \to 0$. In this case, the integration interval in Eqs. (16.40) and (16.41) will decrease. Since the bulk velocity $\tilde{u}$ cannot increase infinitely because it is restricted by the wall speed amplitude, the integrals will vanish at $L \to 0$. The first term in the right-hand side of (16.40) tends to unity. Then, Eqs. (16.40) and (16.41) are reduced to

$$\Phi(x, c_1) = \begin{cases} 1 & \text{at} \quad c_1 > 0, \\ 0 & \text{at} \quad c_1 \leq 0. \end{cases} \tag{16.42}$$

Substituting this solution into the definitions (16.39), we obtain

$$\tilde{u}(x) = \frac{1}{2}, \quad \Pi(x) = \frac{1}{\sqrt{\pi}} \quad \text{at} \quad \delta \to 0 \tag{16.43}$$

so that the expression for Π coincides with (11.9).

Now consider the solution corresponding to the high-frequency oscillation when θ tends to zero, while the rarefaction parameter δ is arbitrary. In this situation, Eqs. (16.40) and (16.41) are reduced to

$$\Phi(x, c_1) = \begin{cases} \exp\left(i\frac{x}{c_1}\right) & \text{at} \quad c_1 > 0, \\ 0 & \text{at} \quad c_1 \leq 0. \end{cases} \tag{16.44}$$

The bulk velocity and shear stress are obtained by substituting this solution into Eq. (16.39)

$$\tilde{u}(x) = \frac{1}{\sqrt{\pi}} I_0(-ix), \quad \Pi(x) = \frac{2}{\sqrt{\pi}} I_1(-ix) \quad \text{at} \quad \theta \to 0, \tag{16.45}$$

where the special functions I_n are defined by (A.29). At the oscillating plate ($x = 0$), we have

$$\tilde{u}|_{x=0} = \frac{1}{2}, \quad \Pi|_{x=0} = \frac{1}{\sqrt{\pi}}, \tag{16.46}$$

that is the same as in the limit $\delta \to 0$. However, the solution (16.45) is different from (16.43) in the gap between the plates. Note that the solution (16.45) is not harmonic so that it cannot be written in the form (16.1).

Thus, there are two free-molecular regimes for the oscillating Couette flow. First, when $\delta \to 0$, θ is arbitrary. Physically, it means that the intermolecular collisions are neglected because they are rare in comparison with the gas–surface collisions. Second, when $\theta \to 0$, δ is arbitrary. Under these conditions, the frequency of the intermolecular collisions is negligibly low in comparison with the oscillation frequency. In both cases, the bulk velocity and shear stress at the oscillating plate are the same and given by (16.46).

16.2.5 Numerical Scheme

The numerical scheme to solve Eq. (16.37) is very similar to that described in Section 11.1.5. First, the nodes in the physical space are defined as

$$x_k = \frac{kL}{N_x}, \quad 0 \le k \le N_x. \tag{16.47}$$

The nodes in the velocity space should be chosen considering the behavior of the perturbation Φ as function of the velocity c_1. A typical function $\Phi(c_1)$ is plotted by the solid line in Figure 16.2, representing the case of $\delta = 0.1$ and $\theta = 0.1$ in the point $x = 0.05$. Such a behavior requires that the velocity nodes should be denser near $c_1 = 0$ and rarer for larger values of c_1. For instance, the nodes (11.131) and weights (11.132) proposed for the planar Poiseuille flow could be used here. Unlike the planar stationary flows, the perturbation considered here is not symmetric so

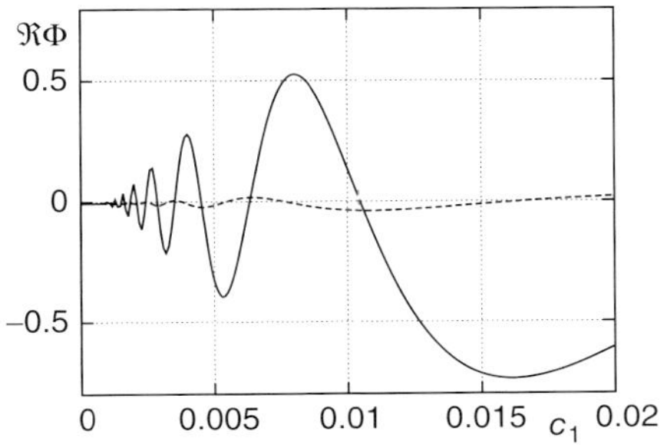

Figure 16.2 Perturbation Φ at $x = 0.05$, $\delta = 0.1$, and $\theta = 0.1$: solid line – nonsplitting scheme; dashed line – splitting scheme.

that both positive and negative values of c_1 must be considered in the numerical scheme, that is, the index i should vary from $-N_c$ to N_c in Eqs. (11.131) and (11.132).

The high oscillation of Φ shown in Figure 16.2 requires many velocity nodes. To reduce their number, the splitting scheme similar to that described in Section 11.3.6 can be used here. The idea is to split the perturbation as

$$\Phi = \Phi_0 + \Phi', \tag{16.48}$$

where the first term Φ_0 obeys the equation

$$(\theta - i)\Phi_0 + c_1 \frac{\partial \Phi_0}{\partial x} = 0 \tag{16.49}$$

and subject to the boundary condition (16.38). The second term Φ' obeys the equation

$$(\theta - i)\Phi' + c_1 \frac{\partial \Phi'}{\partial x} = \theta \tilde{u}, \tag{16.50}$$

and it is subject to the zero boundary conditions

$$\Phi' = \begin{cases} 0 & \text{at} \quad x = 0 \quad \text{and} \quad c_1 > 0, \\ 0 & \text{at} \quad x = L \quad \text{and} \quad c_1 \leq 0. \end{cases} \tag{16.51}$$

As a result, the moments are decomposed too

$$\tilde{u} = \tilde{u}_0 + \tilde{u}', \quad \Pi = \Pi_0 + \Pi', \tag{16.52}$$

where $\tilde{u}_0$ and Π_0 are calculated by (16.39) via Φ_0, while $\tilde{u}'$ and Π' are calculated via Φ' in the same way.

The differential equation (16.49) is integrated analytically and provides Φ_0 in the form

$$\Phi_0(x, c_1) = \begin{cases} \exp\left[-(\theta - i)\frac{x}{c_1}\right] & \text{for} \quad c_1 > 0, \\ 0 & \text{for} \quad c_1 \leq 0. \end{cases} \tag{16.53}$$

Substituting this perturbation into (16.39), the first terms of (16.52) are obtained as

$$\tilde{u}_0(x) = \frac{1}{\sqrt{\pi}} I_0[(\theta - i)x], \quad \Pi_0(x) = \frac{2}{\sqrt{\pi}} I_1[(\theta - i)x]. \tag{16.54}$$

Equation (16.50) is solved numerically. If we choose the centered approximation, then Eq. (9.40) takes the following form for this specific problem

$$\Phi'_k = \frac{(2|c_1|N_x/L - \theta + i)\Phi'_{k-l} + \theta(\tilde{u}_k + \tilde{u}_{k-l})}{2|c_1|N_x/L + \theta - i}, \quad l = \mathrm{sgn}(c_1), \tag{16.55}$$

where the notation

$$\Phi'_k = \Phi'(x_k, c_1), \quad 0 \leq k \leq N_x, \tag{16.56}$$

has been introduced. The sweeping direction is determined by l, namely, the index k varies from 1 to N_x for positive c_1 and from $N_x - 1$ to 0 when $c_1 < 0$. Since $\Phi = 0$

at $c_1 = 0$, this node is skipped in the calculations. When the values Φ'_k are obtained for all values of the velocity nodes, the moments $\tilde{u}'$ and Π' are calculated as

$$\tilde{u}'_k = \sum_{i=1}^{N_c} [\Phi'_k(c_i) + \Phi'_k(-c_i)] W_i, \tag{16.57}$$

$$\Pi'_k = 2 \sum_{i=1}^{N_c} [\Phi'_k(c_i) - \Phi'_k(-c_i)] c_i W_i \tag{16.58}$$

that follow from (16.39). Here, the velocity nodes c_i and weights W_i are given by (11.131) and (11.132), respectively.

The perturbation function Φ' obtained by the splitting scheme is plotted in Figure 16.2 by the dashed line. This graph shows that the oscillatory behavior is taken into account by the known function Φ_0, while the function Φ' to be calculated numerically is quite smoother than the initial function Φ. As a result, the splitting scheme requires less computational effort than the original scheme.

A listing of the numerical program "couette_osc.for" based on the scheme described here is given in Section B.9.

16.2.6 Numerical Results

The numerical scheme parameters and values of the amplitudes A_u, A_Π and phases φ_u, φ_Π calculated by the code "couette_osc.for" are given in Table 16.1. These data coincide with those reported in Ref. [215] where the dimensionless shear stress differs from that defined here by the factor 1/2. The number of the nodes for x is determined basically by the rarefaction parameter δ. It is relatively small for the

Table 16.1 Numerical scheme parameters, amplitudes A_u, A_Π, and phases φ_u, φ_Π at oscillating plate ($x = 0$) for oscillating Couette flow based on the code given in Section B.9.

θ	δ	N_x	N_c	Iters.	A_u	φ_u	A_Π	φ_Π
0[a]					0.5000	0.0000	0.5642	0.0000
0.1	0.1	20	60	10	0.5008	0.0283	0.5625	−0.0246
	1	40	40	10	0.5014	0.0241	0.5634	−0.0181
1	0.01	20	40	8	0.5122	0.0045	0.5594	−0.0004
	0.1	20	40	13	0.5571	0.0510	0.5268	−0.0221
	1	20	40	34	0.5577	0.1905	0.5322	−0.1865
	10	200	40	50	0.5544	0.1779	0.5370	−0.1660
10	0.01	20	40	8	0.5130	0.0005	0.5593	0.0000
	0.1	20	40	14	0.5723	0.0068	0.5225	−0.0026
	1	20	40	46	0.7446	0.0605	0.3483	−0.1146
	10	500	60	496	0.7977	0.1684	0.3247	−0.5015
10[b]	10				0.7349	0.2384	0.3292	−0.5457

a) Equation (16.46).
b) Equations (16.25)–(16.28).

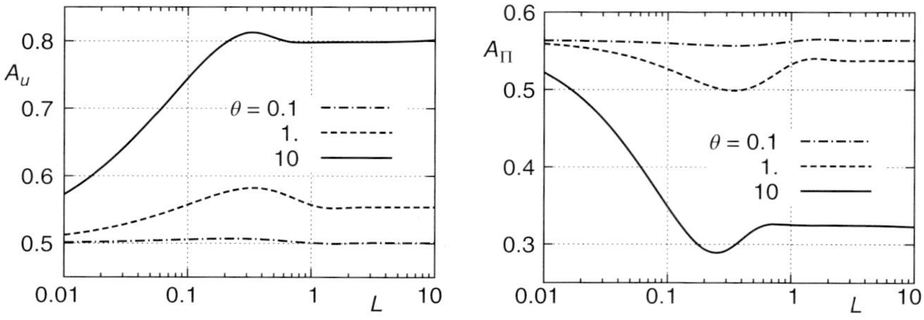

Figure 16.3 Amplitudes of bulk velocity A_u and shear stress A_Π at the surface ($x = 0$) for oscillating Couette flow versus dimensionless distance L. Reproduced from Ref. [215] with permission.

small ($\delta = 0.1$) and transitional ($\delta = 1$) rarefaction parameters, and it drastically increases at $\delta = 10$. The number of the nodes for the velocity c_1 varies weakly. The number of iterations given in the fifth column of Table 16.1 is large when both θ and δ are large, $\theta = \delta = 10$. When the frequency parameter is small $\theta = 0.1$, the amplitude and phase values are close to those calculated by Eq. (16.46) for any rarefaction parameter δ. The amplitudes and phases also tend to those obtained from Eq. (16.46) by decreasing the rarefaction parameter δ when the frequency parameter θ is fixed at any value. The values of amplitudes and phases calculated from Eqs. (16.25)–(16.28) are close to the numerical data obtained when both parameters are large, $\delta = \theta = 10$.

The dependence of the amplitudes A_u and A_Π at the oscillating surface ($x = 0$) on the dimensionless distance L is curious. They are non-monotonic function of the distance L. To see better such a behavior, the amplitudes obtained in Ref. [215] are plotted versus L in Figure 16.3, which shows that the velocity A_u and shear stress A_Π amplitudes undergo the maximum and the minimum, respectively. Then constant values of A_u and A_Π are established when the distance L becomes large.

Additional details about the oscillating Couette flow including the amplitude and phase distribution in the gap between the plates are reported in the papers [215, 216]. The work by Doi [217] reports numerical results on the same topic based on the linearized BE for the HS potential. These results are quite close to those obtained on the BGK model equation.

16.3 Longitudinal Waves

16.3.1 Definitions

Usually, a longitudinal (or sound) wave is generated by a plate oscillating in the direction normal to its plane. Although there are other ways to generate a sound

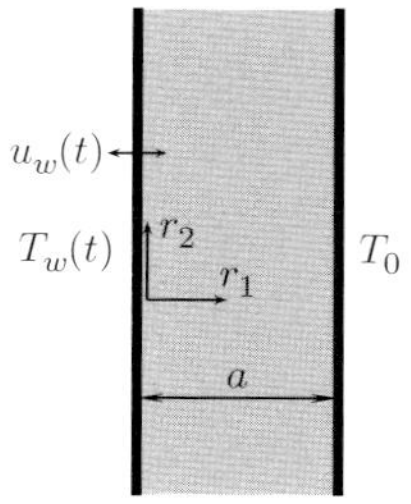

Figure 16.4 Scheme of acoustic wave generation.

wave, for example, a temperature oscillation of a solid surface also causes a longitudinal wave even when the plate is fixed. Such thermally induced waves are usually referred as thermoacoustic waves. This section describes sound waves generated by these two ways.

As in the case of transversal wave, here we consider a monatomic gas confined between two plane, infinite, and parallel plates fixed at $r_1 = 0$ and $r_1 = a$ as is shown in Figure 16.4. The left plate ($r_1 = 0$) is subject to two driving forces. First, it oscillates harmonically in its normal direction with a frequency ω so that its velocity $\boldsymbol{u}_w$ has only the first component depending on the time as

$$\boldsymbol{u}_w = [u_w, 0, 0], \quad u_w = u_m \cos(\omega t) = u_m \Re e^{-i\omega t}, \tag{16.59}$$

where u_m is the amplitude of the surface velocity. The second driving force consists of a periodical heating and cooling of the left plate ($r_1 = 0$) so that its temperature varies as

$$T_w(t) = T_0 + \Delta T_m \cos(\omega t) = T_0 + \Delta T_m \Re e^{-i\omega t}, \tag{16.60}$$

where ΔT_m is a maximum deviation of the temperature from its equilibrium value. Both driving forces generate the sound waves propagating through the gas in the r_1 direction and disturbing its density n, temperature T, and pressure tensor $\mathbf{P}$. In practice, only the diagonal term P_{11} of this tensor is measured; therefore, we will pay more attention to it. Moreover, the bulk velocity u_1 and heat flux q_1 in the r_1 direction appear in the perturbed gas. The waves are determined by the rarefaction δ and oscillation θ parameters defined by (7.46) and (16.11), respectively.

It is assumed that the velocity amplitude u_m is small in comparison with the MPS v_0, and the temperature deviation is small compared with its equilibrium value

$$\xi_u = \frac{u_m}{v_0}, \quad \xi_u \ll 1, \quad \xi_\mathrm{T} = \frac{\Delta T_m}{T_0}, \quad \xi_\mathrm{T} \ll 1. \tag{16.61}$$

Thus, the two quantities ξ_u and ξ_T are used as the small parameters for the linearization. As a result, we have two perturbation functions $h^{(u)}$ and $h^{(\mathrm{T})}$ corresponding to ξ_u and ξ_T, respectively. Then all the moments of our interest are linearized and decomposed into two terms as

$$\begin{bmatrix} n(t,r_1)/n_0 - 1 \\ u_1(t,r_1)/v_0 \\ T(t,r_1)/T_0 - 1 \\ P_{11}(t,r_1)/p_0 - 1 \\ q_1(r_1)/p_0 v_0 \end{bmatrix} = \Re \left\{ \begin{bmatrix} \xi_u \varrho^{(u)}(x) & + & \xi_T \varrho^{(T)}(x) \\ \xi_u \tilde{u}^{(u)}(x) & + & \xi_T \tilde{u}^{(T)}(x) \\ \xi_u \tau^{(u)}(x) & + & \xi_T \tau^{(T)}(x) \\ \xi_u \Pi^{(u)}(x) & + & \xi_T \Pi^{(T)}(x) \\ \xi_u \tilde{q}^{(u)}(x) & + & \xi_T \tilde{q}^{(T)}(x) \end{bmatrix} e^{-i\omega t} \right\}, \tag{16.62}$$

where the subscripts "1" have been omitted at $\tilde{u}$ and $\tilde{q}$. The subscript "11" has also been omitted at Π. Most of the equations given below are the same for the moments with both superscripts (u) and (T). Therefore, the omitted superscript will mean that an expression is valid for both (u) and (T). The coordinate x is related to r_1 by (16.10).

Some relations between the moments can be obtained from the conservation laws (3.26)–(3.28). Substituting (16.62) into Eqs. (3.26)–(3.28), the following relations between the dimensionless moments are obtained

$$-i\varrho + \frac{\mathrm{d}\tilde{u}}{\mathrm{d}x} = 0, \tag{16.63}$$

$$-2i\tilde{u} + \frac{\mathrm{d}\Pi}{\mathrm{d}x} = 0, \tag{16.64}$$

$$-\frac{3}{2}i\tau + \frac{\mathrm{d}}{\mathrm{d}x}(\tilde{q} + \tilde{u}) = 0, \tag{16.65}$$

where the energy equation (1.14) has been used. Since the conservation laws are valid for any gas rarefaction, they can be used as an additional criterion of numerical solution of the kinetic equation given later.

16.3.2
Hydrodynamic Regime

When both parameters δ and θ are large $\theta \gg 1$ and $\delta \gg 1$, the constitutive equations (6.2) and (6.3) are valid. In terms of the dimensionless moments (16.62), they are reduced to

$$\Pi(x) = \varrho(x) + \tau(x) - \frac{4}{3\theta}\frac{\mathrm{d}\tilde{u}}{\mathrm{d}x}, \tag{16.66}$$

$$\tilde{q}(x) = -\frac{15}{8\theta}\frac{\mathrm{d}\tau}{\mathrm{d}x}, \tag{16.67}$$

where the Prandtl number Pr defined by (6.4) is assumed to be 2/3. These differential equations of the first order form the closed system of differential equations together with the conservation laws (16.63)–(16.65). The system is equivalent to the unique differential equation of the fourth order for the velocity $\tilde{u}$

$$A\frac{\mathrm{d}^4\tilde{u}}{\mathrm{d}x^4} + B\frac{\mathrm{d}^2\tilde{u}}{\mathrm{d}x^2} + \tilde{u} = 0, \tag{16.68}$$

where the constants A and B read

$$A = -\frac{5}{8\theta}\left(\frac{4}{3\theta} + i\right), \quad B = \frac{5}{6}\left(1 - \frac{23}{10}\frac{i}{\theta}\right). \tag{16.69}$$

A general solution of this equation can be represented as

$$\tilde{u}(x) = \sum_{j=1}^{2}(a_j e^{i\gamma_j x} + b_j e^{-i\gamma_j x}), \tag{16.70}$$

where γ_1 and γ_2 are the dimensionless wave numbers given as

$$\gamma_1 = \sqrt{\frac{B - \sqrt{B^2 - 4A}}{2A}}, \quad \gamma_2 = \sqrt{\frac{B + \sqrt{B^2 - 4A}}{2A}}. \tag{16.71}$$

Since we already have assumed the parameter θ to be large, these expressions can be simplified if the asymptotes at $\theta \to \infty$ are obtained

$$\gamma_1 = \sqrt{\frac{6}{5}}\left(1 + \frac{7i}{10\theta}\right) + O(\theta^{-2}), \tag{16.72}$$

$$\gamma_2 = (1+i)\sqrt{\frac{2}{3}\theta}\left(1 - \frac{i}{30\theta} + O\left(\theta^{-2}\right)\right), \tag{16.73}$$

where $O(\theta^{-2})$ denotes the terms of the order θ^{-2} so that Eqs. (16.72) and (16.73) provide the wave numbers with the error less than 1% for $\theta > 10$.

Physically, the expression (16.70) represents a linear combination of four harmonic waves. The two opposite waves corresponding to the number γ_1 are the typical ones far from the solid surfaces. If we calculate the dimension wave number k introduced by (16.1) and related to γ as

$$k = \frac{\omega}{v_0}\gamma, \tag{16.74}$$

its real and imaginary parts read

$$k_{R1} = \sqrt{\frac{6}{5}}\frac{\omega}{v_0}, \quad k_{I1} = \frac{7}{10}\sqrt{\frac{6}{5}}\frac{\mu}{v_0 p_0}\omega^2. \tag{16.75}$$

The real part k_{R1} along (16.3) leads to the adiabatic sound speed $v_s = \sqrt{5/6}\, v_0$, while the imaginary term k_{I1} represents the classical attenuation based on the continuous mechanics [23]. The other two waves having the number γ_2 vanish rapidly far from the surfaces because the imaginary part of γ_2 increases by increasing the parameter θ so that the attenuation of these waves is very strong.

Once the velocity is written in the form (16.70), the quantity Π is obtained from (16.64)

$$\Pi = 2\sum_{j=1}^{2}\frac{1}{\gamma_j}\left(a_j e^{i\gamma_j x} - b_j e^{-i\gamma_j x}\right). \tag{16.76}$$

The temperature is calculated as the linear combination of Eqs. (16.63) and (16.66)

$$\tau = \sum_{j=1}^{2} C_j \left(a_j e^{i\gamma_j x} - b_j e^{-i\gamma_j x}\right), \quad C_j = \frac{2}{\gamma_j} - \left(1 - \frac{4i}{3\theta}\right)\gamma_j. \tag{16.77}$$

To calculate the constant a_j and b_j, four boundary conditions are needed. Two of them are for the gas velocities $u^{(m)}$

$$\tilde{u}^{(m)} = \begin{cases} \delta_{mu} & \text{at} \quad x = 0, \\ 0 & \text{at} \quad x = L, \end{cases} \tag{16.78}$$

where $m = u$ or T, δ_{mu} is the Kronecker delta defined by (A.14). Physically, the condition (16.78) means that the gas velocity $\tilde{u}^{(u)}$ oscillates together with the left plate, while it is zero on the right plate. At the same time, the velocity $\tilde{u}^{(\mathrm{T})}$ is zero on both plates. The other two conditions are for the temperatures on both surfaces. To obtain the solution for a larger range of the parameter θ, the temperature jump condition (10.39) is used here, which in the dimensionless form reads

$$\tau^{(m)} = \begin{cases} \delta_{m\mathrm{T}} + \frac{\zeta_\mathrm{T}}{\theta}\frac{\mathrm{d}\tau}{\mathrm{d}x} & \text{at} \quad x = 0, \\ -\frac{\zeta_\mathrm{T}}{\theta}\frac{\mathrm{d}\tau}{\mathrm{d}x} & \text{at} \quad x = L, \end{cases} \tag{16.79}$$

where $m = u$ or T and ζ_T is the TJC (see Section 10.5). Substituting Eq. (16.70) into (16.78) and Eq. (16.77) into (16.79), the system of algebraic equations for the constants $a_j^{(n)}$ and $b_j^{(n)}$ is obtained in the following form:

$$\sum_{j=1}^{2} \left(a_j^{(m)} + b_j^{(m)}\right) = \delta_{mu}, \tag{16.80}$$

$$\sum_{j=1}^{2} \left(a_j^{(m)} e^{i\gamma_j L} + b_j^{(m)} e^{-i\gamma_j L}\right) = 0, \tag{16.81}$$

$$\sum_{j=1}^{2} C_j \left[\left(1 - \frac{\zeta_\mathrm{T}}{\theta} i\gamma_j\right) a_j^{(m)} - \left(1 + \frac{\zeta_\mathrm{T}}{\theta} i\gamma_j\right) b_j^{(m)}\right] = \delta_{m\mathrm{T}}, \tag{16.82}$$

$$\sum_{j=1}^{2} C_j \left[\left(1 + \frac{\zeta_\mathrm{T}}{\theta} i\gamma_j\right) a_j^{(m)} e^{i\gamma_j L} - \left(1 - \frac{\zeta_\mathrm{T}}{\theta} i\gamma_j\right) b_j^{(m)} e^{-i\gamma_j L}\right] = 0. \tag{16.83}$$

The analytical solution of this system is quite cumbersome, but its numerical solution does not represent any difficulty. An example of such a numerical solution for $\theta = 10$ and $\delta = 10$ obtained in Ref. [218] is reproduced in Table 16.2.

16.3.3
Kinetic Equation

As has been mentioned earlier, we consider two driving forces ξ_u and ξ_T and consequently we have the two perturbation functions $h^{(u)}$ and $h^{(\mathrm{T})}$. Each of them obeys the same kinetic equation but they are subject to different boundary conditions,

Table 16.2 Coefficient $a_j^{(m)}$ and $b_j^{(m)}$ obtained as a solution of Eqs. (16.80)–(16.83) for $\theta = 10$ and $\delta = 10$, Ref. [218].

	$m = u$	$m = \mathrm{T}$
$a_1^{(m)}$	$0.5840 + 0.1919i$	$0.1103 - 0.0108i$
$b_1^{(m)}$	$0.3689 - 0.2063i$	$(0.3345 - 0.6711i) \times 10^{-1}$
$a_2^{(m)}$	$(0.3773 + 0.1054i) \times 10^{-1}$	$-0.1451 + 0.0779i$
$b_2^{(m)}$	$(0.9383 + 0.3817i) \times 10^{-2}$	$(0.1373 + 0.0005i) \times 10^{-2}$

so that the superscripts (u) and (T) are omitted when the expressions for $h^{(u)}$ and $h^{(\mathrm{T})}$ are the same and attached in an opposite case.

Since the sound propagation involves heat transfer, the S-model would be appropriate for its description over wide ranges of the parameters θ and δ, that is, Eq. (16.14) with $A_1 = 1, A_2 = 0$, and $A_3 = 1$

$$-ih + c_1 \frac{\partial h}{\partial x} = \theta(H - h). \tag{16.84}$$

In this case, the function H contains four moments and reads

$$H = \varrho + 2c_1\tilde{u} + \left(c^2 - \frac{3}{2}\right)\tau + \frac{4}{15}\left(c^2 - \frac{5}{2}\right)c_1\tilde{q}. \tag{16.85}$$

The dimensionless moments defined by Eqs. (10.39)–(2.26) are calculated via the perturbation function as

$$\begin{bmatrix} \varrho(x) \\ u(x) \\ \tau(x) \\ q(x) \\ \Pi(x) \end{bmatrix} = \frac{1}{\pi^{3/2}} \int \begin{bmatrix} 1 \\ c_1 \\ \frac{2}{3}c^2 - 1 \\ c_1\left(c^2 - \frac{5}{2}\right) \\ 2c_1^2 \end{bmatrix} h(x, \mathbf{c}) e^{-c^2} \, \mathrm{d}\mathbf{c}, \tag{16.86}$$

that follows from (5.12) to (5.16). The surface source terms are derived from the general form (5.54)

$$h_w^{(u)} = \begin{cases} 2c_1, \\ 0, \end{cases} \quad h_w^{(\mathrm{T})} = \begin{cases} c^2 - \frac{3}{2} & \text{at} \quad x = 0, \\ 0 & \text{at} \quad x = L. \end{cases} \tag{16.87}$$

The boundary perturbation under the assumption of the diffuse gas–surface interaction is derived from (5.70)

$$h^{(u)} = \begin{cases} \varrho_{r0}^{(u)} + \sqrt{\pi} + 2c_1 & \text{at} \quad x = 0 \quad \text{and} \quad c_1 > 0, \\ \varrho_{rL}^{(u)} & \text{at} \quad x = L \quad \text{and} \quad c_1 < 0, \end{cases} \tag{16.88}$$

$$h^{(\mathrm{T})} = \begin{cases} \varrho_{r0}^{(\mathrm{T})} + (c^2 - 2) & \text{at} \quad x = 0 \quad \text{and} \quad c_1 > 0, \\ \varrho_{rL}^{(\mathrm{T})} & \text{at} \quad x = L \quad \text{and} \quad c_1 < 0, \end{cases} \tag{16.89}$$

where the quantities ϱ_{r0} and ϱ_{rL} are given by (5.71) and for the specific problem considered here reads

$$\varrho_{r0} = -\frac{2}{\pi}\int_{c_1\leq 0} c_1 e^{-c^2} h(0,\boldsymbol{c})\,\mathrm{d}^3\boldsymbol{c}, \quad \varrho_{rL} = \frac{2}{\pi}\int_{c_1\geq 0} c_1 e^{-c^2} h(L,\boldsymbol{c})\,\mathrm{d}^3\boldsymbol{c}. \tag{16.90}$$

Thus, the kinetic equation applied here is very similarly to that used for the planar heat transfer (11.46). Following the formalism described in Section 11.2, two new perturbation functions $\Phi^{(1)}$ and $\Phi^{(2)}$ are introduced by (11.53) and (11.54), which obey the following kinetic equations:

$$-i\Phi^{(\alpha)} + c_1\frac{\partial \Phi^{(\alpha)}}{\partial x} = \theta(H^{(\alpha)} - \Phi^{(\alpha)}). \tag{16.91}$$

The functions $H^{(\alpha)}$ take the forms

$$H^{(1)} = \varrho + 2c_1\tilde{u} + \left(c_1^2 - \frac{1}{2}\right)\tau + \frac{4}{15}\left(c_1^2 - \frac{3}{2}\right)c_1\tilde{q}, \tag{16.92}$$

$$H^{(2)} = \tau + \frac{4}{15}c_1\tilde{q}. \tag{16.93}$$

In contrast to the planar heat transfer problem, the considered flow is not symmetric though the moments are calculated directly by integrating the perturbation functions with respect to c_1 through the whole interval

$$\begin{bmatrix}\varrho(x)\\ u(x)\\ \tau(x)\\ q(x)\\ \Pi(x)\end{bmatrix} = \frac{1}{\sqrt{\pi}}\int_{-\infty}^{\infty}\left\{\begin{bmatrix}1\\ c_1\\ \frac{2}{3}c_1^2 - \frac{1}{3}\\ c_1\left(c_1^2 - \frac{3}{2}\right)\\ 2c_1^2\end{bmatrix}\Phi^{(1)}(x,c_1) + \begin{bmatrix}0\\ 0\\ \frac{2}{3}\\ c_1\\ 0\end{bmatrix}\Phi^{(2)}(x,c_1)\right\}e^{-c_1^2}\,\mathrm{d}c_1. \tag{16.94}$$

The boundary conditions (16.88) and (16.89) are transformed into

$$\Phi^{(1,u)} = \begin{cases}\varrho_{r0}^{(u)} + \sqrt{\pi} + 2c_1,\\ \varrho_{rL}^{(u)},\end{cases} \quad \Phi^{(2,u)} = \begin{cases}0, & \text{at} \quad x = 0 \quad \text{and} \quad c_1 > 0,\\ 0, & \text{at} \quad x = L \quad \text{and} \quad c_1 < 0,\end{cases} \tag{16.95}$$

$$\Phi^{(1,\mathrm{T})} = \begin{cases}\varrho_{r0}^{(\mathrm{T})} + c_1^2 - 1,\\ \varrho_{rL}^{(\mathrm{T})},\end{cases} \quad \Phi^{(2,\mathrm{T})} = \begin{cases}1 & \text{at} \quad x = 0 \quad \text{and} \quad c_1 > 0,\\ 0 & \text{at} \quad x = L \quad \text{and} \quad c_1 < 0,\end{cases} \tag{16.96}$$

Finally, the expressions (16.90) are reduced to

$$\varrho_{r0} = -2\int_{-\infty}^{0} c_1 e^{-c_1^2}\Phi^{(1)}(0,c_1)\,\mathrm{d}c_1, \quad \varrho_{rL} = 2\int_{0}^{\infty} c_1 e^{-c_1^2}\Phi^{(1)}(L,c_1)\,\mathrm{d}c_1. \tag{16.97}$$

16.3.4 Reciprocal Relation

Since we consider two thermodynamic forces ξ_u and ξ_T perturbing the equilibrium, it is possible to obtain the reciprocal relation between some quantities obtained from each solution $h^{(u)}$ and $h^{(\mathrm{T})}$. The general form of the time-reversed kinetic equation (5.85) is reduced to the following form for the problem in question

$$\begin{aligned}\Lambda_{u\mathrm{T}}^{t} &= (\hat{T}c_1 h_w^{(u)}, h^{(\mathrm{T})})|_{x=0} - \int_0^L \left(\hat{T}\frac{\partial h^{(u)}}{\partial t}, h^{(\mathrm{T})}\right)\,\mathrm{d}x \\ &= \Pi^{(\mathrm{T})}(0) + i\omega \int_0^L (\hat{T}h^{(u)}, h^{(\mathrm{T})})\,\mathrm{d}x,\end{aligned} \tag{16.98}$$

$$\begin{aligned}\Lambda_{\mathrm{T}u}^{t} &= (\hat{T}c_1 h_w^{(\mathrm{T})}, h^{(u)})|_{x=0} - \int_0^L \left(\hat{T}\frac{\partial h^{(\mathrm{T})}}{\partial t}, h^{(u)}\right)\,\mathrm{d}x \\ &= -\tilde{q}^{(u)}(0) + i\omega \int_0^L (\hat{T}h^{(\mathrm{T})}, h^{(u)})\,\mathrm{d}x,\end{aligned} \tag{16.99}$$

where the property

$$(\hat{T}c_1 h_w^{(u)}, h_w^{(\mathrm{T})})|_{x=C} = (\hat{T}c_1 h_w^{(\mathrm{T})}, h_w^{(u)})|_{x=0} = 0, \tag{16.100}$$

has been taken into account. Considering the expressions of the surface source terms $h_w^{(u)}$, $h_w^{(\mathrm{P})}$ given by Eq. (16.87), the definitions (16.86) of $\tilde{q}$ and Π, and the fact that the second terms in Eqs. (16.98) and (16.99) are equal to each other, the reciprocal relation (5.83) leads to the coupling

$$\Pi^{(\mathrm{T})} = -\tilde{q}^{(u)} \quad \text{at} \quad x = 0, \tag{16.101}$$

that is, the pressure tensor $\Pi^{(\mathrm{T})}$ caused by the temperature variation is related to the heat flux $q^{(u)}$ caused by the plate oscillation when both quantities in Eq. (16.101) are calculated at the left plate $x = 0$.

If we consider an auxiliary problem, it is possible to obtain the relation between $\Pi^{(\mathrm{T})}$ and $\tilde{q}^{(u)}$ at the right plate $x = L$. The auxiliary problem consists of the temperature oscillation at the right plate. Then the surface source term (16.87) is replaced by

$$h_w^{\prime(\mathrm{T})} = \begin{cases} 0 & \text{at} \quad x = 0, \\ c^2 - \frac{3}{2} & \text{at} \quad x = L. \end{cases} \tag{16.102}$$

Hence, the time-reversed kinetic coefficients take the form

$$\begin{aligned}\Lambda_{u\mathrm{T}}^{\prime t} &= \left(\hat{T}c_1 h_w^{(u)}, h^{\prime(\mathrm{T})}\right)|_{x=0} - \int_0^L \left(\hat{T}\frac{\partial h^{(u)}}{\partial t}, h^{\prime(\mathrm{T})}\right)\,\mathrm{d}x \\ &= \Pi^{\prime(\mathrm{T})}|_{x=0} + i\omega \int_0^L \left(\hat{T}h^{(u)}, h^{\prime(\mathrm{T})}\right)\,\mathrm{d}x,\end{aligned} \tag{16.103}$$

$$\Lambda'^{t}_{\mathrm{T}u} = -\left(\hat{T}c_1 h'^{(\mathrm{T})}_w, h^{(u)}\right)|_{x=L} - \int_0^L \left(\hat{T}\frac{\partial h'^{(\mathrm{T})}}{\partial t}, h^{(u)}\right)\,\mathrm{d}x$$
$$= \tilde{q}^{(u)}|_{x=L} + i\omega \int_0^1 \left(\hat{T}h'^{(\mathrm{T})}, h^{(u)}\right)\,\mathrm{d}x, \tag{16.104}$$

where the solution $h'^{(\mathrm{T})}$ corresponds to the source term (16.102). The reciprocal relation (5.83) for the auxiliary problem leads to the coupling

$$\Pi'^{(\mathrm{T})}|_{x=0} = \tilde{q}^{(u)}|_{x=L}. \tag{16.105}$$

Since $\Pi'^{(\mathrm{T})}|_{x=0} = \Pi^{(\mathrm{T})}|_{x=L}$, the following coupling for the main problems is obtained as

$$\Pi^{(\mathrm{T})} = q^{(u)} \quad \text{at} \quad x = L. \tag{16.106}$$

Thus, if one is interested only in the pressure tensors $\Pi^{(u)}$ and $\Pi^{(\mathrm{T})}$ at the plates ($x = 0$ and $x = L$), it is enough to solve the kinetic equation (16.84) only for the boundary condition (16.88) corresponding to the source term (16.87). Such a solution provides $\Pi^{(u)}$ and $q^{(u)}$. The quantity $\Pi^{(\mathrm{T})}$ is calculated from Eqs. (16.101) and (16.106). Numerical results reported in the paper [218] confirm the relations (16.101) and (16.106) so that further only the data on the quantity $\Pi^{(\mathrm{T})}$ will be giving omitting those for $\tilde{q}^{(u)}$.

16.3.5 High-Frequency Regime

In case of a high-frequency oscillation when $\theta \to 0$, the kinetic equations (16.91) subject to the boundary conditions (16.95) and (16.96) can be solved analytically. Neglecting the term in the right-hand side of Eq. (16.91), the perturbations are obtained as

$$\Phi^{(1,u)} = \begin{cases} \left(\varrho^{(u)}_{r0} + \sqrt{\pi} + 2c_1\right)\exp\left(i\frac{x}{c_1}\right)0 & \text{at} \quad c_1 > 0, \\ \varrho^{(u)}_{rL}\exp\left(i\frac{x-L}{c_1}\right) & \text{at} \quad c_1 < 0, \end{cases} \tag{16.107}$$

$$\Phi^{(2,u)} = \begin{cases} 0 & \text{at} \quad c_1 > 0, \\ 0 & \text{at} \quad c_1 < 0, \end{cases} \tag{16.108}$$

$$\Phi^{(1,\mathrm{T})} = \begin{cases} \left(\varrho^{(\mathrm{T})}_{r0} + c_1^2 - 1\right)\exp\left(i\frac{x}{c_1}\right) & \text{at} \quad c_1 > 0, \\ \varrho^{(\mathrm{T})}_{rL}\exp\left(i\frac{x-L}{c_1}\right) & \text{at} \quad c_1 < 0, \end{cases} \tag{16.109}$$

$$\Phi^{(2,\mathrm{T})} = \begin{cases} \exp\left(i\frac{x}{c_1}\right) & \text{at} \quad c_1 > 0, \\ 0 & \text{at} \quad c_1 < 0. \end{cases} \tag{16.110}$$

A substitution of these expressions into the definitions of ϱ_{r0} and ϱ_{rL} (16.97) leads to the following expressions:

$$\varrho_{r0}^{(u)} = 2\varrho_{rL} I_1(-iL), \quad \varrho_{rL}^{(u)} = 2\frac{\sqrt{\pi} I_1(-iL) + 2I_2(-iL)}{1 - 4I_1^2(-iL)}, \tag{16.111}$$

$$\varrho_{r0}^{(\mathrm{T})} = 2\varrho_{rL} I_1(-iL), \quad \varrho_{rL}^{(\mathrm{T})} = 2\frac{I_3(-iL) - I_1(-iL)}{1 - 4I_1^2(-iL)}. \tag{16.112}$$

The moments $\Pi^{(u)}$ and $\Pi^{(\mathrm{T})}$ are obtained by substituting the expressions (16.107)–(16.110) into their definition (16.94)

$$\Pi^{(u)}(x) = \frac{2}{\sqrt{\pi}}[2I_3(-ix) + (\varrho_{r0}^{(u)} + \sqrt{\pi})I_2(-ix) + \varrho_{r1}^{(u)} I_2(i(x-L))], \tag{16.113}$$

$$\Pi^{(\mathrm{T})}(x) = \frac{2}{\sqrt{\pi}}[I_4(-ix) + (\varrho_{r0}^{(\mathrm{T})} - 1)I_2(-ix) + \varrho_{r1}^{(\mathrm{T})} I_2(i(x-L))]. \tag{16.114}$$

The expressions for all moments corresponding to the plate oscillation (ξ_u) are given in the paper [219]. Those obtained for the temperature oscillation (ξ_{T}) are reported by Doi [220].

Simpler expressions can be obtained in the limit when the plates are close to each other $L \to 0$. In this case, the quantities $\Pi^{(u)}$ and $\Pi^{(\mathrm{T})}$ do not vary in the gap between the plate. Their amplitudes and phases read

$$A_{\Pi}^{(u)} \to \frac{1}{L}, \varphi_{\Pi}^{(u)} \to \frac{\pi}{2}, \quad A_{\Pi}^{(\mathrm{T})} \to \frac{1}{2}, \quad \varphi_{\Pi}^{(\mathrm{T})} \to 0, \tag{16.115}$$

at any x.

In the limit when the plates are far from each other $L \to \infty$, the quantities $\Pi^{(u)}$ and $\Pi^{(\mathrm{T})}$ are different on the sound generator plate ($x = 0$) and on the sound receptor plate ($x = L$). In terms of the amplitude and phases, these quantities are expressed as

$$A_{\Pi}^{(u)}\Big|_{x=0} \to \left(\frac{2}{\sqrt{\pi}} + \frac{\sqrt{\pi}}{2}\right), \quad \varphi_{\Pi}^{(u)}\Big|_{x=0} \to 0, \tag{16.116}$$

$$A_{\Pi}^{(u)}\Big|_{x=L} \to \frac{2L}{\sqrt{3}} \exp\left[-\frac{3}{2}\left(\frac{L}{2}\right)^{2/3}\right]\left[1 + \frac{\sqrt{3}\pi}{2}\left(\frac{2}{L}\right)^{1/3}\right], \tag{16.117}$$

$$\varphi_{\Pi}^{(u)}\Big|_{x=L} \to \frac{3\sqrt{3}}{2}\left(\frac{L}{2}\right)^{2/3} + \frac{3\pi}{2}, \tag{16.118}$$

$$A_{\Pi}^{(\mathrm{T})}\Big|_{x=0} \to \frac{1}{4}, \quad \varphi_{\Pi}^{(\mathrm{T})}\Big|_{x=0} \to 0, \tag{16.119}$$

$$A_{\Pi}^{(\mathrm{T})}\Big|_{x=L} \to \frac{2}{\sqrt{3}}\left(\frac{L}{2}\right)^{4/3} \exp\left[-\frac{3}{2}\left(\frac{L}{2}\right)^{2/3}\right]\left[1 + \frac{\sqrt{3}\pi}{4}\left(\frac{2}{L}\right)^{1/3}\right], \tag{16.120}$$

$$\varphi_{\Pi}^{(\mathrm{T})}\Big|_{x=L} \to \frac{3\sqrt{3}}{2}\left(\frac{L}{2}\right)^{2/3} + \frac{4\pi}{3}. \tag{16.121}$$

These expressions show that even if the receptor plate is far from the oscillating plate, the wave is not harmonic and cannot be presented in the form (16.1).

16.3.6 Numerical Results

The numerical scheme to solve the kinetic equations (16.91) is quite similar to that to solve Eq. (16.37), but the expressions are quite cumbersome. First, the perturbations $\Phi^{(\alpha,m)}$ ($\alpha = 1, 2$ and $m = u$,T) are split into two terms (16.48), where $\Phi_0^{(\alpha,m)}$ satisfy Eq. (16.49) and boundary conditions (16.95) and (16.96) so that it is obtained analytically. The perturbations $\Phi'^{(\alpha,m)}$ obey (16.50) with $H^{(\alpha)}$ given by (16.92) and (16.93). It is obtained numerically by the scheme (16.55) replacing $\tilde{u}_k$ by $H_k^{(\alpha)}$. The details of the scheme for the mechanical wave generation are given in the paper [219]. The scheme for the thermal wave generation differs only in the expression for $\Phi_0^{(\alpha,m)}$ because of the different boundary conditions.

The numerical and analytical results based on the S-model reported in the papers [218, 219] are partially reproduced in Tables 16.3 and 16.4. The numerical values of amplitudes $A_\Pi^{(u)}$, $A_\Pi^{(\mathrm{T})}$ and phases $\varphi_\Pi^{(u)}$, $\varphi_\Pi^{(\mathrm{T})}$ calculated in the high-frequency regime using Eqs. (16.113), (16.114) assuming $L = 1$ and those calculated in the hydrodynamic regime using Eq. (16.76) are also given in Tables 16.3 and 16.4. In all cases, the phases are calculated so that they vary in the range $[-\pi, \pi]$. These data show that the analytical expressions Eqs. (16.113) and (16.114) provide the values close to the numerical results at $\theta = \delta = 0.1$ and the expression (16.76) yields the values close to those obtained from the

Table 16.3 Amplitudes of pressure tensors $\Pi^{(u)}$ and $\Pi^{(\mathrm{T})}$ at the plates ($x = 0$ and $x = L$) for longitudinal wave versus rarefaction δ and oscillation speed parameter θ, Ref. [218].

		$A_\Pi^{(u)}$		$A_\Pi^{(\mathrm{T})}$	
θ	δ	$x = 0$	$x = L$	$x = 0$	$x = L$
0[a)]	0	1.270	2.239	0.3556	0.4187
0.1	0.1	1.251	2.214	0.3514	0.4148
0.1	1	2.015	0.288	0.2546	0.1523
0.1	10	2.013	0.000	0.2500	0.0000
1	0.1	10.47	10.57	0.4900	0.4902
1	1	1.068	2.009	0.3256	0.3974
1	10	1.924	0.170	0.2480	0.0535
10	0.1	100.1	100.1	0.4964	0.4970
10	1	10.08	10.18	0.4964	0.4970
10	10	0.810	1.897	0.2088	0.3286
10[b)]	10	0.830	1.903	0.2169	0.3348

a) Equations (16.113) and (16.114) for $L = 1$.
b) Equations (16.76) and Table (16.2).

Table 16.4 Phases of pressure tensor $\varphi_{\Pi}^{(u)}$ and $\varphi_{\Pi}^{(\mathrm{T})}$ for longitudinal wave at the plates ($x = 0$ and $x = L$) versus rarefaction δ and oscillation speed θ parameter ($-\pi \le \varphi_{\Pi} \le \pi$), Ref. [218].

		$\varphi_{\Pi}^{(u)}/\pi$		$\varphi_{\Pi}^{(\mathrm{T})}/\pi$	
θ	δ	$x = 0$	$x = L$	$x = 0$	$x = L$
0[a)]	0	0.2644	0.3857	0.1962	0.1593
0.1	0.1	0.2581	0.3836	0.1987	0.1627
0.1	1	−0.0018	0.1247	0.0027	−0.1092
0.1	10	−0.0034	0.4278	−0.0011	0.2305
1	0.1	0.4462	0.4466	0.0316	0.0315
1	1	0.2289	0.3807	0.2219	0.1921
1	10	−0.0252	0.7702	−0.0115	0.5376
10	0.1	0.4929	0.4929	0.0298	0.0298
10	1	0.4638	0.4641	0.0299	0.0299
10	10	0.3240	0.4377	0.3117	0.2698
10[b)]	10	0.3141	0.4334	0.3117	0.2756

a) Equations (16.113) and (16.114) for $L = 1$.
b) Equations (16.76) and Table (16.2).

kinetic equation at $\theta = \delta = 10$. Thus, the kinetic equation solved in the ranges $0.1 \le \theta \le 10$ and $0.1 \le \delta \le 10$ with the analytical solutions (16.113), (16.114) and the solution based on the hydrodynamic equations (16.66), (16.67) cover the whole range of the both parameters θ and δ.

The same problems were solved on the basis of the linearized BE by Doi [217, 220]. A comparative analysis of these data with those obtained on the basis of the S-model in Refs [218, 219] shows that the relative difference of the amplitudes $A_{\Pi}^{(u)}$, $A_{\Pi}^{(\mathrm{T})}$ and phases $\varphi_{\Pi}^{(u)}$, $\varphi_{\Pi}^{(\mathrm{T})}$ does not exceed 3%. Considering that a numerical solution of the BE requires significant computational effort, the model equations solved with quite modest effort are appropriate computational tools if the uncertainty of a few percent is acceptable.

A study of the influence of the gas–surface interaction on the wave generation is performed in the paper [218] on the basis of the CL scattering kernel. According to the data reported in this paper, when $\delta \ll \theta$, the influence of the gas–surface interaction on the pressure tensors $\Pi^{(u)}$ and $\Pi^{(\mathrm{T})}$ is very weak. In other situations, this quantity strongly depends on the energy accommodation coefficient α_n and weakly depends on the momentum accommodation coefficient α_t.

The amplitudes $A_{\Pi}^{(u)}$, $A_{\Pi}^{(\mathrm{T})}$ and phases $\varphi_{\Pi}^{(u)}$, $\varphi_{\Pi}^{(\mathrm{T})}$ calculated from the analytical solutions (16.113) and (16.114), from the kinetic equation (16.91) subject to the boundary conditions (16.95), (16.96) and those calculated in the hydrodynamic regime on the basis of (16.76) are plotted against the distance L in Figures 16.5 and 16.6. The phases are plotted so that they are continuous functions of x, that is, the phases can be out of the range $[-\pi, \pi]$. The behaviors of the amplitude

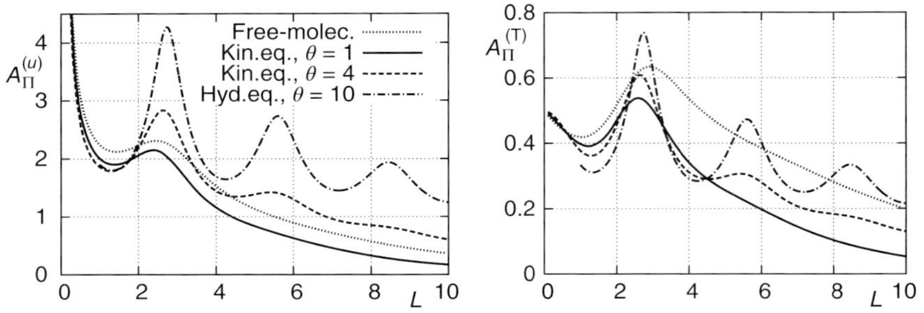

Figure 16.5 Amplitudes of pressure tensors $A_\Pi^{(u)}$ and $A_\Pi^{(T)}$ at receptor ($x = L$) versus L: pointed line – Eqs. (16.113) and (16.114); point-dashed line – Eq. (16.76) for $\theta = 10$.

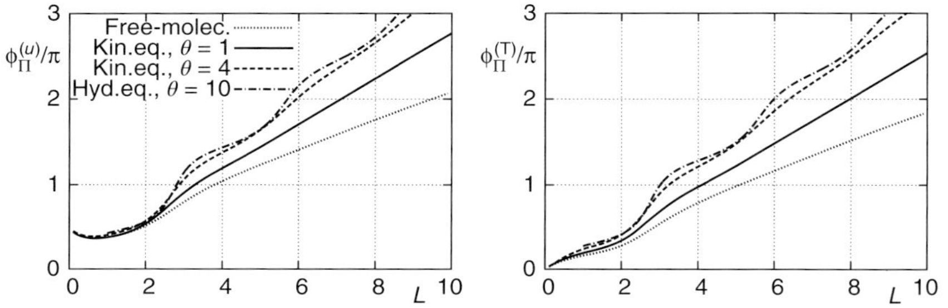

Figure 16.6 Phases of pressure tensors $\varphi_\Pi^{(u)}$ and $\varphi_\Pi^{(T)}$ at receptor ($x = L$) versus L: pointed line – Eqs. (16.113) and (16.114); point-dashed line – Eq. (16.76) for $\theta = 10$.

$A_\Pi^{(u)}$ and phase $\varphi^{(u)}$ of the wave generated by the plate oscillation are qualitatively in agreement with the experimental data reported by Schotter [221]. It is curious that both $A_\Pi^{(u)}$ and $\varphi^{(u)}$ always have a minimum when L is about 1. The behavior of the thermoacoustic wave, $A_\Pi^{(T)}$ and $\varphi^{(T)}$, is quite different from $A_\Pi^{(u)}$ and $\varphi^{(u)}$ for a small distance L. Such a behavior corresponds to the limit values obtained analytically in Eq. (16.115). In case of a large distance L, the waves generated by both mechanical and thermal oscillations are similar to each other and in agreement with the asymptotes (16.117), (16.118), (16.120), and (16.121). One more interesting fact is that both amplitudes $A_\Pi^{(u)}$ and $A_\Pi^{(T)}$ have their minimum value for the intermediate frequency parameter θ when $L > 5$. However, their qualitative behavior is quite similar to that for the high-frequency regime $\theta \to 0$. If we analyze the phases $\varphi_\Pi^{(u)}$ and $\varphi_\Pi^{(T)}$, they smoothly change their behavior in the transition from the high-frequency regime ($\theta \to 0$) to the hydrodynamic one ($\theta \to \infty$).

The DSMC technique was applied to oscillating flows of rarefied gases by Hadjiconstantinou [222, 223]. Some results on axisymmetrical oscillating flow are reported in the papers [224–226].

Exercises

16.1 Check that Eqs. (16.25)–(16.28) obey the differential equations (16.23) subject to the boundary conditions (16.24).

16.2 Prove (16.17).
Hint: Note that all terms contain an expression of the type $\exp\left(\frac{A}{|c_1|}\right)$ where A is always negative.

16.3 Obtain the expression of $\tilde{q}^{(u)}$ in the high-frequency regime and check if it obeys the reciprocal relations (16.101) and (16.106).
Solution:

$$\tilde{q}^{(u)}(x) = \frac{1}{\sqrt{\pi}}\left\{2\left[I_4(-ix) - \frac{3}{2}I_2(-ix)\right] + (\varrho_{r0}^{(u)} + \sqrt{\pi})\left[I_3(-ix) - \frac{3}{2}I_1(-ix)\right]\right.$$
$$\left. -\varrho_{r1}^{(u)}\left[I_3(i(x-L)) - \frac{3}{2}I_1(i(x-L))\right]\right\}, \tag{16.122}$$

$$\Pi^{(\mathrm{T})} = -\tilde{q}^{(u)} = \frac{1}{4} + \frac{\varrho_{rL}^{(\mathrm{T})}}{\sqrt{\pi}}(I_3(-iL) - I_1(-iL)) \quad \text{at} \quad x = 0, \tag{16.123}$$

$$\Pi^{(\mathrm{T})} = \tilde{q}^{(u)} = \frac{2}{\sqrt{\pi}}\left[I_4(-iL) - I_2(-iL)\right.$$
$$\left. +\varrho_{rL}^{(\mathrm{T})}\left(2I_1(-iL)I_2(-iL) + \frac{\sqrt{\pi}}{4}\right)\right] \quad \text{at} \quad x = L. \tag{16.124}$$

16.4 Using the listing in Section B.9, check that the amplitudes A_u, A_Π and phases φ_u, φ_Π at $\theta = 0.1$ and $\delta = 2$ are the practically the same as those at $\theta = 0.1$ and $\delta = 1$.

16.5 Check if the expressions (16.70), (16.76), and (16.77) obey Eqs. (16.63)–(16.65), (16.66), and (16.67).
Hint: First, obtain the expressions for $\varrho(x)$ using (16.63) and $\tilde{q}(x)$ using (16.67). Then substitute τ, $\tilde{q}$, and $\tilde{u}$ into (16.65). To obtain the identity, use Eqs. (16.71).

16.6 Using the data of Table (16.2), the expression (16.76) and that for $\tilde{q}^{(u)}$ obtained in the previous exercise, calculate $\Pi^{(\mathrm{T})}$ and $\tilde{q}^{(u)}$ at $x = 0$ and $x = L$. Check the reciprocal relations (16.101) and (16.106).

A
Constants and Mathematical Expressions

A.1
Physical Constants

Table A.1 summarizes the fundamental physical constants based on the data available through December 31, 2010, and recommended by the Committee on Data for Science and Technology (CODATA) for international use.

Table A.2 contains the atomic weights of the noble gases based on Technical Report [227] of the International Union of Pure and Applied Chemistry (IUPAC) published in 2013.

A.2
Vectors and Tensors

Here, a brief description of some mathematical symbols and notation is given. A more detailed description can be found in Appendix A of the book [4].

Let $\boldsymbol{e}_1$, $\boldsymbol{e}_2$, and $\boldsymbol{e}_3$ be the reference unit vectors. Then a vector $\boldsymbol{a}$ can be represented by its three components a_1, a_2, and a_3 as

$$\boldsymbol{a} = \sum_{i=1}^{3} \boldsymbol{e}_i a_i. \tag{A.1}$$

Einstein's summation convention allows us to achieve more compact form, that is, whenever an index is repeated in a term, such a term is to be summed over all values of that index. With this convention, the expression (A.1) takes the form

$$\boldsymbol{a} = \boldsymbol{e}_i a_i. \tag{A.2}$$

Later in this book, this convention will be adopted.

The scalar (or inner) product of two vectors $\boldsymbol{a}$ and $\boldsymbol{b}$ is defined by

$$\boldsymbol{a} \cdot \boldsymbol{b} = a_i b_i. \tag{A.3}$$

Rarefied Gas Dynamics: Fundamentals for Research and Practice, First Edition. Felix Sharipov.

Table A.1 Fundamental physical constants.

Boltzmann constant	$k_B = 1.3806488(13) \times 10^{-23}$	$J\,K^{-1}$
Molar gas constant	$R_g = 8.3144621(75)$	$J\,K^{-1}\,mol^{-1}$
Avogadro number	$N_A = 6.02214129(27) \times 10^{23}$	mol^{-1}
Loschmidt constant	$N_L = 2.6516462(24) \times 10^{25}$	m^{-3}
Standard atmosphere	1 atm = 101 325	Pa

Table A.2 Standard atomic weights, [227].

Gas	Weight (g mol^{-1})
Helium	4.002602(2)
Neon	20.1797(6)
Argon	39.948(1)
Krypton	83.798(2)
Xenon	131.293(6)

A second-rank tensor $\mathbf{B}$ with components B_{ij} can be represented by the array

$$\mathbf{B} = \begin{pmatrix} B_{11} & B_{12} & B_{13} \\ B_{21} & B_{22} & B_{23} \\ B_{31} & B_{32} & B_{33} \end{pmatrix}. \tag{A.4}$$

A simple product of a vector $\boldsymbol{a}$ and a tensor $\mathbf{B}$ is a vector

$$\boldsymbol{b} = \boldsymbol{a} \cdot \mathbf{B}, \; b_i = a_j B_{ji}. \tag{A.5}$$

A double (or inner) product of two tensors $\mathbf{A}$ and $\mathbf{B}$ produces a scalar

$$\mathbf{A} : \mathbf{B} = A_{ij} B_{ij}. \tag{A.6}$$

Two vectors $\boldsymbol{a}$ and $\boldsymbol{b}$ written together without the dot, like $\boldsymbol{ab}$, form a second-rank tensor defined as

$$\boldsymbol{ab} = \begin{pmatrix} a_1 b_1 & a_1 b_2 & a_1 b_3 \\ a_2 b_1 & a_2 b_2 & a_2 b_3 \\ a_3 b_1 & a_3 b_2 & a_3 b_3 \end{pmatrix}. \tag{A.7}$$

In kinetic theory, it is usual to form a tensor using one vector $\boldsymbol{a}$. Then, its double product with a tensor $\mathbf{B}$ means

$$\mathbf{B} : \boldsymbol{aa} = B_{ij} a_i a_j. \tag{A.8}$$

The unit tensor defined as

$$\mathbf{I} = \begin{pmatrix} 1 & 0 & 0 \\ 0 & 1 & 0 \\ 0 & 0 & 1 \end{pmatrix}. \tag{A.9}$$

The inverse tensor $\mathbf{B}^{-1}$ is defined as

$$\mathbf{B}^{-1} \cdot \mathbf{B} = \mathbf{I}. \tag{A.10}$$

A.3
Nabla Operator

The notation ∇_r or ∇_x is called *nabla operator* and means a vector with the following components

$$\nabla_r = \boldsymbol{e}_i \frac{\partial}{\partial r_i} \quad \text{or} \quad \nabla_x = \boldsymbol{e}_i \frac{\partial}{\partial x_i}, \tag{A.11}$$

where (r_1, r_2, r_3) and (x_1, x_2, x_3) are the components of position vectors $\boldsymbol{r}$ and $\boldsymbol{x}$, respectively. According to Eq. (A.3), the scalar product of the velocity vector $\boldsymbol{v}$ with ∇_r means the following operator:

$$\boldsymbol{v} \cdot \nabla_r = v_i \frac{\partial}{\partial r_i}. \tag{A.12}$$

Again following (A.3), the scalar product of ∇_r with a vector $\boldsymbol{u}$ means the expression

$$\nabla_r \cdot \boldsymbol{u} = \frac{\partial u_i}{\partial r_i}. \tag{A.13}$$

However, the same expression without the dot, that is, $\nabla_r \boldsymbol{u}$, means a tensor with elements equal to $\partial u_j / \partial r_j$.

A.4
Kronecker Delta and Dirac Delta Function

The Kronecker delta is defined as

$$\delta_{ij} = \begin{cases} 1 & \text{if} \quad i = j \\ 0 & \text{if} \quad i \neq j \end{cases}. \tag{A.14}$$

The Dirac delta function is defined as

$$\delta(x) = \begin{cases} \infty & \text{if} \quad x = 0 \\ 0 & \text{if} \quad x \neq 0 \end{cases}, \tag{A.15}$$

It is normalized by unity

$$\int_{-\infty}^{\infty} \delta(x)\, \mathrm{d}x = 1 \tag{A.16}$$

and fulfills the following property

$$\int_{-\infty}^{\infty} f(x)\delta(x - x_0)\, \mathrm{d}x = f(x_0), \tag{A.17}$$

where $f(x)$ is any function continuous in the point x_0.

A.5 Some Integrals

To calculate the moments of the VDF, the following integrals could be useful

$$\int_0^\infty c^{2i}\, e^{-c^2}\, \mathrm{d}c = \frac{(2i)!}{2^{2i+1}(i!)}\sqrt{\pi}, \quad \int_0^\infty c^{2i+1}\, e^{-c^2}\, \mathrm{d}c = \frac{i!}{2}, \tag{A.18}$$

where i is a non-negative integer.

A.6 Taylor Series

Any function $f(x)$ infinitely differentiable at x_0 can be written down in the form of Taylor series, see example, Section 4.10-4 in the handbook [228]. Assuming x_0 and remaining the first-order term, the series is given as

$$f(x) = f(0) + \left.\frac{\mathrm{d}f}{\mathrm{d}x}\right|_{x=0} x + O(x^2), \tag{A.19}$$

where the notation $O(x^2)$ means a term of the order x^2. For some particular functions, Eq. (A.19) takes the form

$$(1+x)^\alpha = 1 + \alpha x + O(x^2), \tag{A.20}$$

$$e^x = 1 + x + O(x^2), \tag{A.21}$$

$$\ln(1+x) = x + O(x^2). \tag{A.22}$$

A.7 Some Functions

The error function is defined as

$$\mathrm{erf}(x) = \frac{2}{\sqrt{\pi}} \int_0^x e^{-c^2}\, \mathrm{d}c. \tag{A.23}$$

The complementary error function is defined as

$$\mathrm{erfc}(x) = \frac{2}{\sqrt{\pi}} \int_x^\infty e^{-c^2}\, \mathrm{d}c = 1 - \mathrm{erf}(x). \tag{A.24}$$

For a large positive argument x, the function erfc(x) has the following asymptotic series

$$\mathrm{erfc}(x) = \frac{e^{-c^2}}{x\sqrt{\pi}} \left(1 - \frac{1}{2x^2} + O\left(\frac{1}{x^4}\right)\right). \tag{A.25}$$

The signum function is defined as

$$\mathrm{sgn}(x) = \begin{cases} 1 & x > 0, \\ 0 & x = 0, \\ -1 & x < 0. \end{cases} \tag{A.26}$$

Modified Bessel function of the first kind is defined as

$$\mathcal{I}_0(x) = \frac{1}{\pi} \int_0^{\pi} \exp(x \cos \varphi)\, \mathrm{d}\varphi. \tag{A.27}$$

Hyperbolic tangent is defined as

$$\tanh(x) = \frac{e^x - e^{-x}}{e^x + e^{-x}}. \tag{A.28}$$

A miscellaneous function $I_n(x)$, see Abramovitz and Stegun [229] (p. 1001), is defined as

$$I_n(x) = \int_0^{\infty} c^n \exp\left(-c^2 - \frac{x}{c}\right)\, \mathrm{d}c. \tag{A.29}$$

The functions of the different order n are related to each other as

$$\frac{\mathrm{d}I_n(x)}{\mathrm{d}x} = -I_{n-1}(x), \quad 2I_n(x) = (n-1)I_{n-2}(x) + xI_{n-3}(x). \tag{A.30}$$

The power series representation has the form

$$I_1(x) = \frac{1}{2} \sum_{k=0}^{\infty} (a_k \ln x + b_k) x^k, \tag{A.31}$$

where

$$a_0 = a_1 = 0, \quad b_0 = -a_2 = 1, \quad b_1 = -\sqrt{\pi}, b_2 = 0.6341754927, \tag{A.32}$$

and

$$a_k = \frac{-2a_{k-2}}{k(k-1)(k-2)}, \quad b_k = \frac{-2b_{k-2} - (3k^2 - 6k + 2)a_k}{k(k-1)(k-2)} \tag{A.33}$$

for $k > 2$. The same representations can be obtained for other I_n using (A.30). The asymptotic representation of $I_n(x)$ at $x \to \infty$ reads

$$I_n(x) \sim \sqrt{\frac{\pi}{3}} 3^{-\frac{n}{2}} \upsilon^{\frac{n}{2}} \exp(-\upsilon) \sum_{k=0}^{\infty} \frac{a_k}{\upsilon^k}, \upsilon = 3\left(\frac{x}{2}\right)^{2/3}, \tag{A.34}$$

$$a_0 = 1, \quad a_1 = \frac{1}{12}(3n^2 + 3n - 1), \tag{A.35}$$

$$12(k+2)a_{k+2} = -(12k^2 + 36k - 3n^2 - 3n + 25)a_{k+1}$$
$$+ \frac{1}{2}(n - 2k)(2k + 3 - n)(2k + 3 + 2n)a_k. \tag{A.36}$$

A.8
Gauss–Ostrogradsky's Theorem

Suppose $\boldsymbol{J}$ is a continuously differentiable vector field defined in a three-dimensional volume V, then we have

$$\int_V \nabla_r \cdot \boldsymbol{J}\, \mathrm{d}^3\boldsymbol{r} = -\oint_\Sigma J_n\, \mathrm{d}^2\Sigma. \tag{A.37}$$

The left side is an integral with respect to the position $\boldsymbol{r}$ over the volume V, the right side is the integral over the closed surface Σ of the volume V, and J_n is the component of the vector $\boldsymbol{J}$ normal to the surface Σ pointing toward the volume V.

A.9
Complex Numbers

A complex number z can be expressed in the form

$$z = a + bi, \tag{A.38}$$

where a and b are real numbers and i is the imaginary unit defined so that its square is equal to -1, that is, $i^2 = -1$. The first term of (A.38) is the real part and the second term is the imaginary part of the complex number z. To distinguish these parts, the following notations are used

$$a = \Re(z), \quad b = \Im(z). \tag{A.39}$$

A complex number can be written in terms of the absolute value and argument as

$$z = |z|e^{i\varphi} = |z|(\cos\varphi + i\sin\varphi), \tag{A.40}$$

where $|z|$ and φ are absolute value and argument of the complex number defined as

$$|z| = \sqrt{a^2 + b^2}, \quad \tan\varphi = \frac{b}{a}, \tag{A.41}$$

respectively. If 2π is added to (or subtracted from) the argument φ, the complex number z does not change.

B
Files and Listings

B.1
Files with Nodes and Weights of Gauss Quadrature

B.1.1
Weighting Function (9.16)

B.1.1.1 File cw4.dat, $N_c = 4$

```
0.1337764469D+00    0.1835325639D+00
0.6243246902D+00    0.2375842405D+00
0.1342537826D+01    0.7528686870D-01
0.2262664477D+01    0.3596326915D-02
```

B.1.1.2 File cw6.dat, $N_c = 6$

```
0.7860065997D-01    0.1110605299D+00
0.3867394030D+00    0.1969891638D+00
0.8664294481D+00    0.1451431364D+00
0.1465698004D+01    0.4288582131D-01
0.2172707796D+01    0.3865781007D-02
0.3036820173D+01    0.5555667667D-04
```

B.1.1.3 File cw8.dat, $N_c = 8$

```
0.5297864228D-01    0.7566300035D-01
0.2673983574D+00    0.1513894051D+00
0.6163029075D+00    0.1556900293D+00
0.1064246297D+01    0.8883067220D-01
0.1588855863D+01    0.2528365329D-01
0.2183921099D+01    0.3028533189D-02
0.2863133907D+01    0.1140021996D-03
0.3686007261D+01    0.6728507742D-06
```

Rarefied Gas Dynamics: Fundamentals for Research and Practice, First Edition. Felix Sharipov.

B.1.2
Weighting Function (9.22)

B.1.2.1 File cpw4.dat, $N_c = 4$

```
0.2800995401D+00      0.3628702601D-01
0.8320770658D+00      0.8569932815D-01
0.1556389870D+01      0.3526522019D-01
0.2463284960D+01      0.1903368745D-02
```

B.1.2.2 File cpw6.dat, $N_c = 6$

```
0.1749161821D+00      0.1528005867D-01
0.5436071645D+00      0.5717139479D-01
0.1046322591D+01      0.6192775186D-01
0.1648827020D+01      0.2246704405D-01
0.2349945600D+01      0.2273655749D-02
0.3203500340D+01      0.3503797227D-04
```

B.1.2.3 File cpw8.dat, $N_c = 8$

```
0.1218127676D+00      0.7632680544D-02
0.3882449489D+00      0.3477557210D-01
0.7651497064D+00      0.5722010701D-01
0.1224690624D+01      0.4302759153D-01
0.1751398297D+01      0.1449004513D-01
0.2343383198D+01      0.1930524586D-02
0.3016608850D+01      0.7793933552D-04
0.3831371301D+01      0.4828489441D-06
```

B.2
Files for Planar Couette Flow

B.2.1
Listing of Program "couette_planar.for"

```
! Calculation of planar Couette flow
      implicit double precision (a-h,o-z)
      parameter(
     & delta=1.d0, ! rarefaction parameter
     & Nx=10, ! number of nodes in physical space
     & Nc=6, ! number of nodes in velocity space
     & e=1.d-10, ! criterion to stop iteration
     & maxit=100 000) !maximum number of iterartions
      dimension
     & u(-Nx:Nx), ! velocity in n-th iteration
```

```
     & un(-Nx:Nx),! velocity in (n+1)-th iteration
     & Pi(-Nx:Nx), ! shear stress in n-th iteration
     & c(Nc),w(Nc) ! nodes and weights
      if(Nc==4) open(10,file='cw4.dat')
      if(Nc==6) open(10,file='cw6.dat')
      if(Nc==8) open(10,file='cw8.dat')
         ! node and weight reading
      read(10,*) (c(i),w(i),i=1,Nc)
      u=0.d0 ! initial approximaion
      Pi0=0.d0
      it=0 ! beginning of iterations
 6    it=it+1 !counter of iterations
      un=0.d0!set moment of (n+1)-th iteration to zero
      Pi=0.d0
      un(-Nx)=0.25d0! boundary values Eq.(11.29)
      Pi(-Nx)=0.5d0/sqrt(dacos(-1.d0))
      do i=1,Nc ! loop for speed
        Phi=0.5d0 ! boundary cond., Eq.(11.21)
        do k=-Nx+1,Nx ! loop for x-coordinate
          Phi=((4.d0*Nx*c(i)-delta)*Phi
     &     +delta*(u(k)+u(k-1)))
     &     /(4.d0*Nx*c(i)+delta)!expression Eq.(11.27)
          un(k)=un(k)+Phi*w(i)!half-moments Eq.(11.28)
          Pi(k)=Pi(k)+2.*Phi*c(i)*w(i)
        end do !  end of loop for x-coordinate
      end do !  end of loop for speed
      do k=0,Nx !  summation of half-moments Eq.(11.30)
        u(k)=un(k)-un(-k)
        Pi(k)=Pi(k)+Pi(-k)
        u(-k)=-u(k)
        Pi(-k)=Pi(k)
      end do
      er=abs((Pi0-Pi(0))/Pi(0)) ! residual calculated
      Pi0=Pi(0)
      if(er>e.and.it<maxit) go to 6!convergence verified
      Pimax=maxval(Pi) ! maximum value of Pi calculated
      Pimin=minval(Pi) ! minimum value of Pi calculated
      open(9,file='Res_couette_planar.dat')
      write(9,18)
      do 104 k=0,Nx,2! writing of results
104     write(9,19) 0.5*k/Nx,u(k),Pi(k)
      write(9,20) delta,Pi0, (Pimax-Pimin)/Pimax
   18 format(4x,'velocity␣profile,␣shear␣stress',
     &/6x,'x',9x,'u',9x,'Pi')
```

```
19  format(f10.3,1x,f10.5,1x,f10.5)
20  format(/2x,'delta=',f7.3,',',2x,'Pi=',f10.5,
   &',',2x,'var. ␣Pi',e9.2)
    stop
    end
```

B.2.2
Output File with Results "Res_couette_planar.dat"

```
  velocity profile, shear stress
    x         u         Pi
   0.000    0.00000    0.33892
   0.100   -0.04406    0.33892
   0.200   -0.08874    0.33892
   0.300   -0.13520    0.33892
   0.400   -0.18650    0.33892
   0.500   -0.25174    0.33892
  delta=  1.000,  Pi=   0.33892,  var. Pi 0.39E-07
```

B.3
Files for Planar Heat Transfer

B.3.1
Listing of Program "heat_planar.for"

```
c  calculation of  planar heat flow
      implicit double precision (a-h,o-z)
      parameter(
     & del=1.d0,! rarefaction parameter
     & Nx=10,!number of nodes in physical space
     & Nc=6,!number of nodes in velocity space
     & e=1.d-10, ! criterion to stop iteration
     & maxit=100 000)!maximum number of iterartions
      dimension
     & rh (-Nx:Nx),!density in n-th iteration
     & rhn(-Nx:Nx),!density in (n+1)-th iteration
     & t  (-Nx:Nx),!temperature in n-th iteration
     & tn (-Nx:Nx),!temperature in (n+1)-th iteration
     & q  (-Nx:Nx),!heat flow in n-th iteration
     & qn (-Nx:Nx),!heat flow in (n+1)-th iteration
     & dH(2,-Nx:Nx),
     & Ph(2), ! perturbation
     & c(Nc),w(Nc) ! nodes and weights
      if(Nc==4) open(10,file='cw4.dat')
```

```
 if(Nc==6) open(10,file='cw6.dat')
 if(Nc==8) open(10,file='cw8.dat')
   ! node and weight reading
 read(10,*) (c(i),w(i),i=1,Nc)
 spi=sqrt(dacos(-1.d0)) ! square root of pi
 rh=0.d0  ! initial approximaion
 t=0.d0
 q=0.d0
 rhr=0.d0!term in bound. cond. Eq.(11.51)
 it=0    ! beginning of iterations
 er=1.
 do while (er>e.and.it<maxit)
 it=it+1 ! counter of iterations
 rhn=0.d0!set moments to zero
 tn=0.d0
 qn=0.d0
 rhr0=0.
 rhn(-Nx)=0.5*rhr -0.125d0! Eq.(11.69)
 tn(-Nx)=0.25
 qn(-Nx)=(rhr+1.d0)/(2.d0*spi)
 do i=1,Nc !   loop for speed
  dH(1,:)=del*(rh(:)+(c(i)**2-0.5)*t(:)!Eq.(11.56)
&      +4.d0/15.d0*((c(i)**2-1.5d0))*c(i)*q(:))
  dH(2,:)=del*(t(:)+4.d0/15.d0*c(i)*q(:))
  Ph(1)=rhr+(c(i)**2-1.)/2.d0!Eq.(11.57)
  Ph(2)=0.5d0 ! Eq.(11.58)
  do k=-Nx+1,Nx ! loop for x-coordinate
   Ph(:)=((4.d0*Nx*c(i)-del)*Ph(:)+dH(:,k)
&     +dH(:,k-1))/(4.d0*Nx*c(i)+del) ! Eq.(11.67)
   rhn(k)=rhn(k)+Ph(1)*w(i) ! Eq.(11.68)
   tn(k)=tn(k)+(Ph(1)*(c(i)**2-0.5d0)+Ph(2))
&      *w(i)*2./3.
   qn(k)=qn(k)+(Ph(1)*c(i)**2+Ph(2))*c(i)*w(i)
  end do !  end of loop for x-coordinate
  rhr0=rhr0+2.d0*spi*c(i)*Ph(1)*w(i)!Eq.(11.70)
 end do !  end of loop for speed
 do k=0,Nx ! Eq.(11.71)
  rh(k)=rhn(k)-rhn(-k)
  t(k)=tn(k)-tn(-k)
  q(k)=qn(k)+qn(-k)
  rh(-k)=-rh(k)
  t(-k)=-t(k)
  q(-k)=q(k)
 end do
 rhr=-rhr0
```

```
      er=abs((q(0)-qn0)/q(0)) ! residual calculated
      qn0=q(0)
      end do
      open(9,file="Res_heat_planar.dat")
      write(9,18)
      do 104 k=0,Nx,2 ! writing of results
104    write(9,19) 0.5*k/Nx,rh(k),t(k),q(k)
      write(9,20) del,qn0,(maxval(q)-minval(q))/qn0
 18   format(13x,'density', 4x,'temperature',2x,
     &'heat␣flow'/6x,'x',9x,'n',9x,'tau', 9x , 'q')
 19   format(f10.3,3(1x,f10.5))
 20   format(/'delta=',f7.3,',',2x,'q=',f10.5,',',
     &2x,'var.␣q',e9.2)
      stop
      end
```

B.3.2
Output File with Results "Res_heat_planar.dat"

```
             density    temperature  heat flow
      x         n          tau          q
     0.000    0.00000    0.00000    0.40013
     0.100    0.03141   -0.03508    0.40013
     0.200    0.06348   -0.07090    0.40013
     0.300    0.09747   -0.10883    0.40013
     0.400    0.13667   -0.15239    0.40013
     0.500    0.19024   -0.21185    0.40013
delta=  1.000,  q=   0.40013,  var. q 0.57E-07
```

B.4
Files for Planar Poiseuille and Creep Flows

B.4.1
Listing of Program "poiseuille_creep_planar.for"

```
! calculation of planar Poiseuille and creep flows
      implicit double precision (a-h,o-z)
      parameter(
     & del=1.d0,! rarefaction parameter
     & Nx=20, ! number of nodes in physical space
     & Nc=6, ! number of nodes in velocity space
     & e=1.d-10, ! criterion to stop iteration
     & maxit=100 000)!maximum number of iterartions
      dimension
```

```
     & u  (-Nx:Nx,2), ! velocity in n-th iteration
     & u0(-Nx:Nx,2), ! split part of velocity
     & un(-Nx:Nx),! velocity in (n+1)-th iteration
     & q  (-Nx:Nx,2),! heat flow in n-th iteration
     & q0(-Nx:Nx,2),! split part of heat flow
     & qn(-Nx:Nx),! heat flow in (n+1)-th iteration
     & dH(2,-Nx:Nx), ! function H
     & Ph(2), ! perturbations
     & g(2), ! source term
     & GG(2),QQ(2), ! mass and heat flow rates
     & c(0:Nc),w(0:Nc) ! nodes and weights
      spi=sqrt(acos(-1.d0)) ! square root of pi
      Ig=1 ! no splitting scheme
      if(del<1.)          Ig=0 ! splitting scheme
      if(Nc<=8) then ! Gauss nodes and weights
       if(Nc==4) open(10,file='cw4.dat')
       if(Nc==6) open(10,file='cw6.dat')
       if(Nc==8) open(10,file='cw8.dat')
         ! nodes and weights reading
       read(10,*) (c(i),w(i),i=1,Nc)
      else ! Simpson nodes and weights
       c01=0.1d0 ! upper limit for first interval
       c0=3.0d0 ! upper limit for second interval
       do 100 i=0,Nc/2
        c(i)=2.d0*c01*i/Nc   !Eq.(11.131)
        c(nc/2+i)=c01+2.d0*(c0-c01)*i/Nc
        w(i)=(((i+1)/2)*2-i+1)*exp(-c(i)**2)*c01
     &   /(0.75d0*Nc*spi)   !(11.132)
100     w(nc/2+i)=(((i+1)/2)*2-i+1)*exp(-c(nc/2+i)**2)
     &  *(c0-c01)/(0.75d0*Nc*spi)
        w(0)=w(0)/2.d0
        w(Nc/2)=w(Nc/2)*c0/(2.d0*c01)
        w(Nc)=w(Nc)/2.
      end if
      u0=0.d0
      q0=0.d0
      if(Ig==0) then ! split part of moments,
        do 101 k=-Nx+1,Nx
          !Eq.(11.129)
         call funcI(del*(0.5d0*k/Nx+0.5d0),aI0,aI2,aI4)
         u0(k,1)=-(0.25-aI0/(2.*spi))/del !Eq.(11.126)
         q0(k,1)=(aI2/2.-aI0/4.d0)/(spi*del) !Eq.(11.127)
         u0(k,2)=q0(k,1)
101      q0(k,2)=-(0.625-(aI4-aI2+2.25d0*aI0)
```

```
      &     /(2.*spi))/del!Eq.(11.128)
       end if
       u=0.d0  ! initial approximaion
       q=0.d0
       do 104 m=1,2! driving force, m=1 -> P, m=2 -> T
         u00=0.d0
         q00=0.d0
         it=0   ! beginning of iterations
 6       it=it+1 ! counter of iterations
         un=0.d0 ! set moments to zero
         qn=0.d0
         g(1)=-0.5*Ig ! source terms Eq.(11.101)
         g(2)=0.d0
         if(m==2) g(2)=-1.d0*Ig
         do 102 i=Ig,Nc !   loop for speed
           if(m==2) g(1)=-0.5d0*(c(i)**2-0.5d0)*Ig
           dH(1,:)=del*(u(:,m)+2./15.*((c(i)**2-0.5))
      &     *q(:,m))!Eq.(11.100)
           dH(2,:)=del*4.d0/15.d0*q(:,m)
           Ph=0.d0 ! boundary cond, Eqs.(11.103)
           do 102 k=-Nx+1,Nx ! loop for x-coordinate
            Ph(:)=((4.d0*Nx*c(i)-del)*Ph(:)+dH(:,k)
      &       +dH(:,k-1)+2.*g(:))
      &       /(4.*Nx*c(i)+del)!Eq.(11.112)
             if(c(i)<1.d-5) Ph(:)=dH(:,k)/del
             un(k)=un(k)+Ph(1)*w(i)!Eq.(11.113)
102          qn(k)=qn(k)+(Ph(1)*(c(i)**2-0.5d0)
      &                     +Ph(2))*w(i)
         un(:)=un(:)+u0(:,m)!Eq.(11.123)
         qn(:)=qn(:)+q0(:,m)
         do 103 k=0,Nx ! Eq.(11.114)
           u(k,m)=un(k)+un(-k)
           q(k,m)=qn(k)+qn(-k)
           u(-k,m)=u(k,m)
103        q(-k,m)=q(k,m)
         er=max(abs((u(0,m)-u00)/u(0,m)),
      &      abs((q(0,m)-q00)/q(0,m)))
         u00=u(0,m)
         q00=q(0,m)
         if(er>e.and.it<maxit) go to 6!convergence
           !flow rate are calculated
         GG(m)=2.d0*(u(0,m)-u(Nx,m))/(3.d0*Nx)
         QQ(m)=2.d0*(q(0,m)-q(Nx,m))/(3.d0*Nx)
```

```
        do 104 k=1,Nx-1,2 !Eqs.(11.115), (11.116)
         GG(m)=GG(m)+(8*u(k,m)+4*u(k+1,m))/(3*Nx)
104      QQ(m)=QQ(m)+(8*q(k,m)+4*q(k+1,m))/(3*Nx)
      open(9,file="Res_pois_cr_pl.dat")
      write(9,18)
      do 105 k=0,Nx,4 ! writing of results
105   write(9,19)0.5*k/Nx,u(k,1),q(k,1),
     &           u(k,2),q(k,2)
      write(9,20) del,-GG(1),GG(2),QQ(1),-QQ(2)
 18   format(6x,'x',9x,'uP',9x,'qP',9x,'uT',9x,'qT')
 19   format(f10.3,4(1x,f10.5))
 20   format(/4x,'del=',f7.3,',',2x,'GP=',f10.6,
     &   ',',2x,'QP=',f10.6/
     &   18x,'GT=',f10.6,',',2x,'QT=',f10.6)
      stop
      end
!function defined by Eq.(A.29)
      subroutine funcI(x,aI0,aI2,aI4)
      implicit double precision (a-h,o-z)
      parameter(n=15)
      dimension a(0:n),b(0:n)
      a(0)=0.d0
      a(1)=0.d0
      b(0)=1.d0
      b(1)=-sqrt(acos(-1.d0))
      a(2)=-b(0)
      b(2)=0.6341754927d0
      do 100 k=3,n !Eq.(A.33)
        a(k)=-2.d0*a(k-2)/(k*(k-1.d0)*(k-2.d0))
100     b(k)=(-2.d0*b(k-2)-(3.d0*k**2
     &        -6.*k+2.d0)*a(k))/(k*(k-1)*(k-2))
      aI0=0.d0
      aI1=0.d0
      aI2=-b(1)/4.d0
      do 101 k=0,n
        aI0=aI0-0.5*(a(k)*(1.d0+k*dlog(x))
     &    +k*b(k))*x**(k-1)
        aI1=aI1+0.5*(a(k)*dlog(x)+b(k))*x**k
101     aI2=aI2-0.5*(a(k)*((k+1)*dlog(x)-1.d0)
     &     /(k+1)+b(k))*x**(k+1)/(k+1.d0)
      aI4=1.5d0*aI2+0.5d0*x*aI1 !Eq.(A.30)
      return
      end
```

B.4.2
Output File "Res_pois_cr_pl.dat" with Results

```
   x          uP          qP          uT          qT
 0.000    -0.87950     0.21851     0.21150    -0.97166
 0.100    -0.86919     0.21528     0.20885    -0.96306
 0.200    -0.83726     0.20505     0.20050    -0.93595
 0.300    -0.77954     0.18559     0.18474    -0.88469
 0.400    -0.68337     0.15024     0.15653    -0.79155
 0.500    -0.51062     0.07960     0.10113    -0.59965
del=  1.000,  GP=  1.553611,  QP=  0.365385
              GT=  0.365432,  QT=  1.753667
```

B.5
Files for Cylindrical Couette Flows

B.5.1
Listing of Program "couette_axisym.for"

```
c  calculation of axi-symmetrical Couette flows
      implicit double precision (a-h,o-z)
      parameter (
     & del=1.d0, ! rarefaction parameter
     & bet=2.d0, ! ratio of radii
     & Nx=20, ! number of nodes in physical space
     & Nxb=bet*Nx+1.d-5,
     & Nc=4,! number of nodes in velocity space
     & Nt=200,! number of nodes for angle theta
     & e=1.d-10, ! criterion to stop iteration
     & maxit=100 000)! maximum number of iterartions
      dimension
     & u(Nx:Nxb),! velocity in n-th iteration
     & P(Nx:Nxb),! shear stress
     & u0(Nx:Nxb),! split part of velocity
     & P0(Nx:Nxb),! split part of shear stress
     & un(Nx:Nxb), ! velocity in (n+1)-th iteration
     & dH(Nx:Nxb), ! delta*H
     & Ph1(Nx:Nxb), ! Phi at j+1
     & Phn1(Nx:Nxb),
     & c(nc),w(nc) ! nodes and weights
      pi=dacos(-1.d0) ! pi number
      Is=1  ! no splitting
      if(del<=2.d0) Is=0 ! splitting
      if(Nc==4) open(10,file='cpw4.dat')
      if(Nc==6) open(10,file='cpw6.dat')
```

```
 if(Nc==8) open(10,file='cpw8.dat')
   ! nodes and weights reading
 read(10,*) (c(i),w(i),i=1,Nc)
 u0=0.d0
 P0=0.d0
 if(Is==0) then
   Nt0=500
   u0(Nx)=0.5d0
   P0(Nx)=1.d0/sqrt(Pi)
   do k=Nx+1,Nxb
     x=bet*k/Nxb
     thet0=asin(1.d0/x)
     do j=0,Nt0
       wt=8.d0/3.*(j-(j/2)*2+1)*
&              thet0*x/(Nt0*pi)
       if(j==0.or.j==Nt0)
&              wt=4./3.*thet0*x/(Nt0*pi)
       thet=j*thet0/Nt0
       s0=x*cos(thet0)
       if(j<Nt0) s0=x*cos(thet)
&            -sqrt(1.d0-(x*sin(thet))**2)
       call funcI(del*s0,aI3,aI4)
         ! Eq.(12.33)
       u0(k)=u0(k)+aI3*sin(thet)**2*wt
       P0(k)=P0(k)+aI4*sin(thet)**2
&         *cos(thet)*2.d0*wt
     end do
   end do
 end if
 u=0.d0 ! initial approximaion
 P=0.d0
 P00=1.d0
 er=1.
 it=0 ! beginning of iterations
 do while (er>e.and.it<maxit)
   it=it+1
   un=0.d0 ! set moments to zero
   P=0.d0
   P(Nx)=Is/sqrt(pi)
   un(Nx)=0.5d0*Is
   do i=1,nc ! loop for speed
     Ph1=0.
     Phn1=0.
     do j=Nt-1,1,-1 ! loop for angle
      dH(:)=4.d0*del*c(i)*u(:)*
```

```
     &          sin(pi*(j+0.5)/Nt)!2*delta*H
           Cx=2.*c(i)*Nx*
     &          abs(cos(pi*(j+0.5)/Nt))! Eq.(12.25)
           Ct=2.*Nt*Nx*c(i)*sin(pi*(j+0.5)/Nt)/pi
           wt=4.*pi/(3*Nt)*(j-(j/2)*2+1)*w(i)
           l=(Nt-(2*j+1))/abs(Nt-(2*j+1))! Eq.(9.73)
           Ni=(Nx+1)*(l+1)/2+
     &            (Nxb-1)*(1-l)/2!first node
           Nf=Nxb*(l+1)/2+Nx*(1-l)/2 !last node
           Ph=c(i)*sin(pi*j/Nt)*Is*(l+1)!Eq.(12.16)
           Ph1(Ni-l)=c(i)*sin(pi*(j+1)/Nt)*Is*(l+1)
           do k=Ni,Nf,l ! loop for x variable
            Ph=((Cx-Ct/(k-0.5*l)-del)*Ph!Eq.(12.24)
     &          +(-Cx+Ct/(k-0.5*l)-del)*Ph1(k)
     &          +( Cx+Ct/(k-0.5*l)-del)*Ph1(k-l)
     &          +dH(k)+dH(k-l))/(Cx+Ct/(k-0.5*l)+del)
            Phn1(k)=Ph ! stored for next j
            un(k)=un(k)+Ph*c(i)*sin(pi*j/Nt)
     &              *wt!Eq.(12.27)
            P(k)=P(k)+Ph*c(i)**2*sin(2.*pi*j/Nt)*wt
           end do
           Ph1=Phn1
          end do ! end of loop for angle
        end do! end of loop for speed
        u=u0+un ! new moments Eq.(12.35)
        P=P0+P
        er=abs((P(Nx)-P00)/P00)!residual
        P00=P(Nx)
      end do ! end of iteration
      open(9,file='Res_couet_axi.dat')
      write(9,18)
      do 104 k=Nx,Nxb,4 ! writing of results
104     write(9,19) k*bet/Nxb,u(k),P(k)*(k*bet/Nxb)**2
      write(9,20) del,P(Nx)
 18   format(4x,'velocity profile and shear stress'
     & /6x,'x',9x,'u',9x,'Pi')
 19   format(f10.3,2(1x,f10.5))
 20   format(/2x,'delta=',f7.3,',',2x,'Pi=',f10.5,2x)
      stop
      end
!function defined by !Eq.(A.29)
      subroutine funcI(x,aI3,aI4)
      implicit double precision (a-h,o-z)
      parameter(n=30)
```

```
      dimension a(0:n),b(0:n)
      if(x<1.d-10) then
        aI3=0.5d0
        aI4=3.d0/8.d0*1.7724538509055159d0
        return
      end if
      a(0)=0.d0
      a(1)=0.d0
      b(0)=1.d0
      b(1)=-1.7724538509055159d0
      a(2)=-1.d0
      b(2)=0.6341754927d0
      do k=3,n !Eq.(A.33)
       a(k)=-2.d0*a(k-2)/(k*(k-1.d0)*(k-2.d0))
       b(k)=(-2.*b(k-2)-(3.d0*k**2-6.*k+2.)
     &        *a(k))/(k*(k-1)*(k-2))
      end do
      aI0=0.d0
      aI1=0.d0
      aI2=-b(1)/4.d0
      do k=0,n
        aI0=aI0-0.5*(a(k)*(1.d0+k*dlog(x))
     &      +k*b(k))*x**(k-1)
        aI1=aI1+0.5*(a(k)*dlog(x)+b(k))*x**k
        aI2=aI2-0.5*(a(k)*((k+1)*dlog(x)-1.d0)
     &        /(k+1)+b(k))*x**(k+1)/(k+1.d0)
      end do
      aI3=aI1+0.5d0*x*aI0 !Eq.(A.30)
      aI4=1.5d0*aI2+0.5d0*x*aI1
      return
      end
```

B.5.2
Output File "Res_couet_axi.dat" **with Results**

```
   velocity profile and shear stress
     x          u          Pi
    1.000     0.57693     0.49682
    1.200     0.28213     0.49306
    1.400     0.19564     0.49337
    1.600     0.14073     0.49352
    1.800     0.09964     0.49360
    2.000     0.06160     0.49365
 delta=  1.000,  Pi=   0.49682
```

B.6
Files for Cylindrical Heat Transfer

B.6.1
Listing of Program "heat_axisym.for"

```
c  calculation of axi-symmetrical heat flow
      implicit double precision (a-h,o-z)
      parameter (
     & del=1.d0, ! rarefaction parameter
     & bet=2.d0, ! ratio of radii
     & Nx=20,!number of nodes in physical space
     & Nxb=bet*Nx+1.d-5,
     & Nc=4,!number of nodes in velocity space
     & Nt=80,! number of nodes for angle theta
     & e=1.d-10, ! criterion to stop iteration
     & maxit=100000)!maximum number of iterartions
      dimension
     & rho(Nx:Nxb),!density in n-th iteration
     & tau(Nx:Nxb),!temperature in n-th iteration
     & q(Nx:Nxb),! heat flow in n-th iteration
     & rhon(Nx:Nxb),!density in (n+1)-th iteration
     & taun(Nx:Nxb),!temperature in (n+1)-th iter.
     & qn(Nx:Nxb),! heat flow in (n+1)-th iteration
     & rho0(2,Nx:Nxb),! split part of density
     & tau0(2,Nx:Nxb),! split part of temperature
     & q0(2,Nx:Nxb),! split part of heat flow
     & Ph(2), ! perturbations
     & dH(2,Nx:Nxb), ! delta*H
     & Ph1(2,Nx:Nxb), ! Phi at j+1
     & Phn1(2,Nx:Nxb),
     & c(nc),w(nc) ! nodes and weights
      pi=dacos(-1.d0) ! pi number
      spi=sqrt(pi)
      Is=1  ! no splitting
      if(del<=2.d0) Is=0 ! splitting
      if(Nc==4) open(10,file='cpw4.dat')
      if(Nc==6) open(10,file='cpw6.dat')
      if(Nc==8) open(10,file='cpw8.dat')
        ! nodes and weights reading
      read(10,*) (c(i),w(i),i=1,Nc)
      rho0=0.d0
      tau0=0.d0
      q0=0.d0
      if(Is==0) then
```

```
Nt0=500
rho0(1,Nx)=0.5d0
rho0(2,Nx)=-0.25d0
tau0(1,Nx)=0.d0
tau0(2,Nx)=0.5d0
q0(1,Nx)=-0.25d0/spi
q0(2,Nx)=1.d0/spi
do k=Nx+1,Nxb
  x=float(k)/Nx
  thet0=asin(1.d0/x)
  do j=0,Nt0
    wt=4.d0/3.*(j-(j/2)*2+1)*thet0/(Nt0*pi)
    if(j==0.or.j==Nt0) wt=2./3.*thet0/(Nt0*pi)
    thet=j*thet0/Nt0
    s0=x*cos(thet0)
    if(j<Nt0) s0=x*cos(thet)
&          -sqrt(1.d0-(x*sin(thet))**2)
    call funcI(del*s0,aI1,aI2,aI3,aI4,aI5,aI6)
    rho0(1,k)=rho0(1,k)+aI1*wt!1st term in (12.74)
    rho0(2,k)=rho0(2,k)
&      +(aI3-1.5*aI1)*wt!2nd term in (12.74)
    tau0(1,k)=tau0(1,k)
&       +(aI3-aI1)*wt*2./3.!1st term in (12.75)
    tau0(2,k)=tau0(2,k)+(aI5-2.5d0*aI3
&       +2.d0*aI1)*wt*2./3.!2nd term in (12.75)
    q0(1,k)=q0(1,k)+(aI4-2.d0*aI2)
&          *cos(thet)*wt!1st term in (12.76)
    q0(2,k)=q0(2,k)+(aI6-3.5d0*aI4+3.5*aI2)
&          *cos(thet)*wt!2nd term in (12.76)
    end do
  end do
end if
rho=0.d0 ! initial approximaion
tau=0.d0
q=0.d0
rhor=1.d0
q00=1.d0
er=1.
it=0 ! beginning of iterations
do while (er>e.and.it<maxit)
  it=it+1
  rhon=0.d0
  taun=0.d0
  qn=0.d0
  rhorn=0.
```

```
      rhon(Nx)=(0.5*rhor-0.25)*Is
      taun(Nx)=0.5d0*Is! (12.71)
      qn(Nx)=(1.d0-0.25d0*rhor)/spi*Is
      do i=1,nc ! loop for speed
        cc1=c(i)**2-1.d0
        cc2=(c(i)**2-2.d0)*c(i)
        do j=Nt,0,-1 ! loop for angle
         dH(1,:)=2.*del*(rho(:)+cc1*tau(:)+4./15.
&         *cc2*cos(pi*(j+0.5)/Nt)*q(:))!(12.63)
         dH(2,:)=2.*del*(0.5*tau(:)+2./15.*c(i)
&         *cos(pi*(j+0.5)/Nt)*q(:))!(12.64)
         Cx=2.*c(i)*Nx*
&           abs(cos(pi*(j+0.5)/Nt))!(12.25)
         Ct=2.*c(i)*Nx*Nt/pi*sin(pi*(j+0.5)/Nt)
         wt=4.*pi/(3.*Nt)*(j-(j/2)*2+1)
&             *w(i)!Eq.(12.70)
         l=(Nt-(2*j+1))
&           /abs(Nt-(2*j+1))!Eq.(9.73)
         Ni=(Nx+1)*(l+1)/2+(Nxb-1)
&           *(1-l)/2!first node
         Nf=Nxb*(l+1)/2+Nx*(1-l)/2 !last node
         Ph(1)=(rhor+(c(i)**2-1.5))
&             *(l+1)/2.d0*Is!Eq.(12.59)
         Ph(2)=(l+1.d0)/4.d0*Is ! Eq.(12.59)
         Ph1(:,Ni-l)=Ph(:)
          if(j==0.or.j==Nt) then
            Ph1=0.d0! scheme adapted to Eq.(12.56)
            Ct=0.d0
            Cx=2.*c(i)*Nx ! Eq.(12.67)
            wt=0.5d0*wt ! Eq.(12.70)
            dH(1,:)=del*(rho(:)+cc1*tau(:)! (12.63)
&                  +l*4./15.*cc2*q(:))
            dH(2,:)=del*(0.5*tau(:) ! (12.64)
&                 +l*2./15.*c(i)*q(:))
          end if
          do k=Ni,Nf,l ! loop for x variable
           Ph(:)=((Cx-Ct/(k-0.5*l)-del)*Ph(:)
&              +(-Cx+Ct/(k-0.5*l)-del)*Ph1(:,k)
&              +( Cx+Ct/(k-0.5*l)-del)*Ph1(:,k-l)
&              +dH(:,k)+dH(:,k-l))
&             /(Cx+Ct/(k-0.5*l)+del) !Eq.(12.24)
           Phn1(:,k)=Ph(:) ! stored for next j
           rhon(k)=rhon(k)+Ph(1)*wt !Eq.(12.68)
           taun(k)=taun(k)+2./3.*(cc1*Ph(1)
&             +Ph(2))*wt
```

```
            qn(k)=qn(k)+(cc2*Ph(1)+c(i)*Ph(2))
     &              *cos(pi*j/Nt)*wt
          end do
          if(l==-1)rhorn=rhorn-2.*spi*Ph(1)
     &             *c(i)*cos(pi*j/Nt)*wt ! Eq.(12.69)
          Ph1=Phn1
        end do ! end of loop for angle
      end do! end of loop for speed
      rhor=rhorn ! new moments Eq.(12.72)
      rho(:)=rhon(:)+rho0(1,:)*rhor+rho0(2,:)
      tau(:)=taun(:)+tau0(1,:)*rhor+tau0(2,:)
      q(:)  =qn(:)  +  q0(1,:)*rhor+q0(2,:)
      er=abs((q(Nx)-q00)/q00)!residual calculated
      q00=q(Nx)
    end do ! end of iteration
    open(9,file='Res_heat_axi.dat')
    write(9,18)
    do 104 k=Nx,Nxb,4!writing of results
104   write(9,19) k*bet/Nxb,rho(k),tau(k),q(k)*k/Nx,
    write(9,20) del,q(Nx)
 18 format(4x,'temperature␣profile␣and␣heat␣flow'
   & /6x,'x',9x,'rho',9x,'tau',9x,'q*x')
 19 format(f10.3,6(1x,f10.5))
 20 format(/2x,'delta=',f7.3,',',2x,'q=',f10.5)
    stop
    end
c  function defined by !Eq.(A.29)
    subroutine funcI(x,aI1,aI2,aI3,aI4,aI5,aI6)
    implicit double precision (a-h,o-z)
    parameter(n=30)
    dimension a(0:n),b(0:n)
    if(x<1.d-10) then
      aI0=1.7724538509055159d0/2.d0
      aI1=0.5d0
      aI2=1.7724538509055159d0/4.d0
    else
    a(0)=0.d0
    a(1)=0.d0
    b(0)=1.d0
    b(1)=-1.7724538509055159d0
    a(2)=-1.dC
    b(2)=0.6341754927d0
    do k=3,n  !Eq.(A.33)
     a(k)=-2.d0*a(k-2)/(k*(k-1.d0)*(k-2.d0))
     b(k)=(-2.*b(k-2)-(3.d0*k**2-6.*k+2.)*a(k))
```

```
     &      /(k*(k-1)*(k-2))
      end do
      aI0=0.d0
      aI1=0.d0
      aI2=-b(1)/4.d0
      do k=0,n
        aI0=aI0-0.5*(a(k)*(1.d0+k*dlog(x))+k*b(k))
     &      *x**(k-1)
        aI1=aI1+0.5*(a(k)*dlog(x)+b(k))*x**k
        aI2=aI2-0.5*(a(k)*((k+1)*dlog(x)-1.d0)
     &       /(k+1)+b(k))*x**(k+1)/(k+1.d0)
      end do
      end if
      aI3=      aI1 +0.5d0*x*aI0 !Eq.(A.30)
      aI4=1.5d0*aI2 +0.5d0*x*aI1
      aI5=2.d0* aI3 +0.5d0*x*aI2
      aI6=2.5d0*aI4 +0.5d0*x*aI3
      return
      end
```

B.6.2
Output File “Res_heat_axi.dat” **with Results**

```
    temperature profile and heat flow
      x         rho          tau         q*x
     1.000    -0.42782     0.62483     0.47336
     1.200    -0.29727     0.42140     0.47201
     1.400    -0.23022     0.33211     0.47224
     1.600    -0.17610     0.26439     0.47235
     1.800    -0.12774     0.20627     0.47240
     2.000    -0.07858     0.14358     0.47410
  delta=  1.000,  q=   0.47336
```

B.7
Files for Axi-Symmetric Poiseuille and Creep Flows

B.7.1
Listing of Program “poiseuille_creep_axisym.for”

```
c axi-symmetrical Poiseuille and creep flows
      implicit double precision (a-h,o-z)
      parameter (
     & del=1.d0,!rarefaction parameter
     & Nx=40,!number of nodes in physical space
```

```
     & Nc=6,!number of nodes in velocity space
     & Nt=20,! number of nodes for angle theta
     & e=1.d-10, ! criterion to stop iteration
     & maxit=100000)!maximum number of iterartions
      dimension
     & u(0:Nx,2),! velocity in n-th iteration
     & un(0:Nx), ! velocity in (n+1)-th iteration
     & q(0:Nx,2),! heat flow in n-th iteration
     & qn(0:Nx), ! heat flow in (n+1)-th iteration
     & dH(2,0:Nx), ! delta*H
     & Ph1(2,0:Nx), ! Phi at j+1
     & Phn1(2,0:Nx),
     & Ph(2), ! perturbations Phi
     & Ph0(2), ! perturbations Phi at x=0
     & g(2), ! scurce term
     & GG(2),QQ(2), ! mass and heat flow rates
     & c(nc),w(nc) ! nodes and weights
      pi=dacos(-1.d0) ! pi number
      if(Nc<=8) then ! Gauss nodes and weights
        if(Nc==4) open(10,file='cpw4.dat')
        if(Nc==6) open(10,file='cpw6.dat')
        if(Nc==8) open(10,file='cpw8.dat')
          ! nodes and weights reading
        read(10,*) (c(i),w(i),i=1,Nc)
      else ! Simpson nodes and weights
        c01=0.1d0 ! upper limit for first interval
        c0=4.d0 ! upper limit for second interval
        do 100 i=1,Nc/2
         c(i)=2.d0*c01*i/Nc !Eq.(11.131)
         c(nc/2+i)=c01+2.d0*(c0-c01)*i/Nc
         w(i)=(((i+1)/2)*2-i+1)*c(i)*exp(-c(i)**2)
     &           *c01/(0.75*Nc*pi) !Eq.(12.115)
100      w(nc/2+i)=c(nc/2+i)*(((i+1)/2)*2-i+1)
     &     *exp(-c(nc/2+i)**2)*(c0-c01)/(0.75*Nc*pi)
        w(Nc/2)=exp(-c01**2)*c0*c01/(1.5d0*pi*Nc)
        w(Nc)=w(Nc)/2.
      end if
      u=0.d0 ! initial approximaion
      q=0.d0
      do m=1,2!driving force, m=1 -> P, m=2 -> T
        u00=0.d0
        q00=0.d0
        er=1.
        it=0 ! beginning of iterations
        do while (er>e.and.it<maxit)
```

```
        it=it+1
        un=0.d0 ! set moments to zero
        qn=0.d0
        do i=1,nc ! loop for speed
          cc1=c(i)**2-1.d0
          g(1)=-0.5*(2-m)-0.5d0*cc1*(m-1)
          g(2)=-0.75d0*(m-1) !Eq.(12.94)
          Ph1=0. ! boundary cond, Eqs.(12.96)
          Phn1=0.
          dH(1,:)=del*(u(:,m)+2./15.*cc1*q(:,m))
          dH(2,:)=del*q(:,m)/5.d0 !Eq.(12.93)
          Cx=2.*Nx*c(i) ! Eq.(12.67)
          wt=2.d0/3.*pi/Nt ! Eq.(12.114)
          Ph1(:,Nx)=0.d0 ! bound. cond. Eq.(12.96)
          do k=Nx-1,0,-1 ! loop for x variable
           Ph1(:,k)=((Cx-del)*Ph1(:,k+1)+dH(:,k)+
     &      dH(:,k+1)+2.*g(:))/(Cx+del) !Eq.(12.109)
           un(k)=un(k)+Ph1(1,k)*w(i)*wt
           qn(k)=qn(k)+(Ph1(1,k)*cc1
     &          +Ph1(2,k))*w(i)*wt !Eq.(12.113)
          end do
          Ph0(:)=Ph1(:,0) !bound. cond. at x=0
          un(0)=un(0)+Ph0(1)*(2.*pi-wt)*w(i)
          qn(0)=qn(0)+(Ph0(1)*cc1+Ph0(2))*
     &         (2.*pi-wt)*w(i) !Eq.(12.110)
          Ph=Ph0
          do k=1,Nx ! loop for x variable
           Ph(:)=((Cx-del)*Ph(:)+dH(:,k)+
     &      dH(:,k-1)+2.*g(:))/(Cx+del) !Eq.(12.109)
           un(k)=un(k)+Ph(1)*w(i)*wt ! Eq.(12.113)
           qn(k)=qn(k)+(Ph(1)*cc1+Ph(2))*w(i)*wt
          end do
          dH=2.d0*dH
          do j=Nt-1,1,-1 ! loop for angle
            Cx=2.*c(i)*Nx*
     &        abs(cos(pi*(j+0.5)/Nt)) !Eq.(12.25)
            Ct=2.*c(i)*Nt*Nx*
     &          sin(pi*(j+0.5)/Nt)/pi
            wt=4.d0/3.*(j-(j/2)*2+1)*pi/Nt !Eq.(12.114)
            l=(Nt-(2*j+1))/
     &          abs(Nt-(2*j+1)) !Eq.(9.73)
            Ni=(l+1)/2+(Nx-1)*(1-l)/2 !first node
            Nf=Nx*(l+1)/2+(1-l)/2 !last node
            Ph=Ph0*(l+1)/2.d0
```

```
             Ph1(:,Ni-1)=Ph(:)
             do k=Ni,Nf,1 ! loop for x variable
              Ph(:)=((Cx-Ct/(k-0.5*1)-del)*Ph(:)
     &          +(-Cx+Ct/(k-0.5*1)-del)*Ph1(:,k)
     &          +( Cx+Ct/(k-0.5*1)-del)*Ph1(:,k-1)
     &          -dH(:,k)+dH(:,k-1)+4.*g(:))
     &          /(Cx+Ct/(k-0.5*1)+del)!Eq.(12.111)
              Phn1(:,k)=Ph(:) ! stored for next j
              un(k)=un(k)+Ph(1)*w(i)*wt!Eq.(12.113)
              qn(k)=qn(k)+(Ph(1)*cc1
     &            +Ph(2))*w(i)*wt
             end do
             Ph1=Phn1
           end do
         end do
         u(:,m)=un(:)
         q(:,m)=qn(:)
         er=max(abs((u(0,m)-u00)/u(0,m)),
     &          abs((q(0,m)-q00)/q(0,m)))
         u00=u(0,m)
         q00=q(0,m)
       end do
       GG(m)=u(Nx,m)/Nx*4.d0/3.!flow rates
       QQ(m)=q(Nx,m)/Nx*4.d0/3.
       do k=0,Nx-2,2
        GG(m)=GG(m)+(u(k,m)*k+2.*u(k+1,m)*(k+1))
     &        *8./(3*Nx**2) !Eq.(12.116)
        QQ(m)=QQ(m)+(q(k,m)*k+2.*q(k+1,m)*(k+1))
     &        *8./(3*Nx**2) !Eq.(12.117)
       end do
      end do ! end of loop for m
      open(9,file='Res_pois_cr_axi.dat')
      write(9,18)
      do 104 k=0,Nx,8! writing of results
104     write(9,19) float(k)/Nx,u(k,1),q(k,1),
     &      u(k,2),q(k,2)
      write(9,20) del,-GG(1),QQ(1),GG(2),-QQ(2)
 18   format(6x,'x',9x,'uP',9x,'qP', 9x , 'uT',9x,'qT')
  19  format(f10.3,4(1x,f10.5))
 20   format(/2x,'delta=',f7.3,',',2x,'GP=',f10.5,
     & ',',2x,'QP=',f10.5/18x,'GT=',f10.5,','
     &    ,2x,'QT=',f10.5)
      stop
      end
```

B.7.2
Output File "Res_pois_cr_axi.dat" **with Results**

```
       x          uP          qP          uT          qT
     0.000    -0.98491     0.27307     0.26416    -1.04311
     0.200    -0.96752     0.26844     0.26002    -1.03075
     0.400    -0.91477     0.25400     0.24715    -0.99228
     0.600    -0.82206     0.22738     0.22353    -0.91997
     0.800    -0.67718     0.18184     0.18353    -0.79316
     1.000    -0.41127     0.08168     0.09783    -0.50162
  delta=  1.000,  GP=   1.47606,  QP=   0.39672
                  GT=   0.39669,  QT=   1.67468
```

B.8
Files for Poiseuille and Creep Flows Through Channel

B.8.1
Listing of Program "poiseuille_creep_chan.for"

```
c Poiseuille and creep flows through channel
      implicit double precision (a-h,o-z)
      parameter (
     & del=1.d0,!rarefaction parameter
     & bet=2.d0,!aspect ratio
     & Nx=40,!number of nodes in physical space
     & Nxb=bet*Nx+1.d-10,
     & Nc=8,!number of nodes in velocity space
     & Nt=40,!number of nodes for angle theta
     & e=1.d-10,!criterion to stop iteration
     & maxit=1000000)! maximum number of iter.
      dimension
     & u(-Nx:Nx,-Nxb:Nxb,2),!velocity in n-th iter.
     & un(-Nx:Nx,-Nxb:Nxb), !velocity in (n+1)-th iter.
     & q(-Nx:Nx,-Nxb:Nxb,2),!heat flow in n-th iter.
     & qn(-Nx:Nx,-Nxb:Nxb), !heat flow in (n+1)-th iter.
     & dH(2,-Nx:Nx,-Nxb:Nxb),! delta*H
     & Ph1(2,-Nx:Nx), ! Phi at j+1
     & Phn1(2,-Nx:Nx), ! Phi at j+1
     & Ph(2), ! perturbations Phi
     & g(2), ! source term
     & GG(2),QQ(2), ! mass and heat flow rates
     & c(nc),w(nc) ! nodes and weights
      pi=dacos(-1.d0) ! pi number
      if(Nc<=8) then ! Gauss nodes and weights
```

```
          if(Nc==4) open(10,file='cpw4.dat')
          if(Nc==6) open(10,file='cpw6.dat')
          if(Nc==8) open(10,file='cpw8.dat')
            ! node and weight reading
          read(10,*) (c(i),w(i),i=1,Nc)
        else ! Simpson nodes and weights
          c01=0.1d0 ! upper limit for first interval
          c0=4.d0 ! upper limit for second interval
          do 100 i=1,Nc/2
            c(i)=2.d0*c01*i/Nc !Eq.(11.131)
            c(nc/2+i)=c01+2.d0*(c0-c01)*i/Nc
            w(i)=(((i+1)/2)*2-i+1)*c(i)*exp(-c(i)**2)
     &         *c01/(0.75*Nc*pi) !Eq.(12.115)
100         w(nc/2+i)=c(nc/2+i)*(((i+1)/2)*2-i+1)
     &       *exp(-c(nc/2+i)**2)*(c0-c01)/(0.75*Nc*pi)
          w(Nc/2)=exp(-c01**2)*c0*c01/(1.5d0*pi*Nc)
          w(Nc)=w(Nc)/2.
        end if
        u=0.d0 ! initial approximaion
        q=0.d0
        do m=1,2! driving force, m=1 -> P, m=2 -> T
          u00=0.d0
          q00=0.d0
          er=1.
          it=0 ! beginning of iterations
          do while (er>e.and.it<maxit)
            it=it+1
            un=0.d0 ! set moments to zero
            qn=0.d0
            do i=1,nc ! loop for speed
              cc1=c(i)**2-1.d0
              g(1)=-0.5*(2-m)-0.5d0*cc1*(m-1)
              g(2)=-0.75d0*(m-1) !Eq.(12.94)
              dH(1,:,:)=del*(u(:,:,m)
     &            +2./15.*cc1*q(:,:,m))
              dH(2,:,:)=del*q(:,:,m)/5.d0!Eq.(12.93)
              do j=0,Nt ! loop for angle
                Cx1=4.*Nx*c(i)*
     &             cos(j*pi/(2.d0*Nt)) !Eq.(13.31)
                Cx2=4.*Nx*c(i)*sin(j*pi/(2.d0*Nt))
                wt=(j-(j/2)*2+1)*pi/(3.*Nt)!Eq.(13.33)
                if(j==0.or.j==Nt) wt=wt/2.d0
                Ph1=0.d0 ! bound.cond. (13.24)
                Phn1=0.
                do k2=-Nxb+1,Nxb
```

```
            Ph=0.d0! bound.cond. (13.23)
            do k1=-Nx+1,Nx ! loop for x variable
             Ph(:)=((Cx1-Cx2-del)*Ph(:)
     &         +(Cx2-Cx1-del)*Ph1(:,k1)
     &         +(Cx1+Cx2-del)*Ph1(:,k1-1)
     &         +dH(:,k1,k2)+dH(:,k1-1,k2)
     &         +dH(:,k1,k2-1)+dH(:,k1-1,k2-1)+4.d0*
     &          g(:))/(Cx1+Cx2+del)!Eq.(13.30)
             Phn1(:,k1)=Ph(:) ! stored for next k2
             un(k1,k2)=un(k1,k2)+Ph(1)*w(i)*wt
             qn(k1,k2)=qn(k1,k2)+(Ph(1)*cc1
     &          +Ph(2))*w(i)*wt ! Eq.(13.32)
            end do
            Ph1=Phn1
           end do
         end do
       end do
       u(:,:,m)=un(:,:)+un(Nx:-Nx:-1,:)
     &     +un(:,Nxb:-Nxb:-1)
     &     +un(Nx:-Nx:-1,Nxb:-Nxb:-1) ! Eq.(13.34)
       q(:,:,m)=qn(:,:)+qn(Nx:-Nx:-1,:)
     &     +qn(:,Nxb:-Nxb:-1)
     &     +qn(Nx:-Nx:-1,Nxb:-Nxb:-1)
       er=max(abs((u(0,0,m)-u00)/u(0,0,m)),
     &    abs((q(0,0,m)-q00)/q(0,0,m)))!residual
       u00=u(0,0,m)
       q00=q(0,0,m)
     end do
   end do
     GG=0. ! flow rate calculation
     QQ=0.
     do 111 k1=0,Nx
     w1=4.d0/(3.*Nxb)*(k1-(k1/2)*2+1)
     if(k1==0.or.k1==Nx) w1=w1/2.
     do 111 k2=0,Nxb
       w2=2./(3.*Nx)*(k2-(k2/2)*2+1)
       if(k2==0.or.k2==Nxb) w2=w2/2.
       GG(:)=GG(:)+u(k1,k2,:)*w1*w2 !Eq.(13.35)
111    QQ(:)=QQ(:)+q(k1,k2,:)*w1*w2 !Eq.(13.36)
   open(9,file='Res_pois_cr_ch.dat')
   write(9,18)
   do 104 k1=0,Nx,8! writing of results
104 write(9,19)float(k1)/Nx,u(k1,0,1),q(k1,0,1),
     &    u(k1,0,2),q(k1,0,2)
   write(9,20) del,bet,-GG(1),QQ(1),GG(2),-QQ(2)
```

```
 18    format(4x,'profiles at x2=0'
      &/6x,'x1',8x,'uP',8x,'qP',8x,'uT',8x,'qT')
  19   format(f10.2,4(1x,f9.5))
 20    format(/2x,'delta=',f7.3,',',2x,'beta=',
      &f7.2,/2x,'GP=',f9.5,',',1x,'QP=',f9.5,1x,
      &/2x,'GT=',f9.5,',',1x,'QT=',f9.5)
       stop
       end
```

B.8.2
Output File "Res_pois_cr_ch.dat" **with Results**

```
    profiles at x2=0
      x1         uP         qP         uT         qT
      0.00   -0.68283    0.20882    0.20370   -0.85027
      0.20   -0.67383    0.20578    0.20095   -0.84138
      0.40   -0.64641    0.19626    0.19242   -0.81401
      0.60   -0.59779    0.17884    0.17687   -0.76370
      0.80   -0.52056    0.14965    0.15107   -0.67836
      1.00   -0.37827    0.08914    0.09883   -0.50030
  delta=   1.000,   beta=    2.00
  GP=  1.05929, QP=  0.31392
  GT=  0.31391, QT=  1.39612
```

B.9
Files for Oscillating Couette Flow

B.9.1
Listing of Program "couette_osc.for"

```
! oscillating Couette flow
      implicit double precision (a-h,o-z)
      parameter(
     & theta=1.d0,!frequency parameter
     & delta=1.d0,!rarefaction parameter
     & Nx=20,!number of nodes in physical space
     & Nc=40,!number of nodes in velocity space
     & e=1.d-10,! criterion to stop iteration
     & maxit=100000)!maximum number of iterartions
      complex*16
     & Phi, ! pertubation
     & u(0:Nx), ! velocity in n-th iteration
     & un(0:Nx),! velocity in (n+1)-th iteration
     & u0(0:Nx), ! split part of velocity
```

```
     & Pi(0:Nx), ! shear stress in n-th iteration
     & Pi0(0:Nx),! split part of  shear stress
     & Pi00,aI0,aI1
      dimension c(Nc),w(Nc) ! nodes and weights
      spi=sqrt(acos(-1.d0))! square root of pi
      u=0.d0 ! initial approximaion
      Pi00=0.d0
! velocity points and weights
      c01=0.1d0 ! upper limit for 1st interval
      c0=3.0d0 ! upper limit for 2nd interval
      do 100 i=1,Nc/2
        c(i)=2.d0*c01*i/Nc !Eq.(11.131)
        c(nc/2+i)=c01+2.d0*(c0-c01)*i/Nc
        w(i)=(((i+1)/2)*2-i+1)*exp(-c(i)**2)*c01/
     &   (0.75d0*Nc*spi) !Eq.(11.132)
100     w(nc/2+i)=(((i+1)/2)*2-i+1)*exp(-c(nc/2+i)**2)
     &   *(c0-c01)/(0.75d0*Nc*spi)
       w(Nc/2)=w(Nc/2)*c0/(2.d0*c01)
       w(Nc)=w(Nc)/2.
      u0(0)=cmplx(0.5d0,0.d0)! split part of moments
      Pi0(0)=cmplx(1.d0/spi,0.d0)
      do 106 k=1,Nx
      call funcI(k*delta/(theta*Nx)*cmplx(theta,-1.d0)
     &   ,AI0,AI1)
       u0(k)=AI0/spi !Eq.(16.54)
106    Pi0(k)=2.*AI1/spi
      it=0 ! beginning of iterations
 6    it=it+1 ! counter of iterations
      un=0.d0 ! set moment of to zero
      Pi=0.d0
      do 101 l=-1,1,2 ! negative and positive speed
      do 101 i=1,Nc !   loop for speed
        CC=2.d0*c(i)*Nx*theta/delta
        Phi=0.d0 ! ! boundary cond., Eq.(16.51)
        Ni=(l+1)/2+(Nx-1)*(1-l)/2 ! initial node
        Nf=Nx*(l+1)/2 ! final node
        do 101 k=Ni,Nf,l ! loop for x-coordinate
         Phi=((CC-cmplx(theta,-1.d0))*Phi+
     &      theta*(u(k)+u(k-l)))
     &     /(CC+cmplx(theta,-1.d0)) ! Eq.(16.55)
         un(k)=un(k)+Phi*w(i) ! Eq.(16.57)
101      Pi(k)=Pi(k)+l*2.*Phi*c(i)*w(i)!Eq.(16.58)
      u=un+u0 ! addition of split part Eq.(16.52)
      Pi=Pi+Pi0
      er=cdabs((Pi00-Pi(0))/Pi(0))!residual calculated
```

```
      Pi00=Pi(0)
      if(er>e.and.it<maxit) go to 6 ! convergence
      open(9,file='Res_couette_osc.dat')
         !writing of results
      write(9,18) delta,theta,Nx,Nc
      write(9,19) cdabs(u(0)),
     &  atan(aimag(u(0)/real(u(0)))),
     &  cdabs(Pi(0)),atan(aimag(Pi(0)/real(Pi(0))))
   18 format(4x,'delta=',f7.3,2x,'theta=',f7.3,2x,
     &    'Nx=',I3,2x,'Nc=',I3)
   19 format(1x,'A_u=',f8.4,1x,'phi_u=',f8.4,1x,
     &1x/'A_Pi=',f8.4,1x,'phi_Pi=',f8.4)
      stop
      end
! function defined by !Eq.(A.29)
      subroutine funcI(x,aI0,aI1)
      implicit double precision (a-h,o-z)
      complex*16 x,aI0,aI1,v
      parameter(n=30,n0=30)
      dimension a(0:n),b(0:n),a0(0:n0),a1(0:n0)
      pi=3.141592653589793d0
      if(cdabs(x)<7.d0) then
        a(0)=0.d0
        a(1)=0.d0
        a(2)=-1.d0
        b(0)=1.d0
        b(1)=-sqrt(pi)
        b(2)=0.6341754927d0
        do 100 k=3, n !Eq.(A.33)
        a(k)=-2.d0*a(k-2)/(k*(k-1.d0)*(k-2.d0))
100     b(k)=(-2.d0*b(k-2)-(3*k**2-6*k+2.d0)*a(k))
     &            /(k*(k-1)*(k-2))
        aI0=0.d0
        aI1=0.d0
        do 101 k=0,n
          aI0=aI0-0.5d0*(a(k)*
     &      (1.d0+k*cdlog(x))+k*b(k))*x**(k-1)
101       aI1=aI1+0.5d0*(a(k)*cdlog(x)+b(k))*x**k
      else
       v=3.d0*(x/2.d0)**(2.d0/3.d0)!Eq.(A.34)
       a0(0)=1.d0 !Eq.(A.35)
       a0(1)=-1.d0/12.d0
       a1(0)=1.d0
       a1(1)=5.d0/12.d0
       aI0=(a0(0)+a0(1)/v)*sqrt(pi/3.d0)*cdexp(-v)
```

```
      aI1=(a1(0)+a1(1)/v)*cdsqrt(pi*v)/
     &     3.d0*cdexp(-v)
      do 102 k=0,n0-2 !Eq.(A.36)
        a0(k+2)=(-(12*k**2+36*k+25)*a0(k+1)
     &        -k*(2*k+3)**2*a0(k))/(12*(k+2))
        a1(k+2)=(-(12*k**2+36*k+19)*a1(k+1)+
     &   (1-2*k)*(k+1)*(2*k+5)*a1(k))/(12*(k+2))
        aI0=aI0+a0(k+2)/(v**(k+2))*sqrt(pi/3.d0)
     &   *cdexp(-v)
102     aI1=aI1+a1(k+2)/(v**(k+2))*cdsqrt(pi*v)
     &    /3.d0*cdexp(-v)
      end if
      return
      end
```

B.9.2
Output File "Res_couette_osc.dat" **with Results**

```
    delta=  1.000  theta=  1.000  Nx= 20  Nc= 40
 A_u=  0.5577 phi_u=  0.1905
A_Pi=  0.5322 phi_Pi= -0.1865
```

References

1. Sears, F. and Salinger, G. (1975) *Thermodynamics, Kinetic Theory and Statistical Thermodynamics*, Addison-Wesley.
2. Kestin, J., Knierim, K., Mason, E.A., Najafi, B., Ro, S.T., and Waldman, M. (1984) Equilibrium and transport properties of the noble gases and their mixture at low densities. *J. Phys. Chem. Ref. Data*, **13** (1), 229–303.
3. Hirschfelder, J.O., Curtiss, C.F., and Bird, R.B. (1954) *The Molecular Theory of Gases and Liquids*, John Wiley & Sons, Inc., New York.
4. Ferziger, J.H. and Kaper, H.G. (1972) *Mathematical Theory of Transport Processes in Gases*, North-Holland Publishing Company, Amsterdam.
5. Chapman, S. and Cowling, T.G. (1952) *The Mathematical Theory of Non-Uniform Gases*, University Press, Cambridge.
6. Hellmann, R., Bich, E., and Vogel, E. (2007) Ab initio potential energy curve for the helium atom pair and thermophysical properties of dilute helium gas. I. Helium-helium interatomic potential. *Mol. Phys.*, **105** (23–24), 3013–3023.
7. Hellmann, R., Bich, E., and Vogel, E. (2008) Ab initio potential energy curve for the neon atom pair and thermophysical properties of the dilute neon gas. I. Neon-neon interatomic potential and rovibrational spectra. *Mol. Phys.*, **106** (1), 133–140.
8. Jäger, B., Hellmann, R., Bich, E., and Vogel, E. (2009) Ab initio pair potential energy curve for the argon atom pair and thermophysical properties of the dilute argon gas. I. Argon-argon interatomic potential and rovibrational spectra. *Mol. Phys.*, **107** (20), 2181–2188. Correction in **108**, 105–(2010).
9. Nasrabad, A. and Deiters, U. (2003) Prediction of thermodynamic properties of krypton by Monte Carlo simulation using ab initio interaction potentials. *J. Chem. Phys.*, **119** (2), 947–952.
10. Janzen, A.R. and Aziz, R.A. (1997) An accurate potential energy curve for helium based on ab initio calculations. *J. Chem. Phys.*, **107** (3), 914–919.
11. Korona, T., Williams, H.L., Bukowski, R., Jeziorski, B., and Szalewicz, K. (1997) Helium dimer potential from symmetry-adapted perturbation theory calculations using large Gaussian geminal and orbital basis sets. *J. Chem. Phys.*, **106**, 5109–5122.
12. Cybulski, S.M. and Toczylowski, R.R. (1999) Ground state potential energy curves for He_2, Ne_2, Ar_2, He-Ne, He-Ar, and Ne-Ar: a coupled-cluster study. *J. Chem. Phys.*, **111** (23), 10 520–10 528.
13. Haley, T.P. and Cybulski, S.M. (2003) Ground state potential energy curves for He-Kr, Ne-Kr, Ar-Kr, and Kr_2: coupled-cluster calculations and comparison with experiment. *J. Chem. Phys.*, **119**, 5487–5496.
14. Slavíček, P., Kalus, R., Paška, P., Odvárková, I., Hobza, P., and Malijevský, A. (2003) State-of-the-art correlated ab initio potential energy curves for heavy rare gas dimers: Ar_2,

Rarefied Gas Dynamics: Fundamentals for Research and Practice, First Edition. Felix Sharipov.

Kr_2, and Xe_2. *J. Chem. Phys.*, **119** (4), 2102–2119.

15. Goldstein, H. (1980) *Classical Mechanics*, Addison-Wesley.
16. Bird, G.A. (1976) *Molecular Gas Dynamics*, Clarendon Press, Oxford.
17. Bird, G.A. (1994) *Molecular Gas Dynamics and the Direct Simulation of Gas Flows*, Oxford University Press, Oxford.
18. Cercignani, C. (1975) *Theory and Application of the Boltzmann Equation*, Scottish Academic Press, Edinburgh.
19. Lifshitz, E.M. and Pitaevskii, L.P. (1981) *Physical Kinetics*, Butterworth-Heinemann, Oxford.
20. Kogan, M.N. (1969) *Rarefied Gas Dynamics*, Plenum Publishing Corporation, New York.
21. Sharipov, F. and Bertoldo, G. (2009) Numerical solution of the linearized Boltzmann equation for an arbitrary intermolecular potential. *J. Comput. Phys.*, **228** (9), 3345–3357.
22. Saksaganskii, G.L. (1988) *Molecular Flow in Complex Vacuum Systems*, Grodon and Breach Science Publishers, New York.
23. Landau, L.D. and Lifshitz, E.M. (1989) *Fluid Mechanics*, Pergamon, New York.
24. Happel, J. and Brenner, H. (1973) *Low Reynolds Number Hydrodynamics*, Noordhoff International Publishing, Leyden, MA.
25. McCourt, F.R.W., Beenakker, J.J.M., Köhler, W.E., and Kuščer, I. (1990) *Nonequilibrium Phenomena in Polyatomic Gases*, Clarendon Press, Oxford.
26. Reichl, L. (1998) *A Modern Course in Statistical Physics*, 2nd edn, John Wiley & Sons, Inc., New York.
27. Landau, L.D. and Lifshitz, E.M. (1980) *Statistical Physics*, 3rd edn, Butterworth-Heinemann (UK), Oxford.
28. Pathria, R.K. and Beale, P.D. (2011) *Statistical Mechanics*, 3rd edn, Elsevier, Amsterdam.
29. Trevena, D.H. (1993) *Statistical Mechanics. An Introduction*, Ellis Horwood, New York.
30. Kuščer, I. (1971) Reciprocity in scattering of gas molecules by surface. *Surf. Sci.*, **25**, 225–237.
31. Cercignani, C. and Lampis, M. (1971) Kinetic model for gas-surface interaction. *Transp. Theory Stat. Phys.*, **1**, 101–114.
32. Sazhin, O.V., Borisov, S.F., and Sharipov, F. (2001) Accommodation coefficient of tangential momentum on atomically clean and contaminated surfaces. *J. Vac. Sci. Technol., A*, **19** (5), 2499–2503. Erratum: **20** (3), 957–(2002).
33. Sharipov, F. (2003) Application of the Cercignani-Lampis scattering kernel to calculations of rarefied gas flows. II. Slip and jump coefficients. *Eur. J. Mech. B/Fluids*, **22**, 133–143.
34. Sharipov, F. (2003) Application of the Cercignani-Lampis scattering kernel to calculations of rarefied gas flows. III. Poiseuille flow and thermal creep through a long tube. *Eur. J. Mech. B/Fluids*, **22**, 145–154.
35. Sharipov, F. and Bertoldo, G. (2006) Heat transfer through a rarefied gas confined between two coaxial cylinders with high radius ratio. *J. Vac. Sci. Technol., A*, **24** (6), 2087–2093.
36. Porodnov, B.T., Suetin, P.E., Borisov, S.F., and Akinshin, V.D. (1974) Experimental investigation of rarefied gas flow in different channels. *J. Fluid Mech.*, **64** (3), 417–437.
37. Semyonov, Y.G., Borisov, S.F., and Suetin, P.E. (1984) Investigation of heat transfer in rarefied gases over a wide range of Knudsen numbers. *Int. J. Heat Mass Transfer*, **27** (10), 1789–1799.
38. Sharipov, F. (1994) Onsager-Casimir reciprocity relations for open gaseous systems at arbitrary rarefaction. I. General theory for single gas. *Physica A*, **203**, 437–456.
39. Sharipov, F. (2012) Power series expansion of the Boltzmann equation and reciprocal relations for non-linear irreversible phenomena. *Phys. Rev. E*, **84** (6), 061 137.
40. Onsager, L. (1931) Reciprocal relations in irreversible processes. I. *Phys. Rev.*, **37**, 405–426.
41. Onsager, L. (1931) Reciprocal relations in irreversible processes. II. *Phys. Rev.*, **38**, 2265–2279.

42. Casimir, H.B.G. (1945) On Onsager's principle of microscopic reversibility. *Rev. Mod. Phys.*, **17**, 343.
43. Sharipov, F. (2006) Onsager-Casimir reciprocal relations based on the Boltzmann equation and gas-surface interaction law. Single gas. *Phys. Rev. E*, **73**, 026 110.
44. Sharipov, F. (2012) Reciprocal relations based on the non-stationary Boltzmann equation. *Physica A*, **391** (5), 1972–1983.
45. Sharipov, F. (1994) Onsager-Casimir reciprocity relations for open gaseous systems at arbitrary rarefaction. II. Application of the theory for single gas. *Physica A*, **203**, 457–485.
46. Sharipov, F. (2010) The reciprocal relations between cross phenomena in boundless gaseous systems. *Physica A*, **389**, 3743–3760.
47. Sharipov, F. (1994) Onsager-Casimir reciprocity relations for open gaseous systems at arbitrary rarefaction. III. Theory and its application for gaseous mixtures. *Physica A*, **209**, 457–476.
48. Sharipov, F. and Kalempa, D. (2006) Onsager-Casimir reciprocal relations based on the Boltzmann equation and gas-surface interaction. Gaseous mixtures. *J. Stat. Phys.*, **125** (3), 661–675.
49. Sharipov, F. (1998) Onsager-Casimir reciprocity relations for open gaseous systems at arbitrary rarefaction. IV Rotating systems. *Physica A*, **260** (3/5), 499–509.
50. Sharipov, F. (1999) Onsager-Casimir reciprocity relation for gyrothermal effect with polyatomic gases. *Phys. Rev. E*, **59** (5), 5128–5132.
51. Sharipov, F. (1995) Onsager-Casimir reciprocity relations for a mixture of rarefied gases interacting with a laser radiation. *J. Stat. Phys.*, **78** (1/2), 413–430.
52. De Groot, S.R. and Mazur, P. (1984) *Non-Equilibrium Thermodynamics*, Dover Publications, New York.
53. Kuščer, I. (1985) Irreversible thermodynamics of rarefied gases. *Physica A*, **133**, 397–412.
54. Loyalka, S.K. (1971) Kinetic theory of thermal transpiration and mechanocaloric effect. I. *J. Chem. Phys.*, **55** (9), 4497–4503.
55. ten Bosch, B.I.M., Beenakker, J.J.M., and Kuščer, I. (1984) Onsager symmetries in field-dependent flow of rarefied molecular gases. *Physica A*, **123**, 443–462.
56. Pekeris, C.L. and Alterman, Z. (1957) Solution of the Boltzmann-Hilbert integral equation. II. The coefficients of viscosity and heat conduction. *Proc. Natl. Acad. Sci. U.S.A.*, **43**, 998–1007.
57. Tipton, E.L., Tompson, R.V., and Loyalka, S.K. (2009) Chapman-Enskog solutions to arbitrary order in Sonine polynomials II: viscosity in a binary, rigid-sphere, gas mixture. *Eur. J. Mech. B/Fluids*, **28** (3), 335–352.
58. Tipton, E.L., Tompson, R.V., and Loyalka, S.K. (2009) Chapman-Enskog solutions to arbitrary order in Sonine polynomials III: diffusion, thermal diffusion, and thermal conductivity in a binary, rigid-sphere, gas mixture. *Eur. J. Mech. B/Fluids*, **28** (3), 353–386.
59. Bich, E., Hellmann, R., and Vogel, E. (2007) Ab initio potential energy curve for the helium atom pair and thermophysical properties of the dilute helium gas. II. Thermophysical standard values for low-density helium. *Mol. Phys.*, **105** (23-24), 3035–3049.
60. Bich, E., Hellmann, R., and Vogel, E. (2008) Ab initio potential energy curve for the neon atom pair and thermophysical properties for the dilute neon gas. II. Thermophysical properties for low-density neon. *Mol. Phys.*, **106** (6), 813–825.
61. Vogel, E., Jaeger, B., Hellmann, R., and Bich, E. (2010) Ab initio pair potential energy curve for the argon atom pair and thermophysical properties for the dilute argon gas. II. Thermophysical properties for low-density argon. *Mol. Phys.*, **108** (24), 3335–3352.
62. Berg, R.F. and Burton, W.C. (2013) Noble gas viscosities at 25°c. *Mol. Phys.*, **111** (2), 195–199.
63. Berg, R. and Moldover, M. (2012) Recommended viscosities of 11 dilute gases at 25 degrees c. *J. Phys. Chem. Ref. Data*, **41** (4), 043104-1–043104-10.

64. Bhatnagar, P.L., Gross, E.P., and Krook, M.A. (1954) A model for collision processes in gases. *Phys. Rev.*, **94**, 511–525.
65. Welander, P. (1954) On the temperature jump in a rarefied gas. *Ark. Fys.*, **7** (5), 507–553.
66. Shakhov, E.M. (1968) Generalization of the Krook kinetic relaxation equation. *Fluid Dyn.*, **3** (5), 95–96.
67. Holway, L.H. (1966) New statistical models for kinetic theory: method of construction. *Phys. Fluids*, **9** (9), 1658–1673.
68. Bird, G.A. (2013) The DSMC method.
69. Lord, R.G. (1991) Some extensions to the Cercignani-Lampis gas-surface scattering kernel. *Phys. Fluids A*, **3** (4), 706–710.
70. Ivanov, M.S. and Rogasinsky, S.V. (1988) Analysis of numerical techniques of the direct simulation Monte Carlo method in rarefied gas dynamics. *Sov. J. Numer. Anal. Math. Modell.*, **3** (6), 453–465.
71. Fedosov, D.A., Rogasinsky, S.V., Zeifman, M.I., Ivanov, M.S., Alexeenko, A.A., and Levin, D.A. (2005) Analysis of numerical errors in the DSMC method, in *Rarefied Gas Dynamics*, vol. **762** (ed. M. Capitelli), AIP Conference Proceedings, pp. 589–594. 24th International Symposium, Italy, 2004.
72. Venkattraman, A., Alexeenko, A.A., Gallis, M.A., and Ivanov, M.S. (2012) A comparative study of no-time-counter and majorant collision frequency numerical schemes in DSMC, in *Rarefied Gas Dynamics*, vol. **1501** (eds M. Mareschal and A. Santos), AIP Conference Proceedings, pp. 489–495. 28th International Symposium, Zaragoza, Spain, 2012.
73. Sharipov, F. and Strapasson, J.L. (2012) Direct simulation Monte Carlo method for an arbitrary intermolecular potential. *Phys. Fluids*, **24** (1), 011 703.
74. Sharipov, F. and Strapasson, J.L. (2012) Ab initio simulation of transport phenomena in rarefied gases. *Phys. Rev. E*, **86** (3), 031 130.
75. Sharipov, F. and Strapasson, J.L. (2013) Benchmark problems for mixtures of rarefied gases. I. Couette flow. *Phys. Fluids*, **25**, 027 101.
76. Strapasson, J.L. and Sharipov, F. (2014) Ab initio simulation of heat transfer through a mixture of rarefied gases. *Int. J. Heat Mass Transfer*, **71**, 91–97.
77. Krylov, V.I. (2005) *Approximate Calculation of Integrals*, Dover Publications, Mineola, NY.
78. Sharipov, F.M. and Subbotin, E.A. (1993) On optimization of the discrete velocity method used in rarefied gas dynamics. *Z. Angew. Math. Phys. (ZAMP)*, **44**, 572–577.
79. Sharipov, F. (2011) Data on the velocity slip and temperature jump on a gas-solid interface. *J. Phys. Chem. Ref. Data*, **40** (2), 023 101.
80. Sharipov, F. (2004) Heat transfer in the Knudsen layer. *Phys. Rev. E*, **69** (6), 061 201.
81. Albertoni, S., Cercignani, C., and Gotusso, L. (1963) Numerical evaluation of the slip coefficient. *Phys. Fluids*, **6** (7), 993–996.
82. Cercignani, C. (1965) The Kramers problem for a not completely diffusion wall. *J. Math. Anal. Appl.*, **10**, 568–586.
83. Cercignani, C., Foresti, P., and Sernagiotto, F. (1968) Dependence of the slip coefficient on the form of the collision frequency. *Nuovo Cimento*, **57B** (2), 297–306.
84. Barichello, L.B., Camargo, M., Rodrigues, P., and Siewert, C.E. (2001) Unified solution to classical flow problem based on the BGK model. *Z. Angew. Math. Phys. (ZAMP)*, **52** (3), 517–534.
85. Siewert, C.E. and Sharipov, F. (2002) Model equations in rarefied gas dynamics: viscous-slip and thermal-slip coefficients. *Phys. Fluids*, **14** (12), 4123–4129.
86. Siewert, C.E. (2001) Kramers' problem for a variable collision frequency model. *Eur. J. Appl. Math.*, **12**, 179–191.
87. Loyalka, S.K., Petrellis, N., and Storvik, T.S. (1975) Some numerical results for the BGK model: thermal creep and viscous slip problems with arbitrary accommodation at the surface. *Phys. Fluids*, **18** (9), 1094–1099.

88. Loyalka, S.K. and Hickey, K.A. (1990) The Kramers problem - velocity slip and defect for a hard-sphere gas with arbitrary accommodation. *Z. Angew. Math. Phys. (ZAMP)*, **41** (2), 245–253.
89. Ohwada, T., Sone, Y., and Aoki, K. (1989) Numerical analysis of the shear and thermal creep flows of a rarefied gas over a plane wall on the basis of the linearized Boltzmann equation for hard-sphere molecules. *Phys. Fluids A*, **1** (9), 1588–1599.
90. Wakabayashi, M., Ohwada, T., and Golse, F. (1996) Numerical analysis of the shear and thermal creep flows of a rarefied gas over the plane wall of a Maxwell-type boundary on the basis of the linearized Boltzmann equation for hard-sphere molecules. *Eur. J. Mech. B/Fluids*, **15** (2), 175–201.
91. Siewert, C.E. (2003) The linearized Boltzmann equation: concise and accurate solutions to basic flow problems. *Z. Angew. Math. Phys. (ZAMP)*, **54**, 273–303.
92. Siewert, C.E. (2003) Viscous-slip, thermal-slip and temperature-jump coefficients as defined by the linearized Boltzmann equation and the Cercignani-Lampis boundary condition. *Phys. Fluids*, **15** (6), 1696–1701.
93. Loyalka, S.K. (1990) Slip and jump coefficients for rarefied gas flows: variational results for Lennard-Jones and *n*(*r*)-6 potential. *Physica A*, **163** (3), 813–821.
94. Sharipov, F. and Bertoldo, G. (2009) Poiseuille flow and thermal creep based on the Boltzmann equation with the Lennard-Jones potential over a wide range of the Knudsen number. *Phys. Fluids*, **21**, 067 101.
95. Klinc, T. and Kuščer, I. (1972) Slip coefficients for general gas-surface interaction. *Phys. Fluids*, **15** (6), 1018–1022.
96. Suetin, P.E. and Chernyak, V.G. (1977) About the dependence of Poiseuille slip and thermal cpeep on interaction law of gaseous molecules with a boundary surface. *Izvestia AN SSSR. Mekhanika Zhidkosti i Gaza*, **6**, 107–114. (in Russian).
97. Millikan, R.A. (1923) Coefficients of slip in gases and the law of reflection of molecules from the surface of solids and liquids. *Phys. Rev.*, **21**, 217–238.
98. Stacy, L. (1923) A determination by the constant deflection method of the value of the coefficient of slip for rough and for smooth surfaces in air. *Phys. Rev.*, **21**, 239–249.
99. Suetin, P.E., Porodnov, B.T., Chernyak, V.G., and Borisov, S.F. (1973) Poiseuille flow at arbitrary Knudsen number and tangential momentum accommodation. *J. Fluid Mech.*, **60** (3), 581–592.
100. Tekasakul, P., Bentz, J.A., Tompson, R.V., and Loyalka, S.K. (1996) The spinning rotor gauge: measurements of viscosity, velocity slip coefficients, and tangential momentum accommodation coefficients. *J. Vac. Sci. Technol., A*, **14** (5), 2946–2952.
101. Bentz, J.A., Thompson, R.V., and Loyalka, S.K. (2001) Measurements of viscosity, velocity slip coefficients, and tangential momentum accommodation coefficients using a modified spinning rotor gauge. *J. Vac. Sci. Technol., A*, **19** (1), 317–324.
102. Ewart, T., Perrier, P., Graur, I.A., and Méolans, J.G. (2007) Mass flow rate measurements in a microchannel, from hydrodynamic to near free molecular regimes. *J. Fluid Mech.*, **584**, 337–356.
103. Ewart, T., Perrier, P., Graur, I., and Méolans, J.G. (2007) Tangential momentum accommodation in microtube. *Microfluid. Nanofluid.*, **3**, 689–695.
104. Sharipov, F. and Kalempa, D. (2003) Velocity slip and temperature jump coefficients for gaseous mixtures. I. Viscous slip coefficient. *Phys. Fluids*, **15** (6), 1800–1806.
105. Loyalka, S.K. (1989) Temperature jump and thermal creep slip: rigid sphere gas. *Phys. Fluids A*, **1**, 403–408.
106. Annis, B.K. (1972) Thermal creep in gases. *J. Chem. Phys.*, **57** (7), 2898–2905.
107. Porodnov, B.T., Kulev, A.N., and Tukhvetov, F.T. (1978) Thermal transpiration in a circular capillary with a small temperature difference. *J. Fluid Mech.*, **88** (4), 609–622.

108. Sharipov, F. and Kalempa, D. (2004) Velocity slip and temperature jump coefficients for gaseous mixtures. II. Thermal slip coefficient. *Phys. Fluids*, **16** (3), 759–764.
109. Barichello, L.B. and Siewert, C.E. (2000) The temperature-jump problem in rarefied-gas dynamics. *Eur. J. Appl. Math.*, **4**, 353–364.
110. Siewert, C.E. (2003) The linearized Boltzmann equation: a concise and accurate solution of the temperature-jump problem. *J. Quant. Spectrosc. Radiat. Transfer*, **77**, 417–432.
111. Sone, Y., Ohwada, T., and Aoki, K. (1989) Temperature jump and Knudsen layer in a rarefied gas over a plane wall: numerical analysis of the linearized Boltzmann equation for hard-sphere molecules. *Phys. Fluids A*, **1** (2), 363–370.
112. Sharipov, F. and Kalempa, D. (2005) Velocity slip and temperature jump coefficients for gaseous mixtures. IV. Temperature jump coefficient. *Int. J. Heat Mass Transfer*, **48** (6), 1076–1083.
113. Sharipov, F., Cumin, L.M.G., and Kalempa, D. (2004) Plane Couette flow of binary gaseous mixture in the whole range of the Knudsen number. *Eur. J. Mech. B/Fluids*, **23**, 899–906.
114. Garcia, R.D.M. and Siewert, C.E. (2008) Couette flow of a binary mixture of rigid-sphere gases described by the linearized Boltzmann equation. *Eur. J. Mech. B/Fluids*, **27**, 823–836.
115. Valougeorgis, D. (1988) Couette flow of a binary gas mixture. *Phys. Fluids*, **31** (3), 521–524.
116. Ohwada, T. (1996) Heat flow and temperature and density distributions in a rarefied gas between parallel plates with different temperature. Finite-difference analysis of the nonlinear Boltzmann equation for hard-sphere molecules. *Phys. Fluids*, **8**, 2153–2160.
117. Graur, I.A. and Polikarpov, A.P. (2009) Comparison of different kinetic models for the heat transfer problem. *Heat Mass Transfer*, **46** (2), 237–244.
118. Pazoki, N. and Loyalka, S.K. (1985) Heat transfer in a rarefied polyatomic gases -I. Plane parallel plates. *Int. J. Heat Mass Transfer*, **28** (11), 2019–2027.
119. Sharipov, F., Cumin, L.M.G., and Kalempa, D. (2007) Heat flux through a binary gaseous mixture over the whole range of the Knudsen number. *Physica A*, **378**, 183–193.
120. Garcia, R.D.M. and Siewert, C.E. (2004) The McCormack model for gas mixtures: heat transfer in a plane channel. *Phys. Fluids*, **16** (9), 3393–3402.
121. Chernyak, V.G., Kalinin, V.V., and Suetin, P.E. (1979) On theory of nonsothermal flow of gas in plane channel. *Inzh.-Fiz. Zh.*, **36** (6), 1059–1065. (in Russian).
122. Cabrera, L.C. and Barichello, L.B. (2005) Unified solutions to some classical problems in rarefied gas dynamics based on the one-dimensional linearized S - model equations. *Z. Angew. Math. Phys. (ZAMP)*, **57**, 285–312.
123. Valougeorgis, D. (2003) An analytical solution of the s-model kinetic equation. *Z. Angew. Math. Phys. (ZAMP)*, **54**, 112–124.
124. Sharipov, F. (2002) Application of the Cercignani-Lampis scattering kernel to calculations of rarefied gas flows. I. Plane flow between two parallel plates. *Eur. J. Mech. B/Fluids*, **21** (1), 113–123.
125. Siewert, C.E. (2002) Generalized boundary condition for the S-model kinetic equations basic to flow in a plane channel. *J. Quant. Spectrosc. Radiat. Transfer*, **72**, 75–88.
126. Loyalka, S.K. and Storvick, T.S. (1979) Kinetic theory of thermal transpiration and mechanocaloric effect. III. Flow of a polyatomic gas between parallel plates. *J. Chem. Phys.*, **71** (1), 339–350.
127. Titarev, V.A. and Shakhov, E.M. (2012) Poiseuille flow and thermal creep in a capillary tube on the basis of the kinetic r-model. *Fluid Dyn.*, **47** (5), 661–672.
128. Naris, S., Valougeorgis, D., Kalempa, D., and Sharipov, F. (2004) Gaseous mixture flow between two parallel plates in the whole range of the gas rarefaction. *Physica A*, **336** (3–4), 294–318.
129. Garcia, R.D.M. and Siewert, C.E. (2007) Channel flow of a binary mixture of

rigid spheres described by the linearized Boltzmann equation and driven by temperature, pressure, and concentration gradients. *SIAM J. Appl. Math.*, **67** (4), 1041–1063.

130. Naris, S. and Valougeorgis, D. (2005) The driven cavity flow over the whole range of the Knudsen number. *Phys. Fluids*, **17** (9), 097 106.

131. Cercignani, C. and Sernagiotto, F. (1967) Cylindrical Couette flow of a rarefied gas. *Phys. Fluids*, **10** (6), 1200–1204.

132. Nanbu, K. (1984) Analysis of cylindrical Couette flows by use of the direction simulation method. *Phys. Fluids*, **27** (11), 2632–2635.

133. Sharipov, F.M. and Kremer, G.M. (1996) Nonlinear Couette flow between two rotating cylinders. *Transp. Theory Stat. Phys.*, **25** (2), 217–229.

134. Sharipov, F.M. and Kremer, G.M. (1996) Linear Couette flow between two rotating cylinders. *Eur. J. Mech. B/Fluids*, **15** (4), 493–505.

135. Tantos, C., Valougeorgis, D., Pannuzzo, M., Frezzotti, A., and Morini, G.L. (2014) Conductive heat transfer in a rarefied polyatomic gas confined between coaxial cylinders. *Int. J. Heat Mass Transfer*, **79**, 378–389.

136. Sharipov, F. (2008) Analytical and numerical calculations of rarefied gas flow, in *Handbook of Vacuum Technology* (ed. K. Jousten), Wiley-VCH Verlag GmbH, Weinheim.

137. Pantazis, S. and Valougeorgis, D. (2010) Heat transfer through rarefied gases between coaxial cylindrical surfaces with arbitrary temperature difference. *Eur. J. Mech. B/Fluids*, **29**, 494–509.

138. Sharipov, F.M. and Kremer, G.M. (1995) On the frame-dependence of constitutive equations. I. Heat transfer through a rarefied gas between two rotating cylinders. *Continuum Mech. Thermodyn.*, **7** (1), 57–72.

139. Sharipov, F.M. and Kremer, G.M. (1995) Heat conduction through a rarefied gas between two rotating cylinders at small temperature difference. *Z. Angew. Math. Phys. (ZAMP)*, **46** (5), 680–692.

140. Sharipov, F.M. and Kremer, G.M. (1999) Non-isothermal Couette flow of a rarefied gas between two cylinders. *Eur. J. Mech. B/Fluids*, **18** (1), 121–130.

141. Biscari, P. and Cercignani, C. (1997) Stress and heat flux in non-inertial reference frames. *Continuum Mech. Thermodyn.*, **9** (1), 1–11.

142. Sharipov, F. (1996) Rarefied gas flow through a long tube at any temperature difference. *J. Vac. Sci. Technol., A*, **14** (4), 2627–2635.

143. Siewert, C.E. and Valougeorgis, D. (2002) An analytical discrete-ordinates solution of the S-model kinetic equation for flow in a cylindrical tube. *J. Quant. Spectrosc. Radiat. Transfer*, **72**, 351–550.

144. Loyalka, S.K. and Hamoodi, S.A. (1990) Poiseuille flow of a rarefied gas in a cylindrical tube: solution of linearized Boltzmann equation. *Phys. Fluids A*, **2** (11), 2061–2065. Erratum. in Phys. Fluids A, **3**, 2825–(1991).

145. Loyalka, S.K., Storvick, T.S., and Lo, S.S. (1982) Thermal transpiration and mechanocaloric effect. IV Flow of a polyatomic gas in a cylindrical tube. *J. Chem. Phys.*, **76** (8), 4157–4170.

146. Sharipov, F. and Kalempa, D. (2002) Gaseous mixture flow through a long tube at arbitrary Knudsen number. *J. Vac. Sci. Technol., A*, **20** (3), 814–822.

147. Sharipov, F. and Kalempa, D. (2005) Separation phenomena for gaseous mixture flowing through a long tube into vacuum. *Phys. Fluids*, **17** (12), 127 102.

148. Sharipov, F. (1999) Rarefied gas flow through a long rectangular channel. *J. Vac. Sci. Technol., A*, **17** (5), 3062–3066.

149. Méolans, J., Nacer, M., Rojas, M., Perrier, P., and Graur, I. (2012) Effects of two transversal finite dimensions in long microchannel: analytical approach in slip regime. *Phys. Fluids*, **24** (11), 112005.

150. Titarev, V.A. and Shakhov, E.M. (2010) Kinetic analysis of the isothermal flow in a long rectangular microchannel. *Comput. Math. Math. Phys.*, **50** (7), 1221–1237.

151. Sharipov, F. (1999) Non-isothermal gas flow through rectangular microchannels. *J. Micromech. Microeng.*, **9** (4), 394–401.
152. Rykov, V.A., Titarev, V.A., and Shakhov, E.M. (2011) Rarefied Poiseuille flow in elliptical and rectangular tubes. *Fluid Dyn.*, **46** (3), 456–466.
153. Graur, I. and Ho, M.T. (2014) Rarefied gas flow through a long rectangular channel of variable cross section. *Vacuum*, **101**, 328–332.
154. Graur, I. and Sharipov, F. (2007) Gas flow through an elliptical tube over the whole range of the gas rarefaction. *Eur. J. Mech. B/Fluids*, **27** (3), 335–345.
155. Graur, I. and Sharipov, F. (2009) Non-isothermal flow of rarefied gas through a long pipe with elliptic cross section. *Microfluid. Nanofluid.*, **6** (2), 267–275.
156. Naris, S., Valougeorgis, D., Sharipov, F., and Kalempa, D. (2004) Discrete velocity modelling of gaseous mixture flows in MEMS. *Superlattices Microstruct.*, **35**, 629–643.
157. Naris, S., Valougeorgis, D., Kalempa, D., and Sharipov, F. (2005) Flow of gaseous mixtures through rectangular microchannels driven by pressure, temperature and concentration gradients. *Phys. Fluids*, **17** (10), 100 607.
158. Sharipov, F. (2012) Benchmark problems in rarefied gas dynamics. *Vacuum*, **86 SI** (11), 1697–1700.
159. Clausing, P. (1932) Uber die stromung sehr verdunntern case durch rohren von beliebiger lange. *Ann. Phys.*, **12**, 961–989.
160. Berman, A.S. (1965) Free molecule transmission probabilities. *J. Appl. Phys.*, **36** (10), 3356.
161. Sharipov, F. (1996) Rarefied gas flow through a slit: influence of the gas-surface interaction. *Phys. Fluids*, **8** (1), 262–268.
162. Roscoe, R. (1949) The flow of viscous fluid round plane obstacles. *Philos. Mag.*, **40**, 338–351.
163. Akinshin, V.D., Makarov, A.M., Seleznev, V.D., and Sharipov, F.M. (1989) Rarefied gas motion in a short planar channel over the entire Knudsen number range. *J. Appl. Mech. Tech. Phys.*, **30** (5), 713–717.
164. Sharipov, F. (1997) Non-isothermal rarefied gas flow through a slit. *Phys. Fluids*, **9** (6), 1804–1810.
165. Sharipov, F.M., Seleznev, V.D., and Makarov, A.M. (1990) Nonisothermal motion of a rarefied gas in a short planar channel over a wide range of Knudsen numbers. *J. Eng. Phys. Thermophys.*, **59** (1), 869–875.
166. Sharipov, F. and Seleznev, V. (1998) Data on internal rarefied gas flows. *J. Phys. Chem. Ref. Data*, **27** (3), 657–706.
167. Titarev, V.A. (2012) Implicit high-order method for calculating rarefied gas flow in a planar microchannel. *J. Comput. Phys.*, **231**, 109–134.
168. Titarev, V.A. (2012) Rarefied gas flow in a planar channel caused by arbitrary pressure and temperature drops. *Int. J. Heat Mass Transfer*, **55** (21-22), 5916–5930.
169. Titarev, V.A. and Shakhov, E.M. (2012) Efficient method for computing rarefied gas flow in a long finite plane channel. *Comput. Math. Math. Phys.*, **52** (2), 269–284.
170. Sharipov, F.M. (1990) Onsager reciprocity relations for rarefied polyatomic gas flow in the presence of an external field. *Zh. Vychisl. Mat. Mat. Fiz.*, **30** (2), 310–318. (in Russian).
171. Sazhin, O. (2008) Gas flow through a slit into a vacuum in a wide range of rarefaction. *J. Exp. Theor. Phys.*, **107** (1), 162–169.
172. Sharipov, F. and Kozak, D.V. (2009) Rarefied gas flow through a thin slit into vacuum simulated by the Monte Carlo method over the whole range of the Knudsen number. *J. Vac. Sci. Technol., A*, **27** (3), 479–484.
173. Sazhin, O. (2009) Rarefied gas flow through a channel of finite length into a vacuum. *J. Exp. Theor. Phys.*, **109** (4), 700–706.
174. Graur, I., Polikarpov, A.P., and Sharipov, F. (2011) Numerical modelling of rarefied gas flow through a slit into vacuum based on the kinetic equation. *Comput. Fluids*, **49**, 87–92.
175. Sharipov, F. and Kozak, D.V. (2011) Rarefied gas flow through a thin slit at

an arbitrary pressure ratio. *Eur. J. Mech. B/Fluids*, **30** (5), 543–549.

176. Sazhin, O. (2012) Pressure-driven flow of rarefied gas through a slit at a various pressure ratios. *J. Vac. Sci. Technol., A*, **30** (2), 021 603.

177. Varoutis, S., Day, C., and Sharipov, F. (2012) Rarefied gas flow through channels of finite length at various pressure ratios. *Vacuum*, **86** (12), 1952–1959.

178. Graur, I., Polikarpov, A.P., and Sharipov, F. (2012) Numerical modelling of rarefied gas flow through a slit at arbitrary gas pressure ratio based on the kinetic equation. *Z. Angew. Math. Phys. (ZAMP)*, **63**, 503–520.

179. Misdanitis, S., Pantazis, S., and Valougeorgis, D. (2012) Pressure driven rarefied gas flow through a slit and an orifice. *Vacuum*, **86** (11), 1701–1708.

180. Aristov, V.V., Frolova, A.A., Zabelok, S.A., Arslanbekov, R.R., and Kolobov, V.I. (2012) Simulations of pressure-driven flows through channels and pipes with unified flow solver. *Vacuum*, **86 SI** (11), 1717–1724.

181. Rovenskaya, O.I., Polikarpov, A.P., and Graur, I.A. (2013) Comparison of the numerical solutions of the full Boltzmann and S-model kinetic equations for gas flow through a slit. *Comput. Fluids*, **80**, 71–78, doi: 10.1016/j.compfluid.2012.05.007.

182. Rovenskaya, O.I. (2013) Comparative analysis of the numerical solution of full boltzmann and BGK model equations for the poiseuille flow in a planar microchannel. *Comput. Fluids*, **81**, 45–56.

183. Rovenskaya, O.I. and Croce, G. (2014) Application a hybrid solver to gas flow through a slit at arbitrary pressure ratio. *Vacuum*, **109**, 266–274.

184. Weissberg, H.L. (1962) End correction for slow viscous flow through long tubes. *Phys. Fluids*, **5** (5), 1033–1036.

185. Akinshin, V.D., Makarov, A.M., Seleznev, V.D., and Sharipov, F.M. (1988) Flow of a rarefied gas in a plane channel of finite length for a wide range of Knudsen numbers. *J. Appl. Mech. Tech. Phys.*, **29** (1), 97–103.

186. Shakhov, E.M. (1999) Linearized two-dimensional problem of rarefied gas flow in a long channel. *Comput. Math. Math. Phys.*, **39** (7), 1192–1200.

187. Pantazis, S., Valougeorgis, D., and Sharipov, F. (2013) End corrections for rarefied gas flows through capillaries of finite length. *Vacuum*, **97**, 26–29.

188. Cercignani, C. and Pagani, C.D. (1966) Variational approach to boundary value problems in kinetic theory. *Phys. Fluids*, **9** (6), 1167–1173.

189. Pantazis, S. and Valougeorgis, D. (2013) Rarefied gas flow through a cylindrical tube due to a small pressure difference. *Eur. J. Mech. B/Fluids*, **38**, 114–127.

190. Titarev, V.A. (2013) Rarefied gas flow in a circular pipe of finite length. *Vacuum*, **94**, 92–103.

191. Titarev, V.A. and Shakhov, E.M. (2012) Computational study of a rarefied gas flow through a long circular pipe into vacuum. *Vacuum*, **86** (11), 1709–1716.

192. Aristov, V.V., Shakhov, E.M., Titarev, V.A., and Zabelok, S. (2014) Comparative study for rarefied gas flow into vacuum through a short circular pipe. *Vacuum*, **103**, 5–8.

193. Sharipov, F. (2002) Rarefied gas flow into vacuum through a thin orifice. Influence of the boundary conditions. *AIAA J.*, **40** (10), 2006–2008.

194. Sharipov, F. (2004) Numerical simulation of rarefied gas flow through a thin orifice. *J. Fluid Mech.*, **518**, 35–60.

195. Sharipov, F. and Strapasson, J.L. (2014) Ab initio simulation of rarefied gas flow through a thin orifice. *Vacuum*, **109**, 246–252.

196. Varoutis, S., Valougeorgis, D., Sazhin, O., and Sharipov, F. (2008) Rarefied gas flow through short tubes into vacuum. *J. Vac. Sci. Technol., A*, **26** (2), 228–238.

197. Varoutis, S., Valougeorgis, D., and Sharipov, F. (2009) Simulation of gas flow through tubes of finite length over the whole range of rarefaction for various pressure drop ratios. *J. Vac. Sci. Technol., A*, **22** (6), 1377–1391.

198. Pantazis, S., Valougeorgis, D., and Sharipov, F. (2014) End corrections for rarefied gas flows through circular tubes of finite length. *Vacuum*, **101**, 306–312.

199. Cercignani, C. and Sernagiotto, F. (1966) Cylindrical Poiseuille flow of a rarefied gas. *Phys. Fluids*, **9** (1), 40–44.
200. Sharipov, F. (2012) Transient flow of rarefied gas through an orifice. *J. Vac. Sci. Technol., A*, **30** (2), 021 602.
201. Sharipov, F. (2013) Transient flow of rarefied gas through a short tube. *Vacuum*, **90**, 25–30.
202. Ho, M.T. and Graur, I. (2014) Numerical study of unsteady rarefied gas flow through an orifice. *Vacuum*, **109**, 253–265.
203. Sharipov, F. and Seleznev, V. (1994) Rarefied gas flow through a long tube at any pressure ratio. *J. Vac. Sci. Technol., A*, **12** (5), 2933–2935.
204. Sharipov, F. (1997) Rarefied gas flow through a long tube at arbitrary pressure and temperature drops. *J. Vac. Sci. Technol., A*, **15** (4), 2434–2436.
205. Marques, W. Jr., Kremer, G.M., and Sharipov, F.M. (2000) Couette flow with slip and jump boundary conditions. *Continuum Mech. Thermodyn.*, **12** (6), 379–386.
206. Sharipov, F. and Bertoldo, G. (2005) Rarefied gas flow through a long tube of variable radius. *J. Vac. Sci. Technol., A*, **23** (3), 531–533.
207. Sharipov, F. and Graur, I. (2014) General approach to transient flows of rarefied gases through long capillaries. *Vacuum*, **100**, 22–25.
208. Lihnaropoulos, J. and Valougeorgis, D. (2011) Unsteady vacuum gas flow in cylindrical tubes. *Fusion Eng. Des.*, **86**, 2139–2142.
209. Maidanik, G., Fox, H.L., and Heckl, M. (1965) Propagation and reflection of sound in rarefied gases. I. Theoretical. *Phys. Fluids*, **8** (2), 259–265.
210. Kahn, D. and Mintzer, D. (1965) Kinetic theory of sound propagation in rarefied gases. *Phys. Fluids*, **8** (6), 1090–1102.
211. Toba, K. (1968) Kinetic theory of sound propagation in a rarefied gas. *Phys. Fluids*, **11** (11), 2495–2497.
212. Hanson, F.B. and Morse, T.F. (1969) Free-molecular expansion polynomials and sound propagation in rarefied gases. *Phys. Fluids*, **12** (8), 1564–1572.
213. Sharipov, F., Marques, W. Jr., and Kremer, G.M. (2002) Free molecular sound propagation. *J. Acoust. Soc. Am.*, **112** (2), 395–401.
214. Sharipov, F. and Kalempa, D. (2008) Numerical modelling of the sound propagation through a rarefied gas in a semi-infinite space on the basis of linearized kinetic equation. *J. Acoust. Soc. Am.*, **124** (4), 1993–2001.
215. Sharipov, F. and Kalempa, D. (2008) Oscillatory Couette flow at arbitrary oscillation frequency over the whole range of the knudsen number. *Microfluid. Nanofluid.*, **4** (5), 363–374.
216. Sharipov, F. and Kalempa, D. (2007) Gas flow near a plate oscillating longitudinally with an arbitrary frequency. *Phys. Fluids*, **19** (1), 017 110.
217. Doi, T. (2010) Numerical analysis of oscillatory Couette flow of a rarefied gas on the basis of the linearized Boltzmann equation for a hard sphere molecular gas. *Z. Angew. Math. Phys.(ZAMP)*, **61** (5), 811–822.
218. Kalempa, D. and Sharipov, F. (2012) Sound propagation through a rarefied gas. Influence of the gas-surface interaction. *Int. J. Heat Fluid Flow*, **30**, 190–199.
219. Kalempa, D. and Sharipov, F. (2009) Sound propagation through a rarefied gas confined between source and receptor at arbitrary Knudsen number and sound frequency. *Phys. Fluids*, **21**, 103 601.1–14.
220. Doi, T. (2011) Numerical analysis of the time-dependent energy and momentum transfers in a rarefied gas between two parallel planes based on the linearized Boltzmann equation. *J. Heat Transfer*, **133**, 022 404.
221. Schotter, R. (1974) Rarefied gas acoustics in the noble gases. *Phys. Fluids*, **17** (6), 1163–1168.
222. Hadjiconstantinou, N.G. (2002) Sound wave propagation in a transition regime micro and nanochannels. *Phys. Fluids*, **14** (2), 802–809.
223. Hadjiconstantinou, N.G. (2005) Oscillatory shear-driven gas flows in the transition and free-molecular flow regimes. *Phys. Fluids*, **17**, 100 611.

224. Emerson, D.R., Gu, X.J., Stefanov, S.K., Yuhong, S., and Barber, R.W. (2007) Nonplanar oscillatory shear flow: from the continuum to the free-molecular regime. *Phys. Fluids*, **19**, 107 105–16.

225. Gospodinov, P., Roussinov, V., and Stefanov, S. (2012) Nonisothermal oscillatory cylindrical couette gas flow in the slip regime: a computational study. *Eur. J. Mech. B/Fluids*, **33**, 14–24.

226. Kalempa, D. and Sharipov, F. (2014) Numerical modelling of thermoacoustic waves in a rarefied gas confined between coaxial cylinders. *Vacuum*, **109**, 326–332.

227. Wieser, M.E., Holden, N., Coplen, T.B., Boehlke, J.K., Berglund, M., Brand, W.A., De Bievre, P., Groening, M., Loss, R.D., Meija, J., Hirata, T., Prohaska, T., Schoenberg, R., O'Connor, G., Walczyk, T., Yoneda, S., and Zhu, X.K. (2013) Atomic weights of the elements 2011 (IUPAC Technical report). *Pure Appl. Chem.*, **85** (5), 1047–1078.

228. Korn, G.A. and Korn, T.M. (1961) *Mathematical Handbook for Scientitsts and Engineers*, McGraw-Hill Book Company, Inc., New York.

229. Abramowitz, M. and Stegun, I.A. (1972) *Handbook of Mathematical Functions with Formulas, Graphs and Mathematical Tables*, 9th edn, Dover Publications, New York.

Index

Rarefied Gas Dynamics: Fundamentals for Research and Practice, First Edition. Felix Sharipov.